Acta Numerica 1993

Acta
Numerica
1993

CAMBRIDGE
UNIVERSITY PRESS

CAMBRIDGE UNIVERSITY PRESS
Cambridge, New York, Melbourne, Madrid, Cape Town, Singapore,
São Paulo, Delhi, Dubai, Tokyo, Mexico City

Cambridge University Press
The Edinburgh Building, Cambridge CB2 8RU, UK

Published in the United States of America by Cambridge University Press, New York

www.cambridge.org
Information on this title: www.cambridge.org/9780521157575

© Cambridge University Press 1993

First published 1993
First paperback edition 2010

A catalogue record for this publication is available from the British Library

ISBN 978-0-521-44356-2 Hardback
ISBN 978-0-521-15757-5 Paperback

Contents

Contents

Acta Numerica (1993), pp. 1–64

Continuation and path following

Eugene L. Allgower* and Kurt Georg*

Department of Mathematics
Colorado State University
Ft. Collins, CO 80523, USA
E-mail: Georg@Math.ColoState.Edu

The main ideas of path following by predictor–corrector and piecewise-linear methods, and their application in the direction of homotopy methods and nonlinear eigenvalue problems are reviewed. Further new applications to areas such as polynomial systems of equations, linear eigenvalue problems, interior methods for linear programming, parametric programming and complex bifurcation are surveyed. Complexity issues and available software are also discussed.

CONTENTS

1. Introduction

Continuation, embedding or homotopy methods have long served as useful theoretical tools in modern mathematics. Their use can be traced back at least to such venerated works as those of Poincaré (1881–1886), Klein (1882–1883) and Bernstein (1910). Leray and Schauder (1934) refined the tool and presented it as a global result in topology, viz. the homotopy invariance of degree. The use of deformations to solve nonlinear systems of equations may be traced back at least to Lahaye (1934). The classical embedding

* Partially supported by the National Science Foundation via grant no. DMS-9104058.

methods were the first deformation methods to be numerically implemented and may be regarded as a forerunner of the predictor–corrector methods for path following which we will discuss here.

Because of their versatility and robustness, numerical continuation or path following methods have now been finding ever wider use in scientific applications. Our aim here is to present some of the recent advances in this subject regarding new adaptations, applications, and analysis of efficiency and complexity. To make the discussion relatively self-contained, we review some of the background of numerical continuation methods. Introductions into aspects of the subject may be found in the books by Garcia and Zangwill (1981), Gould and Tolle (1983), Keller (1987), Rheinboldt (1986), Seydel (1988) and Todd (1976a). The philosophy and notation of the present article will be that of our book Allgower and Georg (1990), which also contains an extensive bibliography up to 1990.

The viewpoint which will be adopted here is that numerical continuation methods are techniques for numerically approximating a solution curve c which is implicitly defined by an underdetermined system of equations. In the literature of numerical analysis, the terms *numerical continuation* and *path following* are used interchangeably.

There are various objectives for which the numerical approximation of c can be used and, depending upon the objective, the approximating technique is adapted accordingly. In fact, continuation is a unifying concept, under which various numerical methods may be subsumed which may otherwise have very little in common. For example, simplicial fixed point methods for solving problems in mathematical economics, the generation of bifurcation diagrams of nonlinear eigenvalue problems involving partial differential equations, and the recently developed interior point methods for solving linear programming problems seem to be quite unrelated. Nevertheless, there is some benefit in considering them as special cases of path following. We personally are struck by the remarkable fact that a technique which was initially developed for solving difficult nonlinear problems now turns out to be extremely useful for treating various problems which are essentially linear: e.g. linear eigenvalue problems, and linear programming and complementarity problems.

The remainder of the article is organized as follows. Section 2 contains the basic ideas of predictor–corrector path following methods. In Section 3 some technical aspects of implementing predictor–corrector methods are addressed, e.g. the numerical linear algebra involved and steplength strategies.

Section 4 deals with various applications of path following methods. We begin with a brief discussion of homotopy methods for fixed point problems and global Newton methods. Then we address the problem of finding multiple solutions. In particular, we discuss recent homotopy methods for finding all solutions of polynomial systems of equations. Next we survey some path following aspects of nonlinear eigenvalue problems, and address the question of handling bifurcations. Finally, three new developments in path following are discussed: (1) The solution of linear eigenvalue problems via special homotopy approaches; (2) the handling of parametric programming problems by following certain branches of critical points via active set strategies; and (3) the path following aspects involved in the interior point methods for solving linear and quadratic programming problems.

Section 5 presents an introduction to the principles of piecewise linear methods. These methods view path following in a different light: instead of approximately following a smooth solution curve, they exactly follow an approximate curve (i.e. a polygonal path). Some instances where these methods are useful are discussed, e.g. linear complementarity problems or homotopy methods where predictor–corrector methods are not implementable, because of lack of smoothness. We also briefly address the related topic of approximating implicitly defined surfaces.

The issue of the computational complexity of path following is considered in Section 6. This issue is related to the Newton–Kantorovich theory and is currently of considerable interest in the context of interior point methods.

We conclude by listing some available software related to path following and indicate how the reader might access these codes. No attempt to compare or evaluate the various codes is offered. In any case, our opinion is that path following codes always need to be considerably adapted to the special purposes for which they are designed. The path following literature offers various tools for accomplishing such tasks. Although there are some general purpose codes, probably none will slay every dragon.

The extensive bibliography contains only cited items. Space considerations prohibited the addressing of some important topics, and consequently some significant recent contributions to the field are not contained in the bibliography.

2. The basics of predictor–corrector path following

The simplest (and most frequently occurring) case of an underdetermined system of nonlinear equations contains just one degree of freedom:

$$H(u) = 0 \text{ where } H : \mathbb{R}^{N+1} \to \mathbb{R}^N \text{ is a smooth map.} \qquad (2.1)$$

When we say that a map is smooth we shall mean that it has as many continuous derivatives as the context of the discussion requires. For convenience, the reader may assume C^∞. In order to apply the Implicit Function Theorem, we need the following standard

Definition 2.1 We call u a *regular point* of H if the Jacobian $H'(u)$ has maximal rank. We call y a *regular value* of H if u is a regular point of

H whenever $H(u) = y$. If a point or value is not regular, then it is called *singular*.

Let $u_0 \in \mathbb{R}^{N+1}$ be a regular point of H such that $H(u_0) = 0$. It follows from the Implicit Function Theorem that the solution set $H^{-1}(0)$ can be locally parametrized about u_0 with respect to some coordinate. By a reparametrization (according to arclength), we obtain a smooth curve $c : J \to \mathbb{R}^{N+1}$ for some open interval J containing zero such that for all $s \in J$:

$$c(0) = u_0 \tag{2.2}$$
$$H'(c(s))\dot{c}(s) = 0, \tag{2.3}$$
$$\|\dot{c}(s)\| = 1, \tag{2.4}$$
$$\det \begin{pmatrix} H'(c(s)) \\ \dot{c}(s)^* \end{pmatrix} > 0. \tag{2.5}$$

These conditions uniquely determine the tangent $\dot{c}(s)$. Here and in the following, $(.)^*$ denotes the Hermitian transpose and $\|.\|$ the Euclidean norm. Condition (2.4) normalizes the parametrization to arclength. This is only for theoretical convenience, and it is not an intrinsic restriction. Condition (2.5) chooses one of the two possible orientations.

The preceding discussion motivates the following

Definition 2.2 Let A be an $(N,N+1)$-matrix with maximal rank. For the purpose of our exposition, the unique vector $t(A) \in \mathbb{R}^{N+1}$ satisfying the conditions

$$At = 0, \tag{2.6}$$
$$\|t\| = 1, \tag{2.7}$$
$$\det \begin{pmatrix} A \\ t^* \end{pmatrix} > 0, \tag{2.8}$$

will be called the *tangent vector induced by A*.

Making use of this definition, solution curve $c(s)$ is characterized as the solution of the initial value problem

$$\dot{u} = t(H'(u)), \quad u(0) = u_0 \tag{2.9}$$

which in this context is occasionally attributed to Davidenko (1953), see also Branin (1972). Note that the domain $\{u \in \mathbb{R}^{N+1} : u \text{ is a regular point}\}$ is open. This differential equation is not used in efficient path following algorithms, but it serves as a useful device in analysing the path. Two examples are:

Lemma 2.3 Let (a, b) be the maximal interval of existence for (2.9). If a

is finite, then $c(s)$ converges to a singular zero point of H as $s \to a$, $s > a$. An analogous statement holds if b is finite.

Lemma 2.4 Let zero be a regular value of H. Then the solution curve c is defined on the real line and satisfies one of the following two conditions:

1. The curve c is diffeomorphic to a circle. More precisely, there is a period $T > 0$ such that $c(s_1) = c(s_2)$ if and only if $s_1 - s_2$ is an integer multiple of T.
2. The curve c is diffeomorphic to the real line. More precisely, c is injective, and $c(s)$ has no accumulation point for $s \to \pm\infty$.

See (2.1.13) and (2.1.14) in Allgower and Georg (1990) for proofs. A more topological and global treatment of the Implicit Function Theorem can be found in the books of Hirsch (1976) or Milnor (1969).

Since the solution curve c is characterized by the initial value problem (2.9), it is evident that the numerical methods for solving initial value problems could immediately be used to numerically trace c. However, in general this is not an efficient approach, since it ignores the contractive properties which the curve c has in view of the fact that it satisfies the equation $H(u) = 0$. Instead, a typical path following method consists of a succession of two different steps:

Predictor step. An approximate step along the curve, usually in the general direction of the tangent of the curve. The initial value problem (2.9) provides motivation for generating predictor steps in the spirit of the technology of numerical solution of initial value problems.

Corrector steps. One or more iterative steps which aim to bring the predicted point back to the curve by an iterative procedure (typically of Newton or gradient type) for solving $H(u) = 0$.

It is usual to call such procedures *predictor–corrector* path following methods. However, let us note that this name should not be confused with the predictor–corrector multistep methods for initial value problems, since the latter do not converge back to the solution curve.

The following pseudocode (in **MATLAB** format) shows the basic steps of a generic predictor–corrector method.

Algorithm 2.5 $u = \text{generic\_pc\_method}(u, h)$

> % *$u \in \mathbb{R}^{N+1}$ such that $H(u) \approx 0$ is an initial point, input*
> % *$h > 0$ is an initial steplength, input*
>> WHILE a stopping criterion is not met
>>> % *predictor step*
>>>> predict v such that $H(v) \approx 0$ and $\|u - v\| \approx h$
>>>> and $v - u$ points in the direction of traversing

% *corrector step*

    let $w \in \mathbb{R}^{N+1}$ approximately solve ...

        $\min_{w} \{ \|v - w\| : H(w) = 0 \}$

% *new point along $H^{-1}(0)$*

    $u = w$

% *steplength adaptation*

    choose a new steplength $h > 0$

END

The predictor–corrector type of algorithms for curve following seem to date to Haselgrove (1961). In contrast to the modern predictor–corrector methods, the classical embedding methods assume that the solution path is parametrized with respect to an explicit parameter which is identified with the last variable in H. Hence, we consider the equation (2.1) in the form

$$H(x, \lambda) = 0. \tag{2.10}$$

If we assume that the partial derivative $H_x(x, \lambda)$ does not vanish, then the solution curve can be parametrized in the form $(x(\lambda), \lambda)$. This assumption has the drawback that *folds* are excluded, i.e. points such that $H(x, \lambda) = 0$ and $H_x(x, \lambda) = 0$. Such points are sometimes called turning points in the literature. The assumption has, however, the advantage that the corrector steps can be more easily handled, in particular if the partial derivative of H with respect to x is sparse. In some applications it is known *a priori* that no folds are present, and then the embedding method is applicable. For purposes of illustration we present an analogous generic embedding method:

Algorithm 2.6 $x = \text{generic\_embedding\_method}(x, \lambda, h)$

% *$(x, \lambda) \in \mathbb{R}^{N+1}$ such that $H(x, \lambda) \approx 0$ is an initial point, input*

% *$h > 0$ is an initial steplength, input*

    WHILE a stopping criterion is not met

        let $y \in \mathbb{R}^N$ approximately solve $H(y, \lambda + h) = 0$

        $(x, \lambda) = (y, \lambda + h)$

        choose a new steplength $h > 0$

    END

The predictor step is hidden; the predictor point would correspond to the starting point of an iterative method for solving $H(y, \lambda + h) = 0$. The most commonly used starting point is the previous point x.

It is common to blend aspects of these two algorithms. A simple example is to use a predictor tangent to the curve $(x(\lambda), \lambda)$ in the embedding algorithm. A more sophisticated example is the use of the bordering algorithm introduced in Keller (1977, 1983) in the corrector phase of the predictor–

corrector method. To avoid dealing with the arclength parameter, one can adopt a strategy of parameter switching, see, e.g., Rheinboldt (1980, 1981).

3. Aspects of implementations

Let us now turn to some of the practical aspects of implementing a predictor–corrector method.

3.1. Newton steps as corrector

A straightforward way of approximating a solution of the minimization problem in the predictor–corrector method (2.5) is given by the Newton step

$$\mathcal{N}_H(v) := v - H'(v)^+ H(v), \tag{3.1}$$

where $H'(v)^+$ denotes the Moore–Penrose inverse of $H'(v)$, see, e.g., Golub and van Loan (1989). Very commonly, an *Euler predictor*, i.e. a predictor step in the direction of the tangent to the curve is used:

$$v = u + ht(H'(u)), \tag{3.2}$$

where $h > 0$ represents the current stepsize.

The following algorithm sketches one version of the predictor–corrector method incorporating an approximate Euler predictor and one Newton-type iteration as a corrector step.

Algorithm 3.1 $u = $ Euler_Newton(u, h)

> WHILE a stopping criterion is not met
>> approximate $A \approx H'(u)$
>> $v = u + ht(A)$ % *predictor step*
>> $u = v - A^+ H(v)$ % *corrector step*
>> choose a new steplength $h > 0$
> END

Discussions of Newton's method using the Moore–Penrose inverse can be found in several text books, e.g. Ortega and Rheinboldt (1970) or Ben-Israel and Greville (1974).

Let us first state a convergence result, see (5.2.1) in Allgower and Georg (1990), which ensures that this algorithm safely follows the solution curve under reasonable assumptions.

Theorem 3.2   Let $H : \mathbb{R}^{N+1} \to \mathbb{R}^N$ be a smooth map having zero as a regular value and let $H(u_0) = 0$. Denote by $c_h(s)$ the polygonal path, starting at u_0, going through all points u generated by Algorithm 3.1 with fixed steplength $h > 0$. Denote by $c(s)$ the corresponding curve in $H^{-1}(0)$ given by the initial value problem (2.9). For definiteness, we assume that

$c_h(0) = c(0) = u_0$, and that both curves are parametrized with respect to arclength. If the estimate $\|A - H'(u)\| = \mathcal{O}(h)$ holds uniformly for the approximation in the loop of the algorithm, then the following quadratic bounds hold uniformly for $0 \leq s \leq s_0$ and s_0 sufficiently small:

$$\|H(c_h(s))\| \leq \mathcal{O}(h^2), \quad \|c_h(s) - c(s)\| \leq \mathcal{O}(h^2).$$

Some major points which remain to be clarified are:

— How do we efficiently handle the numerical linear algebra involved in the calculation of $t(A)$ and $A^+H(v)$?
— How do we formulate efficient steplength strategies?

3.2. The numerical linear algebra involved

A straightforward and simple (but not the most efficient) way to handle the numerical linear algebra would be to use a QR factorization:

$$A^* = Q \begin{pmatrix} R \\ 0^* \end{pmatrix}, \tag{3.3}$$

where Q is an $(N+1, N+1)$ orthogonal matrix, and R is a nonsingular (N, N) upper triangular matrix. We assume that A is an $(N, N+1)$ matrix with maximal rank. If q denotes the last column of Q, then $t(A) = \sigma q$, where the orientation defined in (2.5) leads to the choice

$$\sigma = \text{sign}\left(\det Q \det R \right). \tag{3.4}$$

Hence σ is easy to determine. The Moore–Penrose inverse of A can be obtained from the same decomposition in the following way:

$$A^+ = A^*(AA^*)^{-1} = Q \begin{pmatrix} (R^*)^{-1} \\ 0^* \end{pmatrix}. \tag{3.5}$$

Similar ideas apply if an LU decomposition is given:

$$PA^* = L \begin{pmatrix} U \\ 0^* \end{pmatrix}, \tag{3.6}$$

where L is a lower triangular $(N+1, N+1)$ matrix, U is an (N, N) upper triangular matrix, and P is a permutation matrix corresponding to partial pivoting which is, in general, necessary to improve the numerical stability. Let us first consider the calculation of $t(A)$. If y denotes the last column of $P^*(L^*)^{-1}$, then

$$t(A) = \sigma y/\|y\|, \quad \text{where } \sigma = \text{sign}\left(\det P \det L \det U \right). \tag{3.7}$$

The Moore–Penrose inverse is obtained by

$$A^+ = (I - t(A)t(A)^*)P^*(L^*)^{-1} \begin{pmatrix} (U^*)^{-1} \\ 0^* \end{pmatrix}. \tag{3.8}$$

Hence, a calculation of $w = A^+z$ amounts to essentially one forward-solving with U^*, one back-solving with L^*, and one scalar product with $t(A)$.

These methods are useful for small dense matrices A. However, in many applications of path following methods, the corresponding matrix A is large and sparse, and then this procedure is inefficient. Among such applications are the approximation of branches of nonlinear eigenvalue problems or the central path methods of linear and nonlinear programming. Let us point out some ideas which are useful in dealing with such situations.

In many applications, one encounters matrices A with the following structure:

$$A = (\; L \quad b \;), \tag{3.9}$$

where equations of the form $Lx = y$ permit a fast linear solver. If $(\; c^* \quad d \;)$ denotes an additional row (typically generated via the last predictor direction), then a standard block elimination may be employed via the Schur complement.

Lemma 3.3 Let

$$s = d - c^* L^{-1} b$$

denote the Schur complement of L in the augmented matrix

$$\tilde{A} = \begin{pmatrix} L & b \\ c^* & d \end{pmatrix}.$$

Then

$$\det \tilde{A} = \det L \det s. \tag{3.10}$$

Furthermore, if $\tilde{A}$ is nonsingular, then

$$\tilde{A}^{-1} = \begin{pmatrix} L^{-1} + L^{-1} b s^{-1} c^* L^{-1} & -L^{-1} b s^{-1} \\ -s^{-1} c^* L^{-1} & s^{-1} \end{pmatrix}.$$

As an easy consequence, the tangent $t(A)$ is obtained via

$$t(A) = \sigma y / \|y\|, \tag{3.11}$$

where y denotes the last column of $\tilde{A}^{-1}$. The sign $\sigma \in \{\pm 1\}$ can either be obtained from an angle test with the previous predictor direction or from (3.10), since it can be shown that

$$\sigma = \text{sign}\Big(\det L \det s \Big). \tag{3.12}$$

Note that the computational expense of determining $t(A)$ is roughly one application of the fast solver and a scalar product.

The Moore–Penrose inverse is obtained via

$$A^+ = (I - t(A) t(A)^*)(\tilde{A}^{-1})_N, \tag{3.13}$$

where $(\tilde{A}^{-1})_N$ denotes the submatrix consisting of the first N columns of $\tilde{A}^{-1}$. Hence, a calculation of $w = A^+ z$ amounts to essentially one additional call of the fast solver and two additional scalar products.

Among the fast solvers which are of importance here are direct solvers for sparse linear systems, or preconditioned iterative solvers such as conjugate-gradient or other Krylov methods, see, e.g., Freund, Golub and Nachtigal (1992).

This Schur complement construction is also valid if b, c and d are matrices (of appropriate size). This is of interest in parametric optimization, see Lundberg and Poore (1993). Watson (1986) and deSa, Irani, Ribbens, Watson and Walker (1992) discuss some numerical linear algebra aspects in the context of path following.

The popular bordering algorithm of Keller (1977), see also Chan (1984a), Keller (1983), Menzel and Schwetlick (1978, 1985), is related to these ideas. These approaches are akin to Keller's pseudo arclength method, in which the equation $H(v) = 0$ is extended by an additional parametrization condition $\mathcal{N}(u, v, h) = 0$ which is at least transversal to $H(v) = 0$ for small h, and often models an approximate arclength parametrization. This viewpoint is often convenient, in particular for structured problems.

3.3. Step length control and higher order predictors

The convergence considerations of Theorem 3.2 were carried out under the assumption that the steplength of the Algorithm 3.1 was uniformly constant throughout. This assumption is also typical for complexity studies, see Section 6. Such an approach is inefficient for any practical implementation. An efficient algorithm needs to incorporate an automatic strategy for controlling the steplength. In this respect the predictor–corrector methods are similar to the methods for numerically integrating initial value problems in ordinary differential equations. To some extent, the steplength strategy depends upon the accuracy with which it is desired to numerically trace a solution curve. Path following methods usually split into two categories:

— either the solution curve is to be approximated with some given accuracy, e.g. for plotting purposes; or
— the objective is just to safely follow the curve as fast as possible, until a certain point is reached, e.g. a zero point or critical point with respect to some additional functional defined on the curve.

We briefly sketch some ideas which are used to adjust the steplength.

Steplength control via error models. One method, due to Den Heijer and Rheinboldt (1981), is based upon an error model for the corrector iteration. For Newton corrector steps, such error models can be obtained by analysing the Newton–Kantorovich theory. The steplength is controlled by the number of steps which are taken in the corrector iteration until a given stopping criterion is fulfilled.

We sketch a somewhat modified and simplified version of this steplength strategy. Let us assume that u is a point on the solution curve, and consider, for simplicity, an Euler predictor $v_0(h) = u + ht(H'(u))$. Let $v_0(h)$, $v_1(h)$, ..., $v_k(h)$ be an iterative corrector process for approximating the nearest point to $v_0(h)$ on the curve. Suppose a certain stopping criterion is met after k iterations. The exact nature of the criterion is not important in this context. We assume theoretical convergence to $v_\infty(h)$.

It is assumed that there exists a constant $\gamma > 0$ (which is independent of h) such that the *modified error*

$$\varepsilon_i(h) := \gamma \|v_\infty(h) - v_i(h)\|$$

satisfies inequalities of the following type

$$\varepsilon_{i+1}(h) \leq \psi(\varepsilon_i(h)),$$

where $\psi : \mathbb{R} \to \mathbb{R}$ is a known monotone function such that $\psi(0) = 0$. For example, if Newton's method is employed, Den Heijer and Rheinboldt suggest two models:

$$\psi(\varepsilon) = \frac{\varepsilon^2}{3 - 2\varepsilon}, \quad 0 \leq \varepsilon \leq 1, \tag{3.14}$$

$$\psi(\varepsilon) = \frac{\varepsilon + \sqrt{10 - \varepsilon^2}}{5 - \varepsilon^2} \varepsilon^2, \quad 0 \leq \varepsilon \leq 1. \tag{3.15}$$

We may evaluate *a posteriori* the quotient

$$\omega(h) := \frac{\|v_k(h) - v_{k-1}(h)\|}{\|v_k(h) - v_0(h)\|} \approx \frac{\|v_\infty(h) - v_{k-1}(h)\|}{\|v_\infty(h) - v_0(h)\|} = \frac{\varepsilon_{k-1}(h)}{\varepsilon_0(h)}.$$

Using the estimate $\varepsilon_{k-1}(h) \leq \psi^{k-1}(\varepsilon_0(h))$, we obtain

$$\omega(h) \leq \frac{\psi^{k-1}(\varepsilon_0(h))}{\varepsilon_0(h)}.$$

This motivates taking the solution ε of the equation

$$\omega(h) = \frac{\psi^{k-1}(\varepsilon)}{\varepsilon}$$

as an estimate for $\varepsilon_0(h)$.

We now try to choose the steplength $\tilde{h}$ so that the corrector process satisfies the stopping criterion after a chosen number (say $\tilde{k}$) of iterations. Such a steplength leads to the modified error $\varepsilon_0(\tilde{h})$. Hence, we want the modified error $\varepsilon_{\tilde{k}}(\tilde{h})$ after $\tilde{k}$ iterations to be so small that the stopping criterion is satisfied. Using the inequality $\varepsilon_{\tilde{k}}(\tilde{h}) \leq \psi^{\tilde{k}}(\varepsilon_0(\tilde{h}))$, we accept the solution ε

of the equation

$$\psi^{\tilde{k}}(\varepsilon) = \psi^k(\varepsilon_0(h))$$

as an estimate for $\varepsilon_0(\tilde{h})$. Now we use the asymptotic expansion

$$\|v_\infty(h) - v_0(h)\| = Ch^2 + \mathcal{O}(h^3)$$

to obtain the approximation

$$\left(\frac{h}{\tilde{h}}\right)^2 \approx \frac{\varepsilon_0(h)}{\varepsilon_0(\tilde{h})},$$

which can be used to determine $\tilde{h}$. This steplength $\tilde{h}$ will now be used in the next predictor step. It is usually safeguarded by some additional considerations such as limiting the steplength to some interval $h_{\min} \le \tilde{h} \le h_{\max}$, or limiting the factor $0.5 \le h/\tilde{h} \le 2$, etc.

Steplength control via asymptotic expansion. Another method, based upon asymptotic estimates in the mentality of initial value solvers, is due to Georg (1983). The basic idea in this approach is to observe the performance of the corrector procedure and then to adapt the steplength $h > 0$ accordingly. More precisely, suppose that a point u on the solution curve has been approximated. Suppose further that a steplength $h > 0$ and a predictor point are given. Then a Newton-type iterative corrector process is performed which converges to the next point $z(h)$ on the curve.

The steplength strategy is motivated by the following question: Given the performance of the corrector process, which steplength $\tilde{h}$ would have been 'best' for obtaining $z(\tilde{h})$ from u? This 'ideal' steplength $\tilde{h}$ is determined via asymptotic estimates, and it is then taken as the steplength for the next predictor step. This strategy depends primarily upon two factors: the particular predictor–corrector method being utilized, and the criteria used in deciding what performance is considered 'best'.

Let us illustrate this technique in the case of the following algorithm (cf. Algorithm 3.1):

Algorithm 3.4 $u = \text{Euler\_Newton\_it}(u,\, h)$

> WHILE a stopping criterion is not met
> $\quad v = u + ht(H'(u))$ % *predictor step*
> $\quad A = H'(v)$
> $\quad$ WHILE a convergence criterion is not met
> $\quad\quad\quad v = v - A^+ H(v)$ % *corrector step*
> $\quad$ END
> $\quad u = v$
> $\quad$ choose a new steplength $h > 0$
> END

If $v(h) = u + ht(H'(u))$ denotes the predictor step depending on the steplength h, then the first corrector point is given by

$$w(h) := v(h) - H'(v(h))^+ H(v(h)).$$

Let us call the quotient of the first two successive Newton steps

$$\kappa(u, h) := \frac{\|H'(v(h))^+ H(w(h))\|}{\|H'(v(h))^+ H(v(h))\|}$$

the *contraction rate* of the corrector process. Since Newton's method is locally quadratically convergent, it is plain that $\kappa(u, h)$ will decrease (and hence Newton's method will become faster) as h decreases. The following lemma characterizes the asymptotic behaviour of $\kappa(u, h)$ with respect to h, see (6.1.2) in Allgower and Georg (1990).

Lemma 3.5 Suppose that

$$H''(u)[t(H'(u)), t(H'(u))] \neq 0$$

(i.e. the curve has nonzero curvature at u), then

$$\kappa(u, h) = \kappa_2(u)h^2 + O(h^3)$$

for some constant $\kappa_2(u) \geq 0$ which is independent of h and depends smoothly on u.

In view of this asymptotic relation, the steplength modification $h \to \tilde{h}$ is now easy to explain. Assume that an Euler–Newton step has been performed with steplength h. Then $H'(v(h))^+ H(v(h))$ and $H'(v(h))^+ H(w(h))$ will have been calculated and thus $\kappa(u, h)$ can be obtained without any significant additional cost. Now an *a posteriori* estimate

$$\kappa_2(u) = \frac{\kappa(u, h)}{h^2} + O(h)$$

is available.

In order to have a robust and efficient method we want to continually adapt the steplength h so that a nominal prescribed contraction rate $\tilde{\kappa}$ is maintained. The choice of $\tilde{\kappa}$ will generally depend upon the nature of the problem at hand, and on the desired security with which we want to traverse the curve. That is, the smaller $\tilde{\kappa}$ is chosen, the greater will be the security with which the method will follow the curve. When using the term *securely* or *safely* following the curve we mean that a safeguard prevents the method from jumping to a different part of the curve (at a significantly different arclength value) or to a different connected component of $H^{-1}(0)$. Depending on the structure of the solution manifold $H^{-1}(0)$, this may be an important issue.

Once $\tilde{\kappa}$ has been chosen, we will consider a steplength $\tilde{h}$ to be appropriate if $\kappa(u, \tilde{h}) \approx \tilde{\kappa}$. By using the above equation and neglecting higher order terms we obtain the formula

$$\tilde{h} = h \sqrt{\frac{\tilde{\kappa}}{\kappa(u, h)}}$$

as the steplength for the next predictor step.

In a similar way, other quantities which are important for the performance of the path following method can be taken into account, e.g. the angle of two successive predictor directions, the size of the first Newton step (which gives an approximation of the distance of the predictor point to the curve) or the function value $H(v(h))$. All these quantities admit asymptotic expansions in h (with varying order). For example, Algorithm 6.1.10 and Program 1 in Allgower and Georg (1990) incorporates such features in the steplength strategy.

Kearfott (1989) proposes interval arithmetic techniques to determine a first order predictor which stresses secure path following, see also Kearfott (1990).

The steplength strategies we have discussed up to now have been based upon the Euler predictor, which is only of local order two. This is very often satisfactory since it is usually used in conjunction with rapidly converging correctors such as Newton-type correctors. However, for large systems, often less rapidly convergent iterative methods such as conjugate gradient steps are used. Hence, at least in some cases, one may expect to obtain improved efficiency by using variable order predictors and formulating corresponding steplength strategies. Such strategies could be similar to the ones used in multistep methods for solving initial value problems, see, e.g., Shampine and Gordon (1975). Georg (1982), suggested such a method, see also Georg (1983). Lundberg and Poore (1991) have made an implementation using variable order Adams–Bashforth predictors. Their numerical results show that there is often a definite benefit to be derived by using higher order predictors.

Inexpensive higher order predictors are generally based on polynomial interpolation. In view of the stability of Newton's method as a corrector, it may be advantageous to use more stable predictors. Mackens (1989) has proposed such predictors which are based on Taylor's formula and which are obtained by successive numerical differentiation in a clever way, see also Schwetlick and Cleve (1987) as a predecessor. However, the gain in stability has to be paid for by additional evaluations of the map H and additional applications of the Moore–Penrose inverse of the Jacobian H' (where it may be assumed that H' has already been decomposed).

Let us sketch a general philosophy for higher order predictors which may be useful for implementations. Let u be a point on the solution curve c such that $c(s) = u$. Consider a polynomial predictor of the form

$$c(s + h) \approx p_k(h) = u + \sum_{i=1}^{k} c_i h^i, \tag{3.16}$$

$$c_i \approx \frac{c^{(i)}}{i!}, \tag{3.17}$$

which represents an approximation of the Taylor formula. We see essentially two different ways for obtaining the coefficients c_i: (1) by divided differences or polynomial interpolation making use of previously calculated points on the curve; and (2) by successive numerical differentiation at u. The former is less expensive to calculate, but the latter is more accurate.

We sketch one possible way of determining the next steplength and the next order in the predictor. Let $\varepsilon > 0$ be a given tolerance. The term $\|c_k\| h^k$ can be viewed as a rough estimate for the truncation error of the predictor $p_{k-1}(h)$. Hence, we estimate

$$h_k = \left(\frac{\varepsilon}{\|c_k\|} \right)^{1/k}$$

as the steplength for the predictor p_{k-1} in order to remain within the given tolerance. Due to instabilities of various kinds, we anticipate that

$$h_2 < h_3 \cdots < h_q \geq h_{q+1}$$

will hold for some q. Hence, the predictor p_{q-1} with steplength h_q is our next choice.

This idea can be implemented and modified in various ways, and needs some stabilizing safeguards, such as setting a maximum increase in steplength and in the order. The strategy to be developed depends on the objective of the application at hand.

4. Applications

In this section we present a selection of applications of path following methods. Many more specific examples exist in the literature, some of them are referred to later. Our discussion of applications concentrates to a large extent on cases in which the predictor–corrector methods apply. Applications in which the dimension is relatively low and smoothness does not hold can be handled by the piecewise-linear methods discussed in Section 5.

In many applications of the numerical homotopy methods, it is possible to avoid degeneracies in the solution curve by introducing suitable parameters (perturbations). The theoretical basis of this approach lies in Sard's theorem for maps with additional parameters, see, e.g., Abraham and Robbin (1967) or Hirsch (1976). Yomdin (1990) has given a version of Sard's theorem which is adapted for numerical purposes. We consider the following general form:

Theorem 4.1. (Sard) Let A, B, C be smooth manifolds of finite dimensions with $\dim A \geq \dim C$, and let $F : A \times B \to C$ be a smooth map. Assume that $c \in C$ is a regular value of F, i.e. for $F(a, b) = c$ we have that the total derivative $F'(a, b) : T_a A \times T_b B \to T_c C$ has maximal rank. Here $T_a A$ denotes the tangent space of A at a, etc. Then for almost all $b \in B$ (in the sense of some Lebesgue measure on B) the restricted map $F(\cdot, b) : A \to C$ has c as a regular value.

4.1. Fixed point problems

To illustrate the use of Sard's theorem, let us consider a homotopy arising from a fixed point problem. Let $f : \mathbb{R}^N \to \mathbb{R}^N$ be a smooth map which is bounded. According to the theorem of Brouwer (1912), the map f has at least one fixed point. To simplify the discussion, let us make the assumption that the map $x \mapsto x - f(x)$ has zero as a regular value. This implies that the fixed points of f are isolated, and that Newton's method converges locally. However, the global convergence of Newton's method is by no means guaranteed.

We therefore consider the homotopy

$$H(x, \lambda, p) = x - p - \lambda(f(x) - p). \tag{4.1}$$

For the *trivial level* $\lambda = 0$, we obtain the *trivial map* $H(x, 0, p) = x - p$ which has the unique zero point p, our *starting point*. On the *target level* $\lambda = 1$, we obtain the *target map* $H(x, 1, p) = x - f(x)$ whose zero points are our points of interest, i.e. the fixed points of f.

Let us illustrate by this example how Sard's theorem is typically employed: The Jacobian of H is given by

$$H'(x, \lambda, p) = (\mathrm{Id} - \lambda f'(x), p - f(x), (\lambda - 1)\mathrm{Id}).$$

The first N columns of the Jacobian are linearly independent for $H(x, \lambda, p) = 0$ and $\lambda = 1$ due to our assumptions, and clearly the last N columns are linearly independent for $\lambda \neq 1$. Consequently, by Sard's theorem we can conclude that for almost all $p \in \mathbb{R}^N$ (in the sense of N-dimensional Lebesgue measure) zero is a regular value of the restricted map $H(\cdot, \cdot, p)$.

For such a generic choice of p, the solution manifold $H(\cdot, \cdot, p)^{-1}(0)$ consists of smooth curves which are either diffeomorphic to the circle or to the real line, see Lemma 2.4. Consider the solution curve $c(s) = (x(s), \lambda(s))$ (parametrized for convenience with respect to arclength) such that $c(0) = (p, 0)$. It is easy to see that the initial tangent vector in the direction of increasing λ has the form

$$\dot{c}(0) = (1 + \|f(p) - p\|^2)^{-1/2} \begin{pmatrix} f(p) - p \\ 1 \end{pmatrix},$$

and hence the curve is transversal to the plane $\lambda = 0$.

Since the solution point $(p, 0)$ is unique for $\lambda = 0$, it follows that c is diffeomorphic to the real line. Furthermore, the boundedness of f implies that $x(s)$ is bounded for $0 \leq \lambda(s) \leq 1$. It follows that the curve c reaches the level $\lambda = 1$ after a finite arclength s_0, i.e. $c(s_0) = (x_0, 1)$, and hence x_0 is a fixed point of f which can be approximated by tracing the curve c.

Let us note that

$$(\text{Id} - f'(x_0))\dot{x}(s_0) = \dot{\lambda}(s_0)(f(x_0) - p),$$

and our earlier assumption on f implies that $(\text{Id} - f'(x_0))$ cannot have a nontrivial kernel, and hence $\dot{\lambda}(s_0) \neq 0$, i.e. the curve c is tranversal to the level $\lambda = 1$ at any solution.

This discussion is in the spirit of Chow, Mallet-Paret and Yorke (1978). An earlier approach based on the nonretraction principle of Hirsch (1963) was given by Kellogg, Li and Yorke (1976). General discussions concerning the correspondence between degree arguments and numerical continuation algorithms have been given in Alexander and Yorke (1978), Garcia and Zangwill (1979a, 1981) and Peitgen (1982). Since the appearance of the constructive proofs of the Brouwer fixed point theorem many other constructive existence proofs have been described. Further references may be found in Section 11.1 of Allgower and Georg (1990).

Watson and collaborators have given a great number of engineering applications where an implementation (HOMPACK) of this homotopy method has been employed. As examples, we mention Arun, Reinholtz and Watson (1990), Melville, Trajkovic, Fang and Watson (1990), Vasudevan, Lutze and Watson (1990), Watson (1981), Watson, Li and Wang (1978), Watson and Wang (1981) and Watson and Yang (1980).

4.2. Global Newton methods

Newton's method is a popular method for numerically calculating a zero point of a smooth map $G : \mathbb{R}^N \to \mathbb{R}^N$. As is well known, this method may diverge if the starting point p is not sufficiently near to a zero point $\bar{x}$ of G. Often one would like to determine whether a certain open bounded region $\Omega \subset \mathbb{R}^N$ contains a zero point $\bar{x}$ of G and furthermore, for which starting values p this solution $\bar{x}$ can be obtained by Newton's method. The so-called global Newton methods offer a possibility of answering such questions.

One may interpret Newton's method as the numerical integration of the differential equation

$$\dot{x} = -G'(x)^{-1}G(x)$$

using Euler's method with unit step size. The idea of using this flow to find zero points of G was exploited by Branin (1972). Smale (1976) gave conditions on $\partial \Omega$ under which the flow leads to a zero point of G in Ω. Such numerical methods have been referred to as *global Newton methods*.

Keller (1978) observed that this flow can also be obtained in a numerically stable way from a homotopy equation which he consequently named the *global homotopy* method. Independently, Garcia and Gould (1978, 1980) discussed this flow.

We briefly sketch Keller's approach. The global homotopy method involves tracing the curve defined by the equation $G(x) - (1 - \lambda)G(p) = 0$ starting from $(x, \lambda) = (p, 0) \in \partial\Omega \times \{0\}$ inward into $\Omega \times \mathbb{R}$. If the level $\Omega \times \{1\}$ is encountered, then a zero point of G has been found.

We consider Smale's assumption.

Assumption 4.2 Let the following conditions be satisfied:

1. $\Omega \subset \mathbb{R}^N$ is open and bounded and $\partial\Omega$ is a connected smooth submanifold of $\mathbb{R}^N$;
2. zero is a regular value of G;
3. $G(p) \neq 0$ for $p \in \partial\Omega$;
4. the Jacobian $G'(p)$ is nonsingular for $p \in \partial\Omega$;
5. the Newton direction $-G'(p)^{-1}G(p)$ is not tangent to $\partial\Omega$ at p.

The *global homotopy* $H : \mathbb{R}^N \times \mathbb{R} \times \partial\Omega \longrightarrow \mathbb{R}^N$ is defined by

$$H(x, \lambda, p) := G(x) - (1 - \lambda)G(p).$$

Since p varies over the $(N - 1)$-dimensional surface $\partial\Omega$, it is somewhat difficult to apply Sard's theorem. This task was achieved by Percell (1980). Hence, for almost all $p \in \partial\Omega$ the global homotopy has 0 as a regular value.

Let p be such a generic choice. We consider again the solution curve $c(s) = (x(s), \lambda(s))$ in $H(\cdot, \cdot, p)^{-1}(0)$ such that $c(0) = (p, 0)$ and $\dot{x}(0)$ points into Ω. Keller (1978) showed that the curve hits the target level $\Omega \times \{1\}$ in an odd number of points. This possibility of obtaining more than one solution was first observed by Branin and Hoo (1972).

Given the conditions 1 and 2 of assumption 4.2, the boundary conditions 3–5 can be shown to hold for a sufficiently small ball Ω around a zero point of G. Thus, in a certain sense the global homotopy extends the well known Newton–Kantorovich-type theorems concerning the local convergence of Newton's method, see, e.g., Ortega and Rheinboldt (1970).

4.3. Multiple solutions

In the previous section it was observed that the global homotopy method might actually yield more than one zero point of the map G in a bounded region Ω. This raises the question as to whether one might be able to compute more zero points of G in Ω in addition to those which lie on the global homotopy path. To be more precise, let us suppose that $\Omega \subset \mathbb{R}^N$ is an open bounded region, and that $G : \mathbb{R}^N \to \mathbb{R}^N$ is a smooth map having a zero point $z_0 \in \Omega$. The task is now to find additional zero points of G in Ω, provided they exist. One method which has often been used for handling this problem is deflation, see, e.g., Brown and Gearhart (1971). In this method *a deflated map* $G_1 : \mathbb{R}^N \setminus \{z_0\} \to \mathbb{R}^N$ is defined by $G_1(x) = G(x)/\|x - z_0\|$. One then applies an iterative method to try to find a zero point of G_1. Numerical experience with deflation has shown that it is often a matter of seeming chance whether one obtains an additional solution and if one is obtained, it is very often not the one which is nearest to z_0.

By utilizing homotopy-type methods we can give some conditions which will guarantee the existence of an additional solution and yield insights into the behaviour of deflation. This additional solution will lie on a homotopy path. We illustrate this approach with a discussion of the *d-homotopy*. Let us consider the homotopy map $H_d : \mathbb{R}^N \times \mathbb{R} \to \mathbb{R}$ defined by

$$H_d(x, \lambda) := G(x) - \lambda d$$

where $d \in \mathbb{R}^N$ is some fixed vector with $d \neq 0$. Since we assume that a zero point z_0 is already given, we have $H_d(z_0, 0) = 0$. Let us further assume zero is a regular value of G. Then it follows from Sard's theorem that zero is also a regular value of H_d for almost all $d \in \mathbb{R}^N$. In order to ensure that the solution curve c in $H_d^{-1}(0)$ which contains $(z_0, 0)$ again reaches the level $\lambda = 0$, we need to impose a boundary condition. The following proposition uses a boundary condition which is motivated by a simple degree consideration.

Proposition 4.3 Let the following hypotheses hold:

1. $G : \mathbb{R}^N \to \mathbb{R}^N$ is a smooth map with zero as a regular value;
2. $d \in \mathbb{R}^N \setminus \{0\}$ is a point such that the homotopy H_d also has zero as a regular value;
3. $\Omega \subset \mathbb{R}^N$ is a bounded open set which contains a (known) initial zero point z_0 of G;
4. the *boundary condition* $H_d(x, \lambda) = G(x) - \lambda d \neq 0$ holds for all $x \in \partial\Omega$, $\lambda \in \mathbb{R}$;

Then the curve c in $H_d^{-1}(0)$ which contains $(z_0, 0)$ intersects the level $\Omega \times \{0\}$ an even number of times at points $(z_i, 0)$, $i = 0, \ldots, n$, at which $G(z_i) = 0$.

See (11.5.3) in Allgower and Georg (1990) for a proof. Any two zero points of G which are consecutively obtained by traversing the curve c have opposite index. Allgower and Georg (1983b) have shown that this d-homotopy can be viewed as a continuous version of the deflation technique of Brown and Gearhart.

4.4. Polynomial systems

In the preceding section we considered the task of computing multiple zero points of general smooth maps. In the case of complex polynomial systems it is actually possible to compute (at least in principle) all of the zero points by means of homotopy methods. This subject has received considerable attention in recent years. The book of Morgan (1987) deals exclusively with this topic, using the path following approach. It also contains a number of interesting applications to robotics and other fields.

We consider a system of complex polynomials $P : \mathbb{C}^n \to \mathbb{C}^n$. The task is to find *all* solutions of the equation $P(z) = 0$. If a term of the kth component P_k of P has the form

$$az_1^{r_1} z_2^{r_2} \cdots z_n^{r_n},$$

then its degree is $r_1 + r_2 + \ldots + r_n$. The degree d_k of P_k is the maximum of the degrees of its terms. The *homogeneous part* $\hat{P}$ of P is obtained by deleting in each component P_k all terms having degree less than d_k. The *homogenization* $\tilde{P}$ of P is obtained by multiplying each term of each component P_k with an appropriate power z_0^r such that its degree is d_k. Note that the homogenization $\tilde{P} : \mathbb{C}^{n+1} \to \mathbb{C}^n$ involves one more variable z_0. If

$$(w_0, \ldots, w_n) \neq 0$$

is a zero point of $\tilde{P}$, then the entire ray

$$[w_0 : \cdots : w_n] := \{(\xi w_0, \ldots, \xi w_n) \mid \xi \in \mathbb{C}\}$$

consists of zero points of $\tilde{P}$. Usually, $[w_0 : \cdots : w_n]$ is regarded as a point in the complex projective space $\mathbb{CP}^n$. There are two cases to consider:

1   The solution $[w_0 : \cdots : w_n]$ intersects the hyperplane $z_0 = 0$ transversely, i.e. without loss of generality, $w_0 = 1$. This corresponds to a zero point $(w_1, \ldots, w_n)$ of P. Conversely, each zero point $(w_1, \ldots, w_n)$ of P corresponds to a solution $[1 : w_1 : \cdots : w_n]$ of $\tilde{P}$.

2   The solution $[w_0 : \cdots : w_n]$ lies in the hyperplane $z_0 = 0$, i.e. $w_0 = 0$. This corresponds to a *nontrivial* solution $[w_1 : \cdots : w_n]$ of the homogeneous part $\hat{P}$, and such solutions are called *zero points of P at infinity*.

As in the case of one variable, it is possible to define the multiplicity of a solution. The higher dimensional analogue of the fundamental theorem of algebra is Bezout's theorem, which states that the number of zero points of P (counting their multiplicities and zeros at infinity) equals the product $d = d_1 d_2 \cdots d_n$, provided all solutions are isolated.

Garcia and Zangwill (1979b) and Chow, Mallet-Paret and Yorke (1979) introduced homotopy methods in $\mathbb{C}^n \times \mathbb{R}$ for finding all solutions of the equation $P = 0$. Wright (1985) realized that their approaches could be simplified by going into the complex projective space $\mathbb{CP}^n$. We use his approach to illustrate the homotopy idea for polynomial systems.

Define a homotopy $H = (H_1, \ldots, H_n)$ by involving the homogenization $\tilde{P}$

of P via

$$H_k(z_0, \ldots, z_n, \lambda) = (1 - \lambda)(a_k z_k^{d_k} - b_k z_0^{d_k}) + \lambda \tilde{P}_k(z_0, \ldots, z_n).$$

Wright shows by Sard-type arguments that for almost all coefficients $a_k, b_k \in \mathbb{C}$ the restricted homotopies $H^{(j)}$ which are obtained from H by fixing $z_j = 1$ for $j = 0, \ldots, n$ have zero as a regular value for $\lambda < 1$. He concludes that for $\lambda < 1$, the homogeneous system of polynomials H has exactly d simple zero point curves $c_i(\lambda) \in \mathbb{CP}^n$, $i = 1, \ldots, d$, in complex projective n-space. On the trivial level $\lambda = 0$, the d solutions are obvious, and it is possible to trace the d curves emanating from these solutions into the direction of increasing λ. The solution curves are monotone in λ, and hence all have to reach the target level $\lambda = 1$ on the compact manifold $\mathbb{CP}^n$. Thus, in this approach solutions at infinity are treated no differently than finite solutions. The solution curves are traced in the projective space $\mathbb{CP}^n$, and from the numerical point of view we have the slight drawback that occasionally a chart in $\mathbb{CP}^n$ has to be switched.

Recently, attention has been given to the task of trying to formulate homotopies which eliminate the sometimes wasteful effort involved in tracing paths which go to solutions of $P(z_1, \ldots, z_n) = 0$ at infinity. Work in this direction has been done in Morgan (1986), Li, Sauer and Yorke (1987, 1989) and Li and Wang (1992a,b). Morgan and Sommese (1987) describe the easily implemented 'projective transformation' which allows the user to avoid the drawback of changing coordinate charts on $\mathbb{CP}^n$. Morgan and Sommese (1989) show how to exploit relations among the system coefficients, via 'coefficient parameter continuation'. Such relations occur commonly in engineering problems, as described in Wampler and Morgan (1991), Wampler, Morgan and Sommese (1990, 1992). The papers (Morgan, Sommese and Wampler, 1991–1992) combine a homotopy method with contour integrals to calculate singular solutions to polynomial and nonlinear analytic systems. Morgan, Sommese and Watson (1989) documented that HOMPACK, see Watson, Billups and Morgan (1987), in the case of polynomial systems has some stability issues that CONSOL8, see Morgan (1987), does not have. The path following approach to systems of polynomial equations is particularly suited for parallel processing, see Allison, Harimoto and Watson (1989).

4.5. Nonlinear eigenvalue problems, bifurcation

Path following methods are frequently applied in numerical studies of bifurcation problems. Up to this point we have assumed that zero is a regular value of the smooth mapping $H : \mathbb{R}^{N+1} \to \mathbb{R}^N$. However, bifurcation points are singular points on $H^{-1}(0)$ and hence, if path following algorithms are applied, some special adaptations are required. Generally, bifurcation points are defined in a Banach space context, see for example the book by Chow and Hale (1982). In the case that H represents a mapping arising from a discretization of an operator of the form $\mathcal{H} : E_1 \times \mathbb{R} \to E_2$ where E_1 and E_2 represent appropriate Banach spaces, it is usually of interest to approximate bifurcation points of the operator equation $\mathcal{H} = 0$. Often one can make the discretization H in such a way that the resulting discretized equation $H = 0$ also has a corresponding bifurcation point. Under reasonable assumptions of nondegeneracy it is possible to obtain error estimates for the bifurcation point of the original problem $\mathcal{H} = 0$. Such studies are presented in the papers by Brezzi, Rappaz and Raviart (1980a,b, 1981), Crouzeix and Rappaz (1990), Fink and Rheinboldt (1983, 1984, 1985) and Liu and Rheinboldt (1991).

Since we are primarily concerned with bifurcation in the numerical curve following context, we confine our discussion to the case of the finite dimensional (discretized) equation $H = 0$. However, we note that the theoretical discussion later will essentially extend to the Banach space context if we assume that H is a Fredholm operator of index one. We will discuss how certain types of bifurcation points along a solution curve c can be detected, and having detected a bifurcation point, how one can numerically switch from c onto a bifurcating branch.

Some of the fundamental results on the numerical solution of bifurcation problems are due to Keller (1970), see also Keener and Keller (1974) and Keller (1977). The recent literature on the numerical treatment of bifurcation is very extensive. For an introduction into the field we suggest the lecture notes of Keller (1987). See also the two articles by Doedel, Keller and Kernévez (1991a,b) which discuss the use of the software package AUTO. For surveys and bibliography we suggest the recent book by Seydel (1988) and the recent proceedings (Mittelman and Roose, 1989; Roose, de Dier and Spence, 1990; Seydel, Schneider, Küpper and Troger, 1991). Most authors study bifurcation problems in the context of a nonlinear eigenvalue problem

$$H(x, \lambda) = 0,$$

where λ is the eigenvalue parameter which usually has some physical significance. Conventionally, the solution branches are parametrized according to λ. We have taken the viewpoint that the solution branches c_i are parametrized with respect to the arclength. There is only one essential difference, namely that the former approach also considers folds with respect to λ as singularities.

Such folds are frequently of intrinsic interest, and there are special algorithms for detecting and calculating them. We omit this subject here for reasons of space limitations, and refer the interested reader to, e.g., Bolstad and Keller (1986), Chan (1984b), Fink and Rheinboldt (1986, 1987), Melhem and Rheinboldt (1982), Pönisch and Schwetlick (1981), Schwetlick (1984ab) and Ushida and Chua (1984).

A standard approach to the determination of bifurcation or other singular points is to directly characterize such points by adjoining additional equations to $H = 0$ and handling the resulting new set of equations by some special iterative method. In this context, continuation methods often are used to obtain starting points for these direct methods, see, e.g., Griewank (1985), Moore and Spence (1980) and Yang and Keller (1986). A hybrid method for handling unstable branches has been developed by Shroff and Keller (1991).

Mittelmann and collaborators have made extensive applications of path following and bifurcation methods in the context of minimal surfaces, free boundary problems, obstacle problems and variational inequalities, see, e.g., Hornung and Mittelmann (1991), Maurer and Mittelmann (1991), Miersemann and Mittelmann (1989–1992) and Mittelmann (1990).

In view of the extensive literature we can only touch upon the problem here, and we will confine our discussion to the task of detecting a simple bifurcation point along a solution curve c and effecting a branch switching numerically. We will see that the detection of simple bifurcation points requires only minor modifications of predictor–corrector algorithms. A more detailed discussion along these lines can be found in Chapter 8 of Allgower and Georg (1990). Let us begin by defining a bifurcation point.

Definition 4.4 Suppose that $c : J \to \mathbb{R}^{N+1}$ is a smooth curve, defined on an open interval J containing zero, and parametrized (for reasons of simplicity) with respect to arc length such that $H(c(s)) = 0$ for $s \in J$. The point $c(0)$ is called a *bifurcation point* of the equation $H = 0$ if there exists an $\varepsilon > 0$ such that every neighbourhood of $c(0)$ contains zero points z of H which are not on $c(-\varepsilon, \varepsilon)$.

An immediate consequence of this definition is that a bifurcation point of $H = 0$ must be a singular point of H. Hence the Jacobian $H'(c(0))$ must have a kernel of dimension at least two. We consider the simplest case:

Definition 4.5 A point $\bar{u} \in \mathbb{R}^{N+1}$ is called a *simple bifurcation point* of the equation $H = 0$ if the following conditions hold:

1. $H(\bar{u}) = 0$;
2. $\dim \ker H'(\bar{u}) = 2$;
3. $e^* H''(\bar{u})\big|_{(\ker H'(\bar{u}))^2}$ has one positive and one negative eigenvalue. where e spans $\ker H'(\bar{u})^*$.

Using the well known Liapunov–Schmidt reduction, the following theorem can be shown, which is essentially a restatement of a famous result from Crandall and Rabinowitz (1971).

Theorem 4.6 Let $\bar{u} \in \mathbb{R}^{N+1}$ be a simple bifurcation point of the equation $H = 0$. Then there exist two smooth curves $c_1(s), c_2(s) \in \mathbb{R}^{N+1}$, parametrized with respect to arclength s, defined for $s \in (-\varepsilon, \varepsilon)$ and ε sufficiently small, such that the following holds:

1. $H(c_i(s)) = 0$, $i \in \{1, 2\}$, $s \in (-\varepsilon, \varepsilon)$,;
2. $c_i(0) = \bar{u}$, $i \in \{1, 2\}$,;
3. $\dot{c}_1(0)$, $\dot{c}_2(0)$ are linearly independent;
4. $H^{-1}(0)$ coincides locally with range $(c_1) \cup$ range (c_2), more precisely: $\bar{u}$ is not in the closure of $H^{-1}(0) \setminus (\text{range}(c_1) \cup \text{range}(c_2))$.

By differentiating the equation $e^* H(c_i(s)) = 0$ twice and evaluating the result at $s = 0$, we obtain the following

Lemma 4.7 Let $\bar{u} \in \mathbb{R}^{N+1}$ be a simple bifurcation point of the equation $H = 0$. Then

1. $\ker H'(\bar{u}) = \text{span}\{\dot{c}_1(0), \dot{c}_2(0)\}$,
2. $e^* H''(\bar{u})[\dot{c}_i(0), \dot{c}_i(0)] = 0$ for $i \in \{1, 2\}$.

The following theorem reflects the well known fact, see Krasnosel'skiĭ (1964) or Rabinowitz (1971), that simple bifurcation points cause a switch of orientation along the solution branches. This furnishes a numerically implementable criterion for detecting a simple bifurcation point when traversing one of the curves c_i. For a proof, see, e.g., Theorem (8.1.14) in Allgower and Georg (1990).

Theorem 4.8 Let $\bar{u} \in \mathbb{R}^{N+1}$ be a simple bifurcation point of the equation $H = 0$. Then the determinant of the following augmented Jacobian

$$\det \begin{pmatrix} H'(c_i(s)) \\ \dot{c}_i(s)^* \end{pmatrix}$$

changes sign at $s = 0$ for $i \in \{1, 2\}$.

This theorem implies that when traversing a solution curve c, a simple bifurcation point is detected by a change in orientation. Depending upon the method used to perform the decomposition of the Jacobian during path following, this orientation can often be calculated at very small additional cost. A predictor–corrector algorithm generally has no difficulty in *jumping over*, i.e. proceeding beyond the bifurcation point $\bar{u}$. That is, Keller (1977) has shown that for sufficiently small steplength h, the predictor point will fall into the 'cone of attraction' of the Newton corrector. See Jepson and Decker (1986) for further studies.

Conversely, suppose that a smooth c in $H^{-1}(0)$ is traversed and that $c(0)$ is an isolated singular point of H such that the determinant changes sign at $s = 0$, then using a standard argument in degree theory, see Krasnosel'skiĭ (1964) or Rabinowitz (1971), it can be shown that $c(0)$ is a bifurcation point of $H = 0$. However, $c(0)$ is not necessarily a simple bifurcation point.

Multiple bifurcations often arise from symmetries with respect to certain group actions, i.e. H satisfies an equivariance condition

$$H(\gamma x, \lambda) = \gamma H(x, \lambda)$$

for γ in a group Γ. See the books by Golubitsky and Schaeffer (1985), Golubitsky, Stewart and Schaeffer (1988) and Vanderbauwhede (1982). These symmetries can also be exploited numerically, see, e.g., , Allgower, Böhmer and Mei (1991a,b), Allgower, Böhmer, Georg and Miranda (1992b), Cliffe and Winters (1986), Dellnitz and Werner (1989), Georg and Miranda (1990, 1992), Jepson, Spence and Cliffe (1991), Healey (1988–1989), Healey and Treacy (1991) and Hong (1991); see also the proceedings (Allgower, Böhmer and Golubitsky, 1992a). As this partial list suggests, there is currently very much interest in this topic. However, constraints on our available space prohibits a detailed discussion.

The determinant in Theorem 4.8 is only the simplest example of a so-called *test function*. Such test functions are real functions defined on a neighbourhood of the curve c and are monitored during path following to reveal certain types of singular points by a change of sign. In the case of Hopf bifurcation, the determinant is not an adequate test function. Recently, several authors have proposed and studied classes of test functions for various types of singular points, see, e.g., Dai and Rheinboldt (1990), Garratt, Moore and Spence (1991), Griewank and Reddien (1984), Seydel (1991b) and Werner (1992). A different approach for the prediction of singular points along the path c has been given by Huitfieldt and Ruhe (1990).

Switching branches via perturbation. In the previous section we have seen that it is possible to detect and jump over simple bifurcation points while numerically tracing a solution curve c via a predictor–corrector method. The more difficult task is to numerically branch off onto the second solution curve at the detected bifurcation point $\bar{u}$. The simplest device for branching off numerically rests upon Sard's theorem (4.1). If a small perturbation vector $d \in \mathbb{R}^N$ is chosen at random, then the probability that d is a regular value of H is unity. Of course, in this case $H^{-1}(d)$ has no bifurcation point. Since $d \in \mathbb{R}^N$ is chosen so that $\|d\|$ is small, the solution sets $H^{-1}(0)$ and $H^{-1}(d)$ are close together. On $H^{-1}(d)$, no change of orientation can occur. Therefore, corresponding solution curves in $H^{-1}(d)$ must branch off near the bifurcation point $\bar{u}$. It is easy to implement this idea, see, e.g., Allgower and Chien (1986), Allgower, Chien, Georg and Wang (1991c), Chien (1989), Georg (1981) and Glowinski, Keller and Reinhart (1985).

Recently, an interesting variation on this idea has been proposed by Huitfieldt (1991). He introduces an additional parameter on the perturbation and an additional constraint equation to obtain the *branch connecting equa-*

tion

$$B(u, \tau) := \left(\begin{array}{c} H(u) + \tau d \\ \|u - \hat{u}\|^2 + \tau^2 - \varepsilon^2 \end{array} \right) = 0, \tag{4.2}$$

where $\hat{u}$ is an approximation to the bifurcation point $\bar{u}$. Such approximations are easily obtained via path following together with test function monitoring as described earlier. Note the relationship between this homotopy and the d-homotopy discussed in Section 4.3 in connection with finding multiple solutions.

It is not difficult to see that for almost all d and $\varepsilon > 0$, zero is a regular value of B, provided that $\bar{u}$ is an isolated singular point of H in $H^{-1}(0)$. Let us assume that such a generic choice of d and ε has been made.

Then the solution manifold $B^{-1}(0)$ splits into one or more simple closed curves of the form $(b(s), \tau(s))$. For $\tau(s) = 0$ we obtain $H(b(s)) = 0$. Hence the curves connect points in the intersection of $H^{-1}(0)$ with the sphere $\|u - \hat{u}\|^2 = \varepsilon^2$. Starting points for a path following of $(b(s), \tau(s))$ are available from the tracing of the current solution curve c of $H = 0$. Let $b_i = b(s_i)$, $i = 0, 1, \ldots$, be successively obtained points such that $\tau(s_i) = 0$. It remains to be demonstrated that b_i and b_{i+1} are on different solution branches of the equation $H = 0$.

Since this seems to have been omitted in the paper of Huitfieldt (1991), we sketch a proof. It is easily seen that the determinant of the matrix

$$\left(\begin{array}{cc} H'(b(s)) & d \\ (b(s) - \hat{u})^* & \tau(s) \\ \dot{b}(s)^* & \dot{\tau}(s) \end{array} \right)$$

never changes sign since it never becomes singular. By multiplying this matrix on the right with

$$\left(\begin{array}{cc} \mathrm{Id} & \dot{b}(s) \\ 0^* & \dot{\tau}(s) \end{array} \right)$$

we obtain

$$\left(\begin{array}{ccc} H'(b(s)) & 0 \\ (b(s) - \hat{u})^* & 0 \\ \dot{b}(s)^* & 1 \end{array} \right).$$

Since $\dot{\tau}(s_i)$ changes sign for successive i, we obtain that the determinant of

$$\left(\begin{array}{c} H'(b_i) \\ (b_i - \hat{u})^* \end{array} \right)$$

changes sign for successive i. Under reasonable assumptions this implies that $t(H'(b_{i+1}))$ points out of the sphere $\|u - \hat{u}\|^2 = \varepsilon^2$ if $t(H'(b_i))$ points into it. For a simple bifurcation point (or more generally for a bifurcation point which is detected by a change of determinant in the sense of Theorem 4.8), this means that b_i and b_{i+1} cannot lie on the same solution branch.

Huitfeldt reports very successful numerical tests on some interesting problems of applied mathematics: the Taylor problem, and the von Karman plate equations. In his experiments he succeeded in obtaining all of the bifurcating branches at several multiple bifurcation points, i.e. the 1-manifold $\mathcal{B}^{-1}(0)$ was connected in all cases he considered. However, it does not seem that this should always be the case. Advantages of this approach are that no *a priori* information concerning the multiplicity of the bifurcation is needed, and that it enjoys better numerical stability properties than ordinary perturbation. It should, however, be emphasized that any existing symmetries leading to higher multiplicities ought to be taken into account initially, i.e. by using group actions in the formulation of the problem, see Golubitsky *et al.* (1988) and other references cited earlier.

Branching off via the bifurcation equation. Although the branching off via perturbation techniques works effectively, this approach can have some shortcommings. In general, it cannot be decided in advance which of the two possible directions along the bifurcating branch will be taken. Furthermore, if the perturbation vector d is not chosen correctly (and it is not always clear how this is to be done), one may still have some difficulty in tracing the resulting path. The solution set $H^{-1}(0)$ can be approximated near the bifurcation point $\bar{u}$ only after an additional bifurcating branch has been approximated.

To obtain an approximation of $H^{-1}(0)$ near a simple bifurcation point $\bar{u}$, the alternative is a direct approach. This may consist of two steps, see, e.g., Section 8.3 of Allgower and Georg (1990):

1. Approximation of the bifurcation point $\bar{u}$ by adjoining additional equations to $H = 0$ and handling the resulting new set of equations by some special iterative method.

2. Construct a numerical model for the so-called bifurcation equation in order to approximate all tangents of the bifurcating branches in $\bar{u}$. Lemma 4.7 describes such an equation for the case of a simple bifurcation point. The approaches in Keller (1977, 1987) and Rheinboldt (1978) deal with this idea.

4.6. Complex bifurcation

It has been observed by Allgower (1984) and Allgower and Georg (1983a) that folds in the λ coordinate of solution curves of $H(x, \lambda) = 0$ lead to bifurcation points in a setting of complex extension, see also Section 11.8 of Allgower and Georg (1990). This observation can be used to connect separated real components of $H^{-1}(0)$, and hence may serve as a tool to find additional solutions of the equation $H = 0$. Henderson (1985) and Henderson and Keller (1990) study complex bifurcation in a general Banach space setting. Let us briefly summarize one of their main results.

Let $\mathbf{B}$ be a real Banach space which can be complexified into $\mathbf{B} \oplus i\mathbf{B}$. We use the notation $z = x + iy$ for $z \in \mathbf{B} \oplus i\mathbf{B}$ and $x, y \in \mathbf{B}$. We consider B to be naturally embedded into $\mathbf{B} \oplus i\mathbf{B}$ via $x \mapsto x + i0$. In most cases which occur in applications, e.g. function spaces, the precise meaning of this setting is obvious.

Consider a smooth nonlinear problem of the form

$$H(z, \lambda) = H(x + iy, \lambda) = 0, \quad H : \mathbf{B} \oplus i\mathbf{B} \times \mathbb{R} \to \mathbf{B} \oplus i\mathbf{B}, \qquad (4.3)$$

where H is analytic in the complex variable $z = x + iy$. Furthermore, we assume that H is real for real arguments, and denote the restriction to real arguments by $H_{\mathbf{R}}$, i.e. $H_{\mathbf{R}} : \mathbf{B} \times \mathbb{R} \to \mathbf{B}$.

Let $c(s) = (x(s), \lambda(s))$ be a solution curve of (the real) equation $H_{\mathbf{R}} = 0$ consisting of regular points. We assume that $c(0)$ is a *simple fold*, i.e.

$$\dot{\lambda}(0) = 0, \quad \ddot{\lambda}(0) \neq 0. \qquad (4.4)$$

Then $c(0)$ is a simple bifurcation point of (the complex) equation $H = 0$.

In fact, it can be seen that for the bifurcating curve

$$c_1(s) = (x_1(s) + iy_1(s), \lambda_1(s))$$

the following characterization holds at $s = 0$, see Proposition (11.8.16) of Allgower and Georg (1990):

$$\dot{x}_1(0) = 0, \quad \dot{y}_1(0) = \pm\dot{x}(0), \quad \dot{\lambda}_1(0) = 0, \quad \ddot{\lambda}_1(0) = -\ddot{\lambda}(0).$$

Proposition 2.1 in Li and Wang (1992b) generalizes this result to complex folds.

4.7. Linear eigenvalue problems

In recent years many of the classical problems of numerical linear algebra have been re-examined in the context of homotopies and path following. One of the earliest contributors has been Chu (1984–1991). In these papers iterative processes and matrix factorizations have been studied in the context of flows satisfying various differential equations. A typical example is the Toda flow which has been studied as a continuous analogue of the QR algorithm. A survey of these ideas has been given by Watkins (1984). According to Watkins, although it seems that the Toda flow and related flows yield insight into the workings of algorithms, they do not necessarily directly offer algorithms which are competitive with standard library algorithms that have been developed and polished over numerous years.

Surprisingly, Li and Li (1992), Li and Rhee (1989), Li, Zeng and Cong

(1992) and Li, Zhang and Sun (1991) have been able to construct special implementations of homotopy methods which are now at least competitive with the library routines of EISPACK and IMSL for linear eigenvalue problems.

The versatility of homotopy methods also permits their application to generalized eigenvalue problems, see Chu, Li and Sauer (1988) and non-symmetric matrices, see Li and Zeng (1992) and Li *et al.* (1992). In this case complex eigenvalues are likely to arise, and it is necessary to invoke the idea of complex bifurcation, see Section 4.6.

As an example, let us briefly discuss the homotopy approach given by Li *et al.* (1991). Consider a real symmetric tridiagonal matrix A. We assume that A is irreducible, since otherwise one off-diagonal element $A[i+1, i] = A[i, i+1]$ would vanish and the matrix A would split into two blocks which can be treated independently. We consider a homotopy $H : \mathbb{R}^N \times \mathbb{R} \times [0, 1] \to \mathbb{R}^N \times \mathbb{R}$ defined by

$$H(x, \lambda, s) = \begin{pmatrix} \lambda x - [(1-s)D + sA]x \\ x^*x - 1 \end{pmatrix}.$$

Here D is a real symmetric reducible tridiagonal matrix which is generated from A by setting some of the off-diagonal entries of A to zero. The simplest example for D would be to set all off-diagonal entries to zero. However, it is advantageous to only reduce D to tridiagonal block structure with relatively small blocks, e.g. of size < 50. This technique is referred to as *divide and conquer*.

Since $A(s) := (1 - s)D + sA$ is irreducible for all $s > 0$, the solution set of $H = 0$ consists of $2n$ disjoint smooth curves c (*eigenpaths*) which can be parametrized with respect to s. Note that s is not the arclength, but the homotopy parameter. Hence

$$c(s) = (\pm x(s), \lambda(s)) \quad \text{for } 0 \le s \le 1.$$

The curves obviously occur in pairs, and only one of each pair needs to be traced. At the level $s = 0$, initial values on the curves can be obtained by approximating all eigenvectors and eigenvalues of all small blocks in D. If D is diagonal, this is trivial, and otherwise a QR routine has to be employed.

Let us sketch a typical step of the predictor-corrector method. We note first that it follows from differentiation of $H(c(s)) = 0$ with respect to s that

$$\dot{\lambda}(s) = x(s)^*(A - D)x(s). \tag{4.5}$$

Assume that $(x(s), \lambda(s))$ is (approximately) known. After having decided on a stepsize h (we are not going to discuss this feature), a predicted eigenvalue $\tilde{\lambda}(s+h)$ is obtained from this differential equation by a two-step ODE method. Now a predicted eigenvector $\tilde{x}(s + h)$ is obtained by one step of the inverse power method with shift, i.e. solve

$$(A(s+h) - \tilde{\lambda}(s+h)\mathrm{Id})y = x(s) \quad \text{for } y$$

and set $\tilde{x}(s+h) = y/\|y\|$. Then a Rayleigh quotient iteration is performed as a corrector to approximate $(x(s+h), \lambda(s+h))$.

There are some stability problems for the case that different eigenvalues become close. Sturm sequences are computed to stabilize the procedure.

Let us finally note that this homotopy method has an order-preserving property, i.e. different λ-paths can never cross. Hence the jth eigenvalue of A can be calculated without calculating any other eigenvalues. This is very often an advantageous feature for applications. On the other hand, the homotopy method lends itself conveniently to parallelization, since each solution path can be traced independently of the others and hence also simultaneously.

4.8. Parametric programming problems

Parametric programming problems and sensitivity analysis can also be studied in the context of continuation methods. Consider the problem

$$\min\{f(x, \alpha) : c_i(x, \alpha) = 0, \ i \in E, \ c_i(x, \alpha) \le 0, \ i \in I\}, \qquad (4.6)$$

where $f, c_i : \mathbb{R}^{n+1} \to \mathbb{R}$ are smooth functions. Here

$$E = \{1, \ldots, q\} \quad \text{and} \quad I = \{q+1, \ldots, q+p\}$$

denote the index sets for the equality and inequality constraints, respectively. The local sensitivity of such systems has been analysed, e.g., in Fiacco (1983, 1984) and Robinson (1987). Many authors have used bifurcation and singularity theory to investigate the local behaviour and persistence of minima at the singular points of this system, see, e.g., Bank, Guddat, Klatte, Kummer and Tammer (1983), Gfrerer, Guddat and Wacker (1983), Gfrerer, Guddat, Wacker and Zulehner (1985), Guddat, Guerra Vasquez and Jongen (1990), Guddat, Jongen, Kummer and Nožička (1987), Jongen, Jonker and Twilt (1983, 1986), Jongen and Weber (1990), Kojima and Hirabayashi (1984) and Poore and Tiahrt (1987, 1990). Rakowska, Haftka and Watson (1991) discuss algorithms for tracking paths of optimal solutions. Lundberg and Poore (1993) report on a numerical implementation of a path following method for this problem. Our discussion is motivated by their exposition.

The Fritz John first-order necessary conditions for (4.6) imply the existence of $(\lambda, \nu) \in \mathbb{R}^{p+q} \times \mathbb{R}$ such that

$$\mathcal{L}_x(x, \lambda, \nu, \alpha) = 0, \qquad (4.7)$$

$$c_i(x, \alpha) = 0, \quad i \in E, \qquad (4.8)$$

$$\lambda_i c_i(x, \alpha) = 0, \quad i \in I, \qquad (4.9)$$

$$\nu \ge 0, \ c_i(x, \alpha) \le 0, \quad \lambda_i \ge 0, \ i \in I, \qquad (4.10)$$

where $\mathcal{L}(x, \lambda, \nu, \alpha) = \nu f(x, \alpha) + \sum \lambda_i c_i(x, \alpha)$ is the Lagrangian.

Now an active set strategy is implemented by using the following homotopy equation for a path following algorithm:

$$H\left(x, \{\lambda_i\}_{i \in \mathcal{A}}, \nu, \alpha\right) = \begin{pmatrix} \mathcal{L}_x\left(x, \{\lambda_i\}_{i \in \mathcal{A}}, \nu, \alpha\right) \\ c_i(x, \alpha), \ i \in \mathcal{A} \\ \nu^2 + \sum \lambda_i^2 - 1 \end{pmatrix} = 0, \qquad (4.11)$$

where $\mathcal{A}$ is the set of active constraints. Hence $\mathcal{A}$ includes all of the indices E and some of the indices I. During the path following procedure, this active set is adapted in such a way that the inequalities (4.10) are respected.

There are various technical difficulties (such as handling singularities or efficiently adapting the active set) which have to be overcome in order to create a successful implementation.

4.9. Linear and quadratic programming

Khachiyan (1979) started a new class of polynomial time algorithms for solving the linear programming problem. Karmarkar (1984) subsequently gave a much noted polynomial time algorithm based upon projective rescaling. Gill, Murray, Saunders, Tomlin and Wright (1986) noted that Karmarkar's algorithm is equivalent to a projected Newton barrier method which in turn is closely related to a recent class of polynomial time methods involving a continuation method, namely the tracing of the 'central path'. This last technique can be extended to quadratic programming problems, and both linear and nonlinear complementarity problems. Typically, algorithms of this nature are now referred to as *interior point* methods.

The presentation of a continuous trajectory (central path) of the iterative Karmarkar method was extensively studied by Bayer and Lagarias (1989), see also Sonnevend (1985). Megiddo (1988) related this path to the classical barrier path of nonlinear optimization (Fiacco and McCormick, 1968). Several authors have proposed algorithms that generally follow the central path to a solution, see, e.g., Renegar (1988a), Gonzaga (1988), Vaidya (1990), Kojima, Mizuno and Yoshise (1988, 1989) and Monteiro and Adler (1989).

To make the algorithms more efficient, variable steplength and/or higher order predictor algorithms have been proposed in Adler, Resende, Veiga and Karmarkar (1989), Mizuno, Todd and Ye (1992) and Sonnevend, Stoer and Zhao (1989, 1991). The algorithm of Mizuno *et al.* (1992) has subsequently been shown by Ye, Güler, Tapia and Zhang (1991) to have both polynomial time complexity and quadratic convergence. Kojima, Megiddo and Mizuno (1991a) think that there still remain differences between the theoretical primal-dual algorithms which enjoy global and/or polynomial-time convergence and the efficient implementations of primal-dual algorithms, see, e.g., Marsten, Subramanian, Saltzman, Lustig and Shanno (1990) and McShane, Monma and Shanno (1989).

Adler et al. (1989) report extensive computational experiments for an interior point implementation with solution times being in most cases less than those required by a state-of-the-art simplex method MINOS, see Murtagh and Saunders (1987). Karmarkar and Ramakrishnan (1991) report computational experience on large scale problems which are representative of large classes of applications of current interest. Their interior point implementation incorporates a preconditioned conjugate gradient method as a corrector step and is consistently faster than MINOS by orders of magnitude. Further computational experience comparing an interior point method OB1 and a simplex method CPLEX is reported in technical reports Bixby, Gregory, Lustig, Marsten and Shanno (1991), Carpenter and Shanno (1991) and Lustig, Marsten and Shanno (1991). Polak, Higgins and Mayne (1992) have given an algorithm for solving semi-infinite minimax problems which bears a resemblance to the interior penalty function methods. They report numerical results which show that the algorithm is extremely robust and its performance is at least comparable to that of current first-order minimax algorithms.

There is currently immense activity in studying and developing implementations of interior point algorithms. It is to be expected that our brief account will be outdated in a few years. For further details and literature, we refer to the recent surveys of Gonzaga (1992), Kojima, Megiddo, Noma and Yoshise (1991c), Todd (1989), Wright (1992), and the proceedings edited by Roos and Vial (1991). As an example, we outline the central path approach for a primal-dual linear programming problem, following the introductory parts of Monteiro and Adler (1989) and Mizuno et al. (1992).

Consider the following linear programming problem and its corresponding dual form:

Problem 4.9

$$\min_{x}\{c^*x : Ax = b,\ x \geq 0\}, \tag{4.12}$$

$$\max_{y}\{b^*y : A^*y + z = c,\ z \geq 0\}, \tag{4.13}$$

We make the following standard assumption.

Assumption 4.10 The rank of A equals the number of its rows, and the interior feasible set of the primal-dual problem

$$\mathcal{F}^o := \{(x, z) : x, z > 0,\ Ax = b,\ A^*y + z = c \text{ for some } y\}$$

is not empty.

It is well established that the linear programming problem has a unique solution under these assumptions. The logarithmic barrier function method associated with Problem 4.9 is

$$\min_x \left\{ c^* x - \mu \sum_j \ln x_j : Ax = b, \ x > 0 \right\}, \qquad (4.14)$$

where $\mu > 0$ is the barrier penalty parameter. Under Assumption 4.10, the logarithmic barrier function is strictly convex and has a unique minimal point $x(\mu)$ for all $\mu > 0$. Moreover, $x(\mu)$ tends to the unique solution of Problem 4.9 as μ tends to zero.

The Karush–Kuhn–Tucker optimality condition which characterizes the solution $x(\mu)$ can be expressed in the following way: $(x(\mu), z)$ must belong to the set

$$\mathcal{C} := \{ (x, z) \in \mathcal{F}^o : \operatorname{diag}(x)z = \mu e \}, \qquad (4.15)$$

where e denotes the column of ones. In fact, $\mathcal{C}$ is parametrized by μ and is commonly called the *central path* of the problem. It turns out that μ is related to the so-called *duality gap*: $c^* x - b^* y = x^* z$ via

$$\mu = \frac{x^* z}{n} \qquad (4.16)$$

for $(x, z) \in \mathcal{C}$, where n is the number of columns of A.

From these remarks, it is clear that the objective now is to follow the central path $\mathcal{C}$ as μ tends to zero. In fact, most interior point methods can be viewed, one way or another, as a special path following method along these lines. The methods differ in the choice of predictor step, corrector procedure (usually one or several Newton type iterations) and predictor steplength control. Many papers discussing such methods or introducing new methods also contain a sophisticated complexity analysis, see, for example, Section 6.

These interior point algorithms typically require a phase I in which a feasible starting point is generated. A somewhat different approach is taken by Freund (1991) who introduces a shifted barrier function approach so that the need for phase I is obviated.

Finally, this technique is quite general and can be extended to quadratic programming problems and linear and nonlinear complementarity problems, see, e.g., Kojima, Megiddo and Ye (1992). The literature on interior methods is rapidly increasing, and the subject has become one of the major topics of mathematical programming. In our opinion, it is only a question of time until the venerable simplex methods will be superceded by interior point implementations.

5. Piecewise-linear methods

Up to now we have assumed that the map $H : \mathbb{R}^{n+1} \to \mathbf{R}^n$ was smooth. Next we will discuss piecewise-linear methods which can again be viewed as curve tracing methods, but which can be applied to nonsmooth situations. The piecewise-linear methods trace a polygonal path which is obtained by successively stepping through certain 'transversal' cells of a piecewise-linear manifold. The first and most prominent example of a piecewise-linear algorithm was designed by Lemke and Howson (1964) and Lemke (1965) to calculate a solution of the linear complementarity problem, see Section 5.2. This algorithm played a crucial role in the development of subsequent piecewise-linear algorithms. Scarf (1967) gave a numerically implementable proof of the Brouwer fixed point theorem, based upon Lemke's algorithm. Eaves (1972) observed that a related class of algorithms can be obtained by considering piecewise-linear approximations of homotopy maps. Thus the piecewise-linear continuation methods began to emerge as a parallel to the classical embedding or predictor–corrector methods.

The piecewise-linear methods require no smoothness of the underlying equations and hence have, at least in theory, a more general range of applicability than classical embedding methods. In fact, they can be used to calculate fixed points of set-valued maps. They are more combinatorial in nature and are closely related to the topological degree, see Peitgen and Siegberg (1981). Piecewise-linear continuation methods are usually considered to be less efficient than the predictor–corrector methods when the latter are applicable, especially in higher dimensions. The reasons for this lie in the fact that steplength adaptation and exploitation of special structure are more difficult to implement for piecewise-linear methods.

Eaves (1976) has given a very elegant geometric approach to general piecewise-linear methods, see also Eaves and Scarf (1976). We adopt this point of view and cast the notion of piecewise-linear algorithms into the general setting of subdivided manifolds which we will call *piecewise-linear manifolds*. Our exposition follows the introduction of Georg (1990) to some extent.

Let $\mathbf{E}$ denote some ambient finite dimensional Euclidean space which contains all points arising in the sequel. A *half-space* η and the corresponding *hyperplane* $\partial \eta$ are defined by $\eta = \{y \in \mathbf{E} : x^*y \leq \alpha\}$ and $\partial \eta = \{y \in \mathbf{E} : x^*y = \alpha\}$, respectively, for some $x \in \mathbf{E}$ with $x \neq 0$ and some $\alpha \in \mathbb{R}$. A finite intersection of half-spaces is called a *cell*. If σ is a cell and ξ a half-space such that $\sigma \subset \xi$ and $\tau := \sigma \cap \partial \xi \neq \emptyset$, then the cell τ is called a *face* of σ. For reasons of notation we consider σ also to be a face of itself, and all other faces are *proper* faces of σ. The *dimension* of a cell is the dimension of its affine hull. In particular, the dimension of a singleton is 0 and the dimension of the empty set is -1. If the singleton $\{v\}$

is a face of σ, then v is called a *vertex* of σ. If τ is a face of σ such that $\dim \tau = \dim \sigma - 1$, then τ is called a *facet* of σ.

Definition 5.1 A *piecewise-linear manifold* of dimension n is a system $\mathcal{M} \neq \emptyset$ of cells of dimension n such that the following conditions hold:

1. If $\sigma_1, \sigma_2 \in \mathcal{M}$, then $\sigma_1 \cap \sigma_2$ is a common face of σ_1 and σ_2.
2. A cell τ of dimension $n - 1$ can be a facet of at most two cells in $\mathcal{M}$.
3. The family $\mathcal{M}$ is locally finite, i.e. any relatively compact subset of

$$|\mathcal{M}| := \bigcup_{\sigma \in \mathcal{M}} \sigma \tag{5.1}$$

meets only finitely many cells $\sigma \in M$.

The simplest example of a piecewise-linear manifold is $\mathbb{R}^n$ subdivided into unit cubes with integer vertices.

We introduce the *boundary* $\Delta \mathcal{M}$ of $\mathcal{M}$ as the system of facets which are common to exactly one cell of $\mathcal{M}$. Generally, we cannot expect $\Delta \mathcal{M}$ to again be a piecewise-linear manifold. However, this is true for the case that $|\mathcal{M}|$ is convex. Two cells which have a common facet τ are called *adjacent*. We say that one cell is *pivoted* into the other cell across the facet τ. We will see that piecewise-linear algorithms perform pivoting steps.

Typical for piecewise-linear path following is that only one current cell is stored in the computer, along with some additional data, and the pivoting step is performed by calling a subroutine which makes use of the data to determine an adjacent cell which then becomes the new current cell.

A cell of particular interest is a *simplex* $\sigma = [v_1, v_2, \ldots, v_{n+1}]$ of dimension n which is defined as the convex hull of $n + 1$ affinely independent points $v_1, v_2, \ldots, v_{n+1} \in \mathbf{E}$. These points are the vertices of σ. If a piecewise-linear manifold $\mathcal{M}$ of dimension n consists only of simplices, then $\mathcal{M}$ is called a *pseudo manifold* of dimension n. Such manifolds are of special importance, see, e.g., Gould and Tolle (1983) and Todd (1976a). If a pseudo manifold $\mathcal{T}$ subdivides a set $|\mathcal{T}|$, then we also say that $\mathcal{T}$ *triangulates* $|\mathcal{T}|$. Some triangulations of $\mathbb{R}^n$ of practical importance had been previously considered by Coxeter (1934) and Freudenthal (1942), see also Todd (1976a). Eaves (1984) gave an overview of standard triangulations.

A simple triangulation can be generated by the following pivoting rule, see Allgower and Georg (1979) or Coxeter (1973): if

$$\sigma = [v_1, v_2, \ldots, v_i \ldots, v_{n+1}]$$

is a simplex in $\mathbf{R}^n$, and τ is the facet opposite a vertex v_i, then σ is pivoted across τ into $\tilde{\sigma} = [v_1, v_2, \ldots, \tilde{v}_i \ldots, v_{n+1}]$ by setting

$$\tilde{v}_i = \begin{cases} v_{i+1} + v_{i-1} - v_i & \text{for} \quad 1 < i < n + 1, \\ v_2 + v_{n+1} - v_1 & \text{for} \quad i = 1, \\ v_n + v_1 - v_{n+1} & \text{for} \quad i = n + 1. \end{cases}$$

In fact, a minimal (nonempty) system of n-simplices in $\mathbb{R}^n$ which is closed under this pivoting rule is a triangulation of $\mathbb{R}^n$.

Let $\mathcal{M}$ be a piecewise-linear manifold of dimension $n+1$. We call $H: |\mathcal{M}| \to \mathbb{R}^n$ a *piecewise-linear map* if the restriction $H_\sigma : \sigma \to \mathbb{R}^n$ of H to σ is an affine map for all $\sigma \in \mathcal{M}$. In this case, H_σ can be uniquely extended to an affine map on the affine space spanned by σ. The Jacobian H_σ' has the property $H_\sigma'(x - y) = H_\sigma(x) - H_\sigma(y)$ for x, y in this affine space. Note that under an appropriate choice of basis H_σ' corresponds to an $(n, n+1)$-matrix which has a one-dimensional kernel in case of nondegeneracy, i.e. if its rank is maximal.

A piecewise-linear algorithm is a method for following a polygonal path in $H^{-1}(0)$. To avoid degeneracies, we introduce a concept of regularity, see Eaves (1976). A point $x \in |\mathcal{M}|$ is called a *regular point* of H if x is not contained in any face of dimension $< n$, and if H_τ' has maximal rank n for all facets τ. A value $y \in \mathbb{R}^n$ is a *regular value* of H if all points in $H^{-1}(y)$ are regular. By definition, y is vacuously a regular value if it is not contained in the range of H. If a point or value is not regular it is called *singular*. An analogue of Sard's theorem 4.1 holds, see, e.g., Eaves (1976) or Peitgen and Siegberg (1981) for details. This enables us to confine ourselves to regular values. We note that degeneracies could be handled via the concept of lexicographical ordering, see Dantzig (1963) and Todd (1976a).

Hence, for reasons of simplicity, we assume that all piecewise-linear maps under consideration here have zero as a regular value. This implies that $H^{-1}(0)$ consists of polygonal paths whose vertices are always in the interior of some facet. If σ is a cell, then $\sigma \cap H^{-1}(0)$ is a segment (two endpoints), a ray (one endpoint) or a line (no endpoint). The latter case is not of interest for piecewise-linear path following. A step of the method consists of following the ray or segment from one cell into a uniquely determined adjacent cell. The method is typically started at a point of the boundary or on a ray (coming from infinity), and it is typically terminated at a point of the boundary or in a ray (going to infinity). The numerical linear algebra required to perform one step of the method is typical for linear programming and usually involves n^2 operations for dense matrices (at least in the case that the cells are simplices).

Nearly all piecewise-linear manifolds $\mathcal{M}$ which are of importance for practical implementations, are orientable. If $\mathcal{M}$ is orientable and of dimension $n+1$, and if $H : \mathcal{M} \to \mathbb{R}^n$ is a piecewise-linear map, then it is possible to introduce an index for the piecewise-linear solution manifold $H^{-1}(0)$ which has important invariance properties and occasionally yields some useful information, see Eaves (1976), Eaves and Scarf (1976), Lemke and Grotzinger (1976), Shapley (1974) and Todd (1976c). It should be noted that this index is closely related, see, e.g., Peitgen (1982), to the topological index which is a standard tool in topology and nonlinear analysis. Occasionally, index arguments are used to ensure a certain behaviour of the solution path.

We now give some examples of how the piecewise-linear path following methods are used.

5.1. Piecewise-linear homotopy algorithms

Let us first show how these ideas can be used to approximate a fixed point of a continuous bounded map $f : \mathbb{R}^n \to \mathbb{R}^n$ by applying piecewise-linear path following to an appropriate piecewise-linear homotopy map. Eaves (1972) presented the first such method. A restart method based on somewhat similar ideas was developed by Merrill (1972). A number of authors have studied the efficiency and complexity of piecewise-linear homotopy algorithms, see, e.g., Alexander (1987), Eaves and Yorke (1984), Saigal (1977, 1984), Saigal and Todd (1978), Saupe (1982), Todd (1982) and Todd (1986).

As an example of a piecewise-linear homotopy algorithm, let us sketch the algorithm of Eaves and Saigal (1972). We consider a triangulation T of $\mathbb{R}^n \times (0,1]$ into $(n+1)$-simplices σ such that every simplex is contained in some slab $\mathbb{R}^n \times [2^{-k}, 2^{-k-1}]$ for $k = 0, 1, \ldots$. Let us call the maximum of the last coordinates of all vertices of σ the *level* of σ. We call T a *refining* triangulation if for $\sigma \in T$, the diameter of σ tends to zero as the level of σ tends to zero. The first such triangulation was proposed by Eaves (1972). Todd (1976a) gave a triangulation with refining factor $1/2$. Subsequently, many triangulations with arbitrary refining factors were developed, see Eaves (1984).

Consider the homotopy

$$\tilde{H}(x, \lambda) = x - \lambda x_0 - (1 - \lambda) f(x).$$

The idea is to follow a solution path from $(x_0, 1)$ to $(\bar{x}, 0)$ where x_0 is the starting point of the method and $\bar{x}$ is a fixed point of f we wish to approximate. However, there are no smoothness assumptions on f, and therefore a more subtle path following approach involving piecewise-linear approximations is required.

We denote by H the piecewise-linear map which interpolates $\tilde{H}$ on the vertices of the given refining triangulation T. Then it is possible to follow the polygonal solution path $c(s) = (x(s), \lambda(s))$ in $H^{-1}(0)$ starting at $c(0) = (x_0, 1)$. For convenience we regard c to be parametrized by arclength $0 \leq s < s_0 \leq \infty$. From the boundedness of the map f it follows that $\lambda(s)$ tends to zero as s tends to s_0. Furthermore,

$$\lim_{s \to s_0} \|x(s) - f(x(s))\| = 0.$$

Since $x(s)$ remains bounded as s tends to s_0, this implies that every accumulation point of $x(s)$ is a fixed point of f.

These ideas can be extended to set-valued maps.

5.2. Lemke's algorithm

The first and most prominent example of a piecewise-linear algorithm was designed by Lemke (1965) and Lemke and Howson (1964) in order to calculate a solution of the linear complementarity problem. Subsequently, several authors have studied complementarity problems from the standpoint of piecewise-linear homotopy methods, see, e.g., Kojima (1974, 1979), Kojima, Nishino and Sekine (1976), Saigal (1971, 1976) and Todd (1976b). Complementarity problems can also be considered from an interior point algorithm viewpoint, see Section 4.9, hence by following a smooth path, see, e.g., Kojima, Mizuno and Noma (1990b), Kojima, Mizuno and Yoshise (1991d), Kojima, Megiddo and Noma (1991b), Kojima, Megiddo and Mizuno (1990a) and Mizuno (1992).

We present the Lemke algorithm as an example of a piecewise-linear algorithm since it played a crucial role in the development of subsequent piecewise-linear algorithms. Let us consider the following *linear complementarity problem*: Given an affine map $g : \mathbb{R}^n \to \mathbb{R}^n$, find an $x \in \mathbb{R}^n$ such that

$$x \in \mathbb{R}^n_+; \quad g(x) \in \mathbb{R}^n_+; \quad x^* g(x) = 0.$$

Here $\mathbb{R}_+$ denotes the set of nonnegative real numbers, and in the sequel we also denote the set of positive real numbers by $\mathbb{R}_{++}$. If $g(0) \in \mathbb{R}^n_+$, then $x = 0$ is a trivial solution to the problem. Hence this trivial case is always excluded and the additional assumption

$$g(0) \notin \mathbb{R}^n_+$$

is made. Linear complementarity problems arise in quadratic programming, bimatrix games, variational inequalities and economic equilibria problems, and numerical methods for their solution have been of considerable interest, see, e.g., Cottle (1974), Cottle and Dantzig (1968), Cottle, Golub and Sacher (1978) and Lemke (1980). See also the proceedings (Cottle, Gianessi and Lions, 1980) for further references.

For $x \in \mathbb{R}^n$ we introduce the positive part $x_+ \in \mathbb{R}^n_+$ by setting $e_i^* x_+ := \max\{e_i^* x, 0\}$, $i = 1, \dots, n$ and the negative part $x_- \in \mathbb{R}^n_+$ by $x_- := (-x)_+$. The following formulae are then obvious: $x = x_+ - x_-$, $(x_+)^*(x_-) = 0$.

It is not difficult to show the following: Define $f : \mathbb{R}^n \to \mathbb{R}^n$ by $f(z) := g(z_+) - z_-$. If x is a solution of the linear complementarity problem, then $z := x - g(x)$ is a zero point of f. Conversely, if z is a zero point of f, then $x := z_+$ solves the linear complementarity problem.

The advantage which f provides is that it is obviously a piecewise-linear map if we subdivide $\mathbb{R}^n$ into orthants. This is the basis for our description of Lemke's algorithm. For a fixed $d \in \mathbb{R}^n_{++}$ we define the homotopy H :

$\mathbb{R}^n \times [0, \infty) \to \mathbb{R}^n$ by

$$H(x, \lambda) := f(x) + \lambda d.$$

For a given subset $I \subset \{1, 2, \ldots, n\}$ an orthant can be written in the form

$$\sigma_I := \{ (x, \lambda) : \lambda \geq 0, \; e_i^* x \geq 0 \text{ for } i \in I, \; e_i^* x \leq 0 \text{ for } i \notin I \}.$$

The collection of all such orthants forms a piecewise-linear manifold $\mathcal{M}$ (of dimension $n + 1$) which subdivides $\mathbb{R}^n \times [0, \infty)$. Furthermore it is clear that $H : \mathcal{M} \to \mathbb{R}^n$ is a piecewise-linear map since $x \mapsto x_+$ switches its linearity character only at the coordinate hyperplanes.

Let us assume for simplicity (as usual) that zero is a regular value of H. Lemke's algorithm is started on a ray: if $\lambda > 0$ is sufficiently large, then

$$(- g(0) - \lambda d)_+ = 0 \quad \text{and} \quad (- g(0) - \lambda d)_- = g(0) + \lambda d \in \mathbb{R}_{++}^n,$$

and consequently

$$H(- g(0) - \lambda d, \lambda) = 0.$$

Hence, the ray defined by

$$\lambda \in [\lambda_0, \infty) \longmapsto -g(0) - \lambda d \in \sigma_\emptyset \tag{5.2}$$

$$\text{for} \quad \lambda_0 := \max_{i=1,\ldots,N} \frac{-g(0)[i]}{d[i]} \tag{5.3}$$

is used (for decreasing λ-values) to start the path following. Since the piecewise-linear manifold $\mathcal{M}$ consists of the orthants of $\mathbb{R}^n \times [0, \infty)$, it is finite, and there are only two possibilities:

1. The algorithm terminates on the boundary $|\partial \mathcal{M}| = \mathbb{R}^n \times \{0\}$ at a point $(z, 0)$. Then z is a zero point of f, and hence z_+ solves the linear complementarity problem.
2. The algorithm terminates on a secondary ray. Then it can be shown, see Cottle (1974), that the linear complementarity problem has no solution, at least if the Jacobian g' belongs to a certain class of matrices.

5.3. Variable dimension algorithms

In recent years, a new class of piecewise-linear algorithms has attracted considerable attention. They are called *variable dimension algorithms* since they all start from a single point, a zero-dimensional simplex, and successively generate simplices of varying dimension, until a so-called completely labelled simplex is found. Numerical results from Kojima and Yamamoto (1984) indicate that these algorithms improve the computational efficiency of piecewise-linear homotopy methods. The first variable dimension algorithm is due to Kuhn (1969). However, this algorithm had the disadvantage that it could only be started from a vertex of a large triangulated standard simplex S, and therefore piecewise-linear homotopy algorithms were preferred.

By increasing the sophistication of Kuhn's algorithm considerably, van der Laan and Talman (1979) developed an algorithm which could start from any point inside S. It soon became clear, see Todd (1978), that this algorithm could be interpreted as a homotopy algorithm. Numerous other variable dimension algorithms were developed. Some of the latest are due to Dai, Sekitani and Yamamoto (1992), Dai and Yamamoto (1989), Kamiya and Talman (1990), Talman and Yamamoto (1989). Two unifying approaches have been given, one due to Kojima and Yamamoto (1982), the other due to Freund (1984a,b). A variable dimension algorithm which is easy to comprehend and may serve the reader as a gateway is the octrahedral algorithm of Wright (1981).

5.4. Approximating manifolds

The emphasis of this survey is on path following methods. We should note, however, that the ideas of predictor–corrector and piecewise-linear curve tracing can be extended to the approximation of implicitly defined manifolds $H^{-1}(0)$ where $H : \mathbb{R}^{N+K} \to \mathbb{R}^N$. Limitations of space preclude a detailed discussion.

There are two basic types of algorithms: one is the moving frame algorithm of Rheinboldt (1987), see also Rheinboldt (1988b), which is a higher dimensional analogue of the predictor–corrector method, the other is a piecewise-linear algorithm which has been developed in Allgower and Gnutzmann (1987), Allgower and Schmidt (1985), Gnutzmann (1989), Widmann (1990a,b), see also Chapter 15 of Allgower and Georg (1990).

The moving frame algorithm involves predictors that arise from a local triangulation of the tangent space at a current point. The corrector consists of a Newton-like method for projecting the generated mesh back to the manifold. This method is well-suited for smooth manifolds in which the dimension N is large, such as in multiple parameter nonlinear eigenvalue problems, see, e.g., Rheinboldt (1988b, 1992a). It has been applied to the calculation of fold curves and to differential-algebraic equations, see Dai and Rheinboldt (1990) and Rheinboldt (1986, 1991, 1992b).

So far, it has not been possible to make the moving frame algorithm global in the sense that a compact manifold is approximated (without holes or overlaps) by a piecewise-linear compact manifold. The latter task can be accomplished by the application of piecewise-linear algorithms. However, these algorithms become extremely costly for large N. The piecewise-linear algorithms have been applied to the visualization of body surfaces, see Allgower and Gnutzmann (1991), and to the approximation of surface and body integrals, see Allgower, Georg and Widmann (1991d). They can also be used as automatic mesh generators for boundary element methods, see Georg (1991).

6. Complexity

In modern complexity investigations of continuation-type methods the so-called α-theory of Smale (1986) is a convenient tool. This theory is closely related to the classical Kantorovich estimates for Newton iterations, see, e.g., Ortega and Rheinboldt (1970) and Deuflhard and Heindl (1979). In contrast to the Kantorovich estimates, Smale's estimates are based on information at only one point, involving however all derivatives. The maps under consideration have to be analytic.

On the other hand, an analytic map is characterized by all its derivatives at one point. In fact, Rheinboldt (1988a) showed that Smale's estimates can be derived from the Kantorovich estimates. However, for complexity considerations, it is more convenient to have all the relevant information situated at only one point. Let us briefly present Smale's estimates and show how they are used for complexity discussions. Our presentation is based on the introductory parts of the papers of Shub and Smale (1991) and Renegar and Shub (1992).

Let E, F be complex Banach spaces and $f : E \to F$ an analytic map. It would be possible to assume that f is given only on some open domain, but for reasons of simplicity of exposition we assume f to be defined on all of E. Then for each point $x \in E$ such that $Df(x) : E \to F$ is an isomorphism the following quantities are defined:

$$\beta(f, x) = \|Df(x)^{-1}f(x)\|, \tag{6.1}$$

$$\gamma(f, x) = \sup_{k>1} \frac{1}{k!} \|Df(x)^{-1}D^k f(x)\|^{1/(k-1)}, \tag{6.2}$$

$$\alpha(f, x) = \beta(f, x)\gamma(f, x), \tag{6.3}$$

$$\mathcal{N}_f(x) = x - Df(x)^{-1}f(x). \tag{6.4}$$

Note that $\mathcal{N}_f(x)$ is the Newton iterate of x. It it also convenient to introduce the notation

$$\mathcal{N}_f^\infty(x) = \lim_{i \to \infty} \mathcal{N}_f^i(x) \tag{6.5}$$

provided Newton's method started at x is convergent.

A related one-dimensional 'control' Newton method is occasionally generated from the following family of functions

$$h_{\beta,\gamma}(t) = \beta - t + \frac{\gamma t^2}{1 - \gamma t}. \tag{6.6}$$

For $0 < \alpha < 3 - 2\sqrt{2} \approx 0.1716$, the function $h_{\beta,\gamma}$ has two real positive roots, the smaller one being

$$\frac{\tau(\alpha)}{\gamma} = \frac{(\alpha+1) - \sqrt{(\alpha+1)^2 - 8\alpha}}{4\gamma}. \tag{6.7}$$

Moreover, $h''_{\beta,\gamma} > 0$ on the interval $(0, 1/\gamma)$. Thus, Newton's method starting at zero generates a strictly increasing sequence $t_i(\beta, \gamma) = \mathcal{N}^i_{h_{\beta,\gamma}}(0)$ converging to this root.

Occasionally, a slightly smaller upper bound for α is used, namely $\alpha_0 = \frac{1}{4}(13 - 3\sqrt{17}) \approx 0.1577$.

The following is a modification of Smale's α-theorem.

Theorem 6.1 Let $x_0 \in E$, $\alpha = \alpha(f, x_0)$, $\gamma = \gamma(f, x_0)$. If $\alpha \leq \alpha_0 \approx 0.1577$, then the iterates $x_{i+1} = \mathcal{N}_f(x_i)$ are defined and converge to a zero point $x_\infty = \mathcal{N}_f^\infty(x_0) \in E$ with the rate

$$\|x_{i+1} - x_i\| \leq \left(\frac{1}{2}\right)^{2^i - 1} \|x_1 - x_0\|.$$

Moreover, the following estimates hold:

$$\|x_\infty - x_0\| \leq \frac{\tau(\alpha)}{\gamma}, \quad \|x_\infty - x_1\| \leq \frac{\tau(\alpha) - \alpha}{\gamma}.$$

An easy consequence is

Corollary 6.2 $\|x_\infty - x_i\| \leq \varepsilon$ for $i \geq 1 + \log|\log \tau(\alpha)/\varepsilon\gamma|$.

Furthermore, by using the control Newton iterates $t_i = t_i(\beta, \gamma)$, a stricter estimate can be obtained under the same hypotheses:

Theorem 6.3 $\|x_i - x_{i-1}\| \leq t_i - t_{i-1}$.

Another property which is important for complexity discussions is the fact that α is upper semi-continuous, more precisely:

Proposition 6.4 Let $\psi(u) := 2u^2 - 4u + 1$ and $u := \gamma(f, x_0)\|x_0 - x\|$. Then

$$\alpha(f, x) \leq \frac{\alpha(f, x_0)(1 - u) + u}{\psi(u)^2}.$$

From the previous proposition it is possible to obtain a uniform estimate for Newton steps:

Theorem 6.5 There are universal constants $\bar{\alpha} \approx 0.0802$ and $\bar{u} \approx 0.0221$ with the following property: Let $\bar{\gamma} > 0$ and $x, \zeta \in E$. If $\beta(f, \zeta) \leq \bar{\alpha}/\bar{\gamma}$ and $\|x - \zeta\| \leq \bar{u}/\bar{\gamma}$, then $\|\mathcal{N}_f(x) - \mathcal{N}_f^\infty(\zeta)\| \leq \bar{u}/\bar{\gamma}$.

This theorem is used to investigate the complexity of path following in the following way: Let $H : [0, 1] \times E \to F$ be a continuous (homotopy) map which is analytic in the second argument. We further assume that a continuous solution path $\zeta : [0, 1] \to E$ exists, i.e. $H(t, \zeta(t)) = 0$ for $t \in [0, 1]$, such that the derivative $H_\zeta(t, \zeta(t))$ is an isomorphism. The following crude path-following method can be designed: choose a subdivision $0 = t_0 < t_1 < \cdots < t_k = 1$ and define

$$x_i := \mathcal{N}_{H(t_i, .)}(x_{i-1}) \quad \text{for } i = 1, \ldots, k. \tag{6.8}$$

It is clear that this method follows the solution curve if $\|x_0 - \zeta(0)\|$ and $|t_i - t_{i-1}|$ are small enough. Of course, the crucial number for complexity considerations is the number k of Newton steps involved in this embedding method. If it is wished to obtain some points of the solution curve with high accuracy, then the complexity described in Corollary 6.2 has to be added.

The preceding analysis immediately furnishes a tool to determine the estimates necessary for a successful tracing of the solution curve:

Theorem 6.6 Let $\|x_0 - \zeta(0)\| \leq \bar{u}/\bar{\gamma}$, and let the mesh t_i be so fine that $\beta(H(t_i, .), \zeta(t_{i-1})) \leq \bar{\alpha}/\bar{\gamma}$ and $\gamma(H(t_i, .), \zeta(t_{i-1})) \leq \bar{\gamma}$. Then the embedding method (6.8) follows the solution path ζ. In fact, $\|x_i - \zeta(t_i)\| \leq \bar{u}/\bar{\gamma}$.

To summarize, we have outlined a program for approaching complexity investigations when Newton steps are the primary tool of path following methods. As can be seen from the last theorem, the success of the approach depends heavily on the availability of estimates $\beta(H(t, .), \zeta(s)) \leq C_1|t - s|$ and $\gamma(H(t, .), \zeta(s)) \leq C_2|t - s|$ with explicit constants C_1 and C_2.

This program was carried out by Shub and Smale (1991) for the case of a homotopy method for calculating all solutions of a system of polynomial equations (Bezout's theorem). A previous effort along similar lines was described by Renegar (1987).

Recently, this approach has also been used by Renegar and Shub (1992) for a unified complexity analysis of various interior methods designed for solving linear and convex quadratic programming problems. They obtain and re-derive various 'polynomial time' estimates. The linear programming barrier method was first analysed by Gonzaga (1988). The quadratic programming barrier method was analysed by Goldfarb and Liu (1991). A primal-dual linear programming algorithm was investigated by Kojima *et al.* (1988) and Monteiro and Adler (1989). The algorithm has roots in Megiddo (1988). Primal-dual linear complementarity and quadratic programming algorithms were discussed by Kojima *et al.* (1989) and Monteiro and Adler (1989). All of these algorithms follow the *central trajectory* studied by Bayer and Lagarias (1989) and Megiddo and Shub (1989). For the case of the linear complementarity problem, Mizuno, Yoshise and Kikuchi (1989) present several implementations and report computational experience which confirms the polynomial complexity.

This discussion involved path following methods of Newton type. Renegar

(1985), see also Renegar (1988b), has made complexity investigations for piecewise-linear path following methods.

7. Available software

We conclude the paper by listing some available software related to path following and indicate how the reader might access these codes. No attempt to compare or evaluate the various codes is offered. In any case, our opinion is that path following codes always need to be considerably adapted to the special purposes for which they are designed. The path following literature offers various tools for accomplishing such tasks. Although there are some general purpose codes, probably none will slay every dragon.

Rheinboldt, Roose and Seydel (1990) present a list of features and options that appear to be necessary or desirable for continuation codes. This should be viewed as a guideline for people who want to create a new code.

Several of the codes can be accessed via *netlib*: The best way to obtain them is to ftp into `netlib@research.att.com`, login as netlib, password = your e-mail address. It is also possible to e-mail to netlib by writing *send index*. Information on how to proceed will then be e-mailed back to you.

7.1. ALCON

This sofware package has been written by Deuflhard, Fiedler and Kunkel (1987). ALCON is a continuation method for algebraic equations $f(x, \tau) = 0$, based on QR factorization as a solver for the arising equations in the Gauss–Newton iteration of the corrector step. Turning points and simple bifurcations can be computed on demand. It can be found in the electronic library of the Konrad Zuse Zentrum für Informationstechnik in Berlin. The reader may telnet or ftp to sc.ZIB-Berlin.de (130.73.108.11) and login under the user identification elib, no password is required. The sources can be found in the directory /pub/ELIB/codelib either in unpacked form or as a tar.Z file.

7.2. AUTO

This is a software package written by E. Doedel. It is mainly intended to investigate bifurcation phenomena. There is a charge of $175 for the software, a manual by Doedel and Kernévez (1986) is also available, contact: S. K. Shull, Applied Mathematics, 217-50, California Institute of Technology, Pasadena, CA 91125, USA. Telephone: (818) 356-4560.

7.3. BIFPACK

This package has been written by Seydel (1991a). It is meant primarily for bifurcation analysis of ODEs. This is not a public domain software.

However, as a research tool, it is freely distributed for *noncommercial* use, except for a $20 contribution for handling. Indicate whether you prefer **BIFPACK** on 5.25 in or on 3.5 in diskette (1.4 MB, DOS double-density). Contact: Professor Rüdiger Seydel, Abt. Mathematik VI, Universität Ulm, Postfach 4066, W - 7900 Ulm, Germany.

e-mail: seydel@rz.uni-ulm.dbp.de

7.4. CONKUB

This is an interactive program for continuation and bifurcation of large systems of nonlinear equations written by Mejia (1986), see also Mejia (1990). It is currently available from him via e-mail:

ray@helix.nih.gov.

7.5. DERPAR

This package was written by Kubíček (1976), and Holodniok and Kubíček (1984). This is a Fortran subprogram for the evaluation of the dependence of the solution of a nonlinear system on a parameter. The modified method of Davidenko, which applies the Implicit Function Theorem, is used in combination with Newton's method and Adam's integration formulae. The program can be accessed via netlib, see number 502 in the directory *toms*.

7.6. HOMPACK

This is a suite of FORTRAN 77 subroutines for solving nonlinear systems of equations by homotopy methods, written by L. T. Watson, see Watson *et al.* (1987). There are subroutines for fixed point, zero finding, and general homotopy curve tracking problems, utilizing both dense and sparse Jacobian matrices, and implementing three different algorithms: ODE-based, normal flow and augmented Jacobian. The program can be accessed via netlib under the directory *hompack*. See also number 652 in the directory *toms*.

7.7. LOCBIF

A. Khibnik and collaborators in Moscow have developed several codes for path following and bifurcation analysis. **CYCLE** is a one-parameter continuation program for limit cycles. **LINLBF** has been designed for multi-parameter bifircation analysis of equilibrium points, limit cycles, fixed points of maps, respectively. **LOCBIF** is an interactive program built originally on the top of **LINLBF**. People interested in trying this software should contact A. Khibnik via e-mail:

na.khibnik@na-net.ornl.gov.

7.8. OB1

This interior point method has been written by I. J. Lustig, R. E. Marsten and D. F. Shanno. The version of OB1 that implements a primal-dual algorithm for linear programming is available in source code form to academics from Roy Marsten at Georgia Tech. This is the December 1989 version, also known as the WRIP (Workshop on Research in Programming) version. The current version of OB1 is commercial. It implements a primal-dual predictor–corrector algorithm for linear programming and is available from XMP Software at prices ranging from $15,000 to $100,000: XMP Software, Suite 279, Bldg 802, 930 Tahoe Blvd, Incline Village, NV 89451, phone: (702) 831- 4XMP, e-mail:

tlowe@mcimail.com

7.9. PATH

This software package for dynamical systems was originally coded in FORTRAN 77 by Kaas-Petersen (1989), and is currently modified to include a graphical interface. According to the workers at the Technical University of Denmark, it seems to be able to handle much larger systems of ODE's than AUTO. For more details and availability, readers may contact Michael Rose via e-mail:

lamfmr@lamf.dth.dk.

7.10. PITCON

This is a Fortran subprogram for continuation and limit points, written by Rheinboldt and Burkardt (1983b), see also Rheinboldt and Burkardt (1983a). It is used for computing solutions of a nonlinear system of equations containing a parameter. The location of target points where a given variable has a specified value can be located. Limit points are also identified. It uses a local parameterization based on curvature estimates to control the choice of parameter value. The program can be accessed via netlib under the directory *contin*. See also number 596 in the directory *toms*.

7.11. PLALGO

This is a software for piecewise-linear homotopy methods developed by Todd (1981). It can be obtained from him via e-mail:

miketodd@orie.cornell.edu.

No support is available, and he says that on-line documentation is weak, although he can send a hard copy.

7.12. pla_s_k

This is a C program, written by Widmann (1990a), for triangulating surfaces in $\mathbb{R}^3$ which are implicitly defined, see Section 5.4. It incorporates mesh smoothing and some other features. It is particularly suited for mesh generation (e.g. for boundary element methods) and for visualization purposes. The program can be obtained via e-mail:

Georg@Math.ColoState.Edu.

7.13. PLTMG

This package has been written by R. E. Bank, see also Bank and Chan (1986). It solves elliptic partial differential equations in general regions of the plane. It features adaptive local mesh refinement, multigrid iteration, and a pseudo-arclength continuation option for parameter dependencies. The package includes an initial mesh generator and several graphics packages. Full documentation can be obtained in the PLTMG User's Guide by R. E. Bank, available from SIAM publications via e-mail:

SIAMPUBS@wharton.upenn.edu.

The program can be accessed via netlib under the directory *pltmg*.

7.14. Last and least

The book Allgower and Georg (1990) contains several Fortran codes for path following which are to be regarded primarily as illustrations. The intention was to encourage the readers to experiment and be led to make improvements and adaptations suited to their particular applications. We emphasize that these programs should not be regarded as programs of library quality. They can be obtained via e-mail:

Georg@Math.ColoState.Edu.

Acknowledgment

We are grateful to numerous colleagues for supplying references and other information. Limitations of space, however, forced in many cases abbreviated treatments.

REFERENCES

R. Abraham and J. Robbin (1967), *Transversal Mappings and Flows*, W. A. Benjamin (New York).

I. Adler, G. C. Resende, G. Veiga and N. Karmarkar (1989), 'An implementation of Karmarkar's algorithm for linear programming', *Math. Programming* **44**, 297–335.

J. C. Alexander (1987), 'Average intersection and pivoting densities', *SIAM J. Numer. Anal.* **24**, 129–146.

J. C. Alexander and J. A. Yorke (1978), 'Homotopy continuation method: Numerically implementable topological procedures', *Trans. Amer. Math. Soc.* **242**, 271–284.

E. L. Allgower (1984), 'Bifurcations arising in the calculation of critical points via homotopy methods', in *Numerical Methods for Bifurcation Problems* Vol. 70 of *ISNM* (T. Küpper, H. D. Mittelmann and H. Weber, eds), Birkhäuser (Basel) 15–28.

E. L. Allgower and C.-S. Chien (1986), 'Continuation and local perturbation for multiple bifurcations', *SIAM J. Sci. Statist. Comput.* **7**, 1265–1281.

E. L. Allgower and K. Georg (1979), 'Generation of triangulations by reflections', *Utilitas Math.* **16**, 123–129.

E. L. Allgower and K. Georg (1983a), Predictor–corrector and simplicial methods for approximating fixed points and zero points of nonlinear mappings, in *Mathematical Programming: The State of the Art* (A. Bachem, M. Grötschel and B. Korte, eds), Springer (Berlin, Heidelberg, New York), 15–56.

E. L. Allgower and K. Georg (1983b), Relationships between deflation and global methods in the problem of approximating additional zeros of a system of nonlinear equations, in *Homotopy Methods and Global Convergence* (B. C. Eaves, F. J. Gould, H.-O. Peitgen and M. J. Todd, eds), Plenum (New York), 31–42.

E. L. Allgower and K. Georg (1990), *Numerical Continuation Methods: An Introduction*, Vol. 13 of *Series in Computational Mathematics*, Springer (Berlin, Heidelberg, New York), 388.

E. L. Allgower and S. Gnutzmann (1987), 'An algorithm for piecewise linear approximation of implicitly defined two-dimensional surfaces', *SIAM J. Numer. Anal.* **24**, 452–469.

E. L. Allgower and S. Gnutzmann (1991), 'Simplicial pivoting for mesh generation of implicitly defined surfaces', *Computer Aided Geometric Design* **8**, 305–325.

E. L. Allgower and P. H. Schmidt (1985), 'An algorithm for piecewise-linear approximation of an implicitly defined manifold', *SIAM J. Numer. Anal.* **22**, 322–346.

E. L. Allgower, K. Böhmer and M. Golubitsky, eds (1992a), *Bifurcation and Symmetry*, Vol. 104 of *ISNM*, Birkhäuser (Basel).

E. L. Allgower, K. Böhmer and Z. Mei (1991a), 'A complete bifurcation scenario for the 2-D nonlinear Laplacian with Neumann boundary conditions', in *Bifurcation and Chaos: Analysis, Algorithms, Applications* (R. Seydel, F. W. Schneider, T. Küpper and H. Troger, eds), Vol. 97 of *ISNM*, Birkhäuser (Basel), 1–18.

E. L. Allgower, K. Böhmer and Z. Mei (1991b), 'On new bifurcation results for semi-linear elliptic equations with symmetries', in *The Mathematics of Finite Elements and Applications VII* (J. Whiteman, ed.), MAFELAP 1990, Academic (Brunel), 487–494.

E. L. Allgower, K. Böhmer, K. Georg and R. Miranda (1992b), 'Exploiting symmetry in boundary element methods', *SIAM J. Numer. Anal.* **29**, 534–552.

E. L. Allgower, C.-S. Chien, K. Georg and C.-F. Wang (1991c), 'Conjugate gradient methods for continuation problems', *J. Comput. Appl. Math.* **38**, 1–16.

E. L. Allgower, K. Georg and R. Widmann (1991d), 'Volume integrals for boundary element methods', *J. Comput. Appl. Math.* **38**, 17–29.

D. C. S. Allison, S. Harimoto and L. T. Watson (1989), 'The granularity of parallel homotopy algorithms for polynomial systems of equations', *Int. J. Comput. Math.* **29**, 21–37.

V. Arun, C. F. Reinholtz and L. T. Watson (1990), 'Enumeration and analysis of variable geometry truss manipulators', Technical Report TR 90-10, Virginia Polytechnic Institute, Blacksburg.

B. Bank, J. Guddat, D. Klatte, B. Kummer and K. Tammer (1983), *Non-Linear Parametric Optimization*, Birkhäuser (Basel).

R. E. Bank and T. F. Chan (1986), 'PLTMGC: A multi-grid continuation program for parameterized nonlinear elliptic systems', *SIAM J. Sci. Statist. Comput.* **7**, 540–559.

D. Bayer and J. C. Lagarias (1989), 'The nonlinear geometry of linear programming, I: Affine and projective scaling trajectories, II: Legendre transform coordinates and central trajectories', *Trans. Amer. Math. Soc.* **314**, 499–581.

A. Ben-Israel and T. N. E. Greville (1974), *Generalized Inverses: Theory and Applications*, John Wiley (New York).

S. Bernstein (1910), 'Sur la généralisation du problème de Dirichlet', *Math. Ann.* **69**, 82–136.

R. E. Bixby, J. W. Gregory, I. J. Lustig, R. E. Marsten and D. F. Shanno (1991), 'Very large-scale linear programming: A case study in combining interior point and simplex methods', Technical Report RRR 34-91, Rutgers University, New Brunswick, NJ.

J. H. Bolstad and H. B. Keller (1986), 'A multigrid continuation method for elliptic problems with folds', *SIAM J. Sci. Statist. Comput.* **7**, 1081–1104.

F. H. Branin (1972), 'Widely convergent method for finding multiple solutions of simultaneous nonlinear equations', *IBM J. Res. Develop.* **16**, 504–522.

F. H. Branin and S. K. Hoo (1972), 'A method for finding multiple extrema of a function of n variables', in *Numerical Methods for Non-linear Optimization* (F. A. Lootsma, ed.), Academic (New York, London), 231–237.

F. Brezzi, J. Rappaz and P. A. Raviart (1980a), 'Finite dimensional approximation of nonlinear problems. Part 1: Branches of nonsingular solutions', *Numer. Math.* **36**, 1–25.

F. Brezzi, J. Rappaz and P. A. Raviart (1980b), 'Finite dimensional approximation of nonlinear problems. Part 2: Limit points', *Numer. Math.* **37**, 1–28.

F. Brezzi, J. Rappaz and P. A. Raviart (1981), 'Finite dimensional approximation of nonlinear problems. Part 3: Simple bifurcation points', *Numer. Math.* **38**, 1–30.

L. E. J. Brouwer (1912), 'Über Abbildung von Mannigfaltigkeiten', *Math. Ann.* **71**, 97–115.

K. M. Brown and W. B. Gearhart (1971), 'Deflation techniques for the calculation of further solutions of a nonlinear system', *Numer. Math.* **16**, 334–342.

T. J. Carpenter and D. F. Shanno (1991), 'An interior point method for quadratic programs based on conjugate projected gradients', Technical Report RRR 55-91, Rutgers University, New Brunswick, NJ.

T. F. Chan (1984a), 'Deflation techniques and block-elimination algorithms for solving bordered singular systems', *SIAM J. Sci. Statist. Comput.* **5**, 121–134.

T. F. Chan (1984b), 'Newton-like pseudo-arclength methods for computing simple turning points', *SIAM J. Sci. Statist. Comput.* **5**, 135–148.

C.-S. Chien (1989), 'Secondary bifurcations in the buckling problem', *J. Comput. Appl. Math.* **25**, 277–287.

S. N. Chow and J. K. Hale (1982), *Methods of Bifurcation Theory*, Springer (New York).

S. N. Chow, J. Mallet-Paret and J. A. Yorke (1978), 'Finding zeros of maps: Homotopy methods that are constructive with probability one', *Math. Comput.* **32**, 887–899.

S. N. Chow, J. Mallet-Paret and J. A. Yorke (1979), A homotopy method for locating all zeros of a system of polynomials, in *Functional Differential Equations and Approximation of Fixed Points* (H.-O. Peitgen and H.-O. Walther, eds), Vol. 730 of *Lecture Notes in Mathematics*, Springer (Berlin, Heidelberg, New York), 77–88.

M. T. Chu (1984a), 'The generalized Toda flow, the QR algorithm and the center manifold theory', *SIAM J. Alg. Disc. Meth.* **5**, 187–201.

M. T. Chu (1984b), 'A simple application of the homotopy method to symmetric eigenvalue problems', *Lin. Alg. Appl.* **59**, 85–90.

M. T. Chu (1986), 'A continuous approximation to the generalized Schur's decomposition', *Lin. Alg. Appl.* **78**, 119–132.

M. T. Chu (1988), 'On the continuous realization of iterative processes', *SIAM Rev.* **30**, 375–387.

M. T. Chu (1990), 'Solving additive inverse eigenvalue problems for symmetric matrices by the homotopy method', *IMA J. Numer. Anal.* **9**, 331–342.

M. T. Chu (1991), 'A continuous Jacobi-like approach to the simultaneous reduction of real matrices', *Lin. Alg. Appl.* **147**, 75–96.

M. T. Chu, T.-Y. Li and T. Sauer (1988), 'Homotopy methods for general λ-matrix problems', *SIAM J. Matrix Anal. Appl.* **9**, 528–536.

K. A. Cliffe and K. H. Winters (1986), 'The use of symmetry in bifurcation calculations and its application to the Bénard problem', *J. Comput. Phys.* **67**, 310–326.

R. W. Cottle (1974), 'Solution rays for a class of complementarity problems', *Math. Programming Study* **1**, 58–70.

R. W. Cottle and G. B. Dantzig (1968), 'Complementary pivot theory of mathematical programming', *Lin. Alg. Appl.* **1**, 103–125.

R. W. Cottle, F. Gianessi and J. L. Lions, eds (1980), *Variational Inequalities and Complementarity Problems*, John Wiley (London).

R. W. Cottle, G. H. Golub and R. S. Sacher (1978), 'On the solution of large structured linear complementarity problems: The block partitioned case', *Appl. Math. Optim.* **4**, 347–363.

H. S. M. Coxeter (1934), 'Discrete groups generated by reflections', *Ann. Math.* **6**, 13–29.

H. S. M. Coxeter (1973), *Regular Polytopes*, third edition, Dover (New York).

M. G. Crandall and P. H. Rabinowitz (1971), 'Bifurcation from simple eigenvalues', *J. Funct. Anal.* **8**, 321–340.

M. Crouzeix and J. Rappaz (1990), *On Numerical Approximation in Bifurcation Theory*, RMA, Masson (Paris).

R.-X. Dai and W. C. Rheinboldt (1990), 'On the computation of manifolds of fold points for parameter-dependent problems', *SIAM J. Numer. Anal.* **27**, 437–446.

Y. Dai and Y. Yamamoto (1989), 'The path following algorithm for stationary point problems on polyhedral cones', *J. Op. Res. Soc. Japan* **32**, 286–309.

Y. Dai, K. Sekitani and Y. Yamamoto (1992), 'A variable dimension algorithm with the Dantzig-Wolfe decomposition for structured stationary point problems', *ZOR — Methods and Models of Operations Research* **36**, 23–53.

G. B. Dantzig (1963), *Linear Programming and Extensions*, Princeton University (Princeton, NJ).

D. Davidenko (1953), 'On a new method of numerical solution of systems of nonlinear equations', *Dokl. Akad. Nauk USSR* **88**, 601–602. In Russian.

M. Dellnitz and B. Werner (1989), 'Computational methods for bifurcation problems with symmetries — with special attention to steady state and Hopf bifurcation points', *J. Comput. Appl. Math.* **26**, 97–123.

C. Den Heijer and W. C. Rheinboldt (1981), 'On steplength algorithms for a class of continuation methods', *SIAM J. Numer. Anal.* **18**, 925–948.

C. deSa, K. M. Irani, C. G. Ribbens, L. T. Watson and H. F. Walker (1992), 'Preconditioned iterative methods for homotopy curve tracking', *SIAM J. Sci. Statist. Comput.* **13**, 30–46.

P. Deuflhard and G. Heindl (1979), 'Affine invariant convergence theorems for Newton's method and extensions to related methods', *SIAM Numer. Anal.* **16**, 1–10.

P. Deuflhard, B. Fiedler and P. Kunkel (1987), 'Efficient numerical pathfollowing beyond critical points', *SIAM J. Numer. Anal.* **24**, 912–927.

E. Doedel and J. P. Kernévez (1986), *AUTO: Software for Continuation and Bifurcation Problems in Ordinary Differential Equations*, California Institute of Technology.

E. Doedel, H. B. Keller and J. P. Kernévez (1991a), 'Numerical analysis and control of bifurcation problems. Part I: Bifurcation in finite dimensions', *Int. J. Bifurcation and Chaos*.

E. Doedel, H. B. Keller and J. P. Kernévez (1991b), 'Numerical analysis and control of bifurcation problems. Part II: Bifurcation in infinite dimensions', *Int. J. Bifurcation and Chaos*.

B. C. Eaves (1972), 'Homotopies for the computation of fixed points', *Math. Programming* **3**, 1–22.

B. C. Eaves (1976), 'A short course in solving equations with pl homotopies', in *Nonlinear Programming* (R. W. Cottle and C. E. Lemke, eds), Vol. 9 of *SIAM-AMS Proc.*, American Mathematical Society (Providence, RI), 73–143.

B. C. Eaves (1984), *A Course in Triangulations for Solving Equations with Deformations*, Vol. 234 of *Lecture Notes in Economics and Mathematical Systems*, Springer (Berlin, Heidelberg, New York).

B. C. Eaves and R. Saigal (1972), 'Homotopies for computation of fixed points on unbounded regions', *Math. Programming* **3**, 225–237.

B. C. Eaves and H. Scarf (1976), 'The solution of systems of piecewise linear equations', *Math. Oper. Res.* **1**, 1–27.

B. C. Eaves and J. A. Yorke (1984), 'Equivalence of surface density and average directional density', *Math. Oper. Res.* **9**, 363–375.

A. V. Fiacco (1983), *Introduction to Sensitivity and Stability Analysis in Nonlinear Programming*, Academic (New York).

A. V. Fiacco, ed. (1984), *Sensitivity, Stability and Parametric Analysis*, Vol. 21 of *Mathematical Programming Study*, North-Holland (Amsterdam).

A. V. Fiacco and G. McCormick (1968), *Nonlinear Programming: Sequential Unconstrained Minimization Techniques*, John Wiley (New York, NY).

J. P. Fink and W. C. Rheinboldt (1983), 'On the discretization error of parametrized nonlinear equations', *SIAM J. Numer. Anal.* **20**, 732–746.

J. P. Fink and W. C. Rheinboldt (1984), 'Solution manifolds and submanifolds of parametrized equations and their discretization errors', *Numer. Math.* **45**, 323–343.

J. P. Fink and W. C. Rheinboldt (1985), 'Local error estimates for parametrized nonlinear equations', *SIAM J. Numer. Anal.* **22**, 729–735.

J. P. Fink and W. C. Rheinboldt (1986), 'Folds on the solution manifold of a parametrized equation', *SIAM J. Numer. Anal.* **23**, 693–706.

J. P. Fink and W. C. Rheinboldt (1987), 'A geometric framework for the numerical study of singular points', *SIAM J. Numer. Anal.* **24**, 618–633.

H. Freudenthal (1942), 'Simplizialzerlegungen von beschränkter Flachheit', *Ann. Math.* **43**, 580–582.

R. M. Freund (1984a), 'Variable dimension complexes. Part I: Basic theory', *Math. Oper. Res.* **9**, 479–497.

R. M. Freund (1984b), 'Variable dimension complexes. Part II: A unified approach to some combinatorial lemmas in topology', *Math. Oper. Res.* **9**, 498–509.

R. M. Freund (1991), 'A potential-function reduction algorithm for solving a linear program directly from an infeasible 'warm start' ', in *Interior Point Methods for Linear Programming: Theory and Practice* (C. Roos and J.-P. Vial, eds), Vol. 52 of *Math. Programming, Ser. B*, Mathematical Programming Society (North-Holland, Amsterdam), 441–446.

R. W. Freund, G. H. Golub and N. M. Nachtigal (1992), 'Iterative solution of linear systems', *Acta Numerica* **1**, 57–100.

C. B. Garcia and F. J. Gould (1978), 'A theorem on homotopy paths', *Math. Oper. Res.* **3**, 282–289.

C. B. Garcia and F. J. Gould (1980), 'Relations between several path following algorithms and local and global Newton methods', *SIAM Rev.* **22**, 263–274.

C. B. Garcia and W. I. Zangwill (1979a), 'An approach to homotopy and degree theory', *Math. Oper. Res.* **4**, 390–405.

C. B. Garcia and W. I. Zangwill (1979b), 'Finding all solutions to polynomial systems and other systems of equations', *Math. Programming* **16**, 159–176.

C. B. Garcia and W. I. Zangwill (1981), *Pathways to Solutions, Fixed Points, and Equilibria*, Prentice-Hall (Englewood Cliffs, NJ).

T. J. Garratt, G. Moore and A. Spence (1991), 'Two methods for the numerical detection of Hopf bifurcations', in *Bifurcation and Chaos: Analysis, Algorithms, Applications* (R. Seydel, F. W. Schneider, T. Küpper and H. Troger, eds), Vol. 97 of *ISNM*, Birkhäuser (Basel), 129–134.

K. Georg (1981), 'On tracing an implicitly defined curve by quasi-Newton steps and calculating bifurcation by local perturbation', *SIAM J. Sci. Statist. Comput.* **2**, 35–50.

K. Georg (1982), 'Zur numerischen Realisierung von Kontinuitätsmethoden mit Prädiktor-Korrektor- oder simplizialen Verfahren', Habilitationsschrift, University of Bonn (Germany).

K. Georg (1983), 'A note on stepsize control for numerical curve following', in *Homotopy Methods and Global Convergence* (B. C. Eaves, F. J. Gould, H.-O. Peitgen and M. J. Todd, eds), Plenum (New York), 145–154.

K. Georg (1990), 'An introduction to PL algorithms', in *Computational Solution of Nonlinear Systems of Equations* (E. L. Allgower and K. Georg, eds), Vol. 26 of *Lectures in Applied Mathematics*, American Mathematical Society (Providence, RI), 207–236.

K. Georg (1991), 'Approximation of integrals for boundary element methods', *SIAM J. Sci. Statist. Comput.* **12**, 443–453.

K. Georg and R. Miranda (1990), 'Symmetry aspects in numerical linear algebra with applications to boundary element methods', Preprint, Colorado State University.

K. Georg and R. Miranda (1992), 'Exploiting symmetry in solving linear equations', in *Bifurcation and Symmetry* (E. L. Allgower, K. Böhmer and M. Golubitsky, eds), Vol. 104 of *ISNM*, Birkhäuser (Basel), 157–168.

H. Gfrerer, J. Guddat and H.-J. Wacker (1983), 'A globally convergent algorithm based on imbedding and parametric optimization', *Computing* **30**, 225–252.

H. Gfrerer, J. Guddat, H.-J. Wacker and W. Zulehner (1985), 'Path-following for Kuhn-Tucker curves by an active set strategy', in *Systems and Optimization* (A. Baghi and H. T. Jongen, eds), Vol. 66 of *Lecture Notes Contr. Inf. Sc.*, Springer (Berlin, Heidelberg, New York), 111–132.

P. E. Gill, W. Murray, M. A. Saunders, J. A. Tomlin and M. H. Wright (1986), 'On projected Newton barrier methods for linear programming and an equivalence to Karmarkar's projective method', *Math. Programming* **36**, 183–209.

R. Glowinski, H. B. Keller and L. Reinhart (1985), 'Continuation-conjugate gradient methods for the least squares solution of nonlinear boundary value problems', *SIAM J. Sci. Statist. Comput.* **6**, 793–832.

S. Gnutzmann (1989), 'Stückweise lineare Approximation implizit definierter Mannigfaltigkeiten', PhD thesis, University of Hamburg, Germany.

D. Goldfarb and S. Liu (1991), 'An algorithm for solving linear programming problems in $O(n^3L)$ operations', *Math. Programming* **49**, 325–340.

G. H. Golub and C. F. van Loan (1989), *Matrix Computations*, second edn, J. Hopkins University (Baltimore, London).

M. Golubitsky and D. G. Schaeffer (1985), *Singularities and Groups in Bifurcation Theory*, Vol. 1, Springer (Berlin, Heidelberg, New York).

M. Golubitsky, I. Stewart and D. G. Schaeffer (1988), *Singularities and Groups in Bifurcation Theory*, Vol. 2, Springer (Berlin, Heidelberg, New York).

C. C. Gonzaga (1988), 'An algorithm for solving linear programming problems in $O(n^3 L)$ operations,, in *Progress in Mathematical Programming, Interior Point and Related Methods* (N. Megiddo, ed.), Springer (New York), 1–28.

C. C. Gonzaga (1992), 'Path-following methods for linear programming', *SIAM Rev.* **34**, 167–224.

F. J. Gould and J. W. Tolle (1983), *Complementary Pivoting on a Pseudomanifold Structure with Applications on the Decision Sciences*, Vol. 2 of *Sigma Series in Applied Mathematics*, Heldermann (Berlin).

A. Griewank (1985), 'On solving nonlinear equations with simple singularities or nearly singular solutions', *SIAM Rev.* **27**, 537–563.

A. Griewank and G. W. Reddien (1984), 'Characterization and computation of generalized turning points', *SIAM J. Numer. Anal.* **21**, 176–185.

J. Guddat, F. Guerra Vasquez and H. T. Jongen (1990), *Parametric Optimization: Singularities, Path Following, and Jumps*, John Wiley (Chichester).

J. Guddat, H. T. Jongen, B. Kummer and F. Nožička, eds (1987), *Parametric Optimization and Related Topics*, Akademie (Berlin).

C. B. Haselgrove (1961), 'The solution of nonlinear equations and of differential equations with two-point boundary conditions', *Comput. J.* **4**, 255–259.

T. J. Healey (1988a), 'Global bifurcation and continuation in the presence of symmetry with an application to solid mechanics', *SIAM J. Math. Anal.* **19**, 824–840.

T. J. Healey (1988b), 'A group theoretic approach to computational bifurcation problems with symmetry', *Comput. Meth. Appl. Mech. Eng.* **67**, 257–295.

T. J. Healey (1989), 'Symmetry and equivariance in nonlinear elastostatics. Part I', *Arch. Rational Mech. Anal.* **105**, 205–228.

T. J. Healey and J. A. Treacy (1991), 'Exact block diagonalization of large eigenvalue problems for structures with symmetry', *Int. J. Numer. Methods Engrg.* **31**, 265–285.

M. E. Henderson (1985), 'Complex Bifurcation', PhD thesis, CALTECH, Pasadena.

M. E. Henderson and H. B. Keller (1990), 'Complex bifurcation from real paths', *SIAM J. Appl. Math.* **50**, 460–482.

M. W. Hirsch (1963), 'A proof of the nonretractibility of a cell onto its boundary', *Proc. Amer. Math. Soc.* **14**, 364–365.

M. W. Hirsch (1976), *Differential Topology*, Springer (Berlin, Heidelberg, New York).

M. Holodniok and M. Kubíček (1984), 'DERPER — An algorithm for the continuation of periodic solutions in ordinary differential equations', *J. Comput. Phys.* **55**, 254–267.

B. Hong (1991), 'Computational methods for bifurcation problems with symmertries on the manifold', Technical Report ICMA-91-163, University of Pittsburgh, PA.

U. Hornung and H. D. Mittelmann (1991), 'Bifurcation of axially symmetric capillary surfaces', *J. Coll. Interf. Sci.* **146**, 219–225.

J. Huitfieldt (1991), 'Nonlinear eigenvalue problems — prediction of bifurcation points and branch switching', Technical Report 17, University of Göteborg, Sweden.

J. Huitfieldt and A. Ruhe (1990), 'A new algorithm for numerical path following applied to an example from hydrodynamical flow', *SIAM J. Sci. Statist. Comput.* **11**, 1181–1192.

A. D. Jepson and D. W. Decker (1986), 'Convergence cones near bifurcation', *SIAM J. Numer. Anal.* **23**, 959–975.

A. D. Jepson, A. Spence and K. A. Cliffe (1991), 'The numerical solution of nonlinear equations having several parameters. Part III: Equations with Z_2-symmetry', *SIAM J. Numer. Anal.* **28**, 809–832.

H. T. Jongen and G.-W. Weber (1990), 'On parametric nonlinear programming', *Ann. Oper. Res.* **27**, 253–284.

H. T. Jongen, P. Jonker and F. Twilt (1983), *Nonlinear Optimization in R^N. I. Morse Theory, Chebyshev Approximation*, Peter Lang (New York).

H. T. Jongen, P. Jonker and F. Twilt (1986), *Nonlinear Optimization in R^N. II. Transversality, Flows, Parametric Aspects*, Peter Lang (New York).

C. Kaas-Petersen (1989), *PATH — User's Guide*, University of Leeds, England.

K. Kamiya and A. J. J. Talman (1990), 'Variable dimension simplicial algorithm for balanced games', Technical Report 9025, Tilburg University, The Netherlands.

N. K. Karmarkar (1984), 'A new polynomial-time algorithm for linear programming', *Combinatorica* **4**, 373–395.

N. K. Karmarkar and K. G. Ramakrishnan (1991), 'Computational results of an interior point algorithm for large scale linear programming', in *Interior Point Methods for Linear Programming: Theory and Practice* (C. Roos and J.-P. Vial, eds), Vol. 52 of *Math. Programming, Ser. B*, Mathematical Programming Society, North-Holland (Amsterdam), 555–586.

R. B. Kearfott (1989), 'An interval step control for continuation methods', *Math. Comput.* to appear.

R. B. Kearfott (1990), 'Interval arithmetic techniques in the computational solution of nonlinear systems of equations: Introduction, examples and comparisons', in *Computational Solution of Nonlinear Systems of Equations* (E. L. Allgower and K. Georg, eds), Vol. 26 of *Lectures in Applied Mathematics*, American Mathematical Society (Providence, RI), 337–357.

J. P. Keener and H. B. Keller (1974), 'Perturbed bifurcation theory', *Arch. Rational Mech. Anal.* **50**, 159–175.

H. B. Keller (1970), 'Nonlinear bifurcation', *J. Diff. Eq.* **7**, 417–434.

H. B. Keller (1977), 'Numerical solution of bifurcation and nonlinear eigenvalue problems', in *Applications of Bifurcation Theory* (P. H. Rabinowitz, ed.), Academic (New York, London), 359–384.

H. B. Keller (1978), 'Global homotopies and Newton methods', in *Recent Advances in Numerical Analysis* (C. de Boor and G. H. Golub, eds), Academic (New York), 73–94.

H. B. Keller (1983), 'The bordering algorithm and path following near singular points of higher nullity', *SIAM J. Sci. Statist. Comput.* **4**, 573–582.

H. B. Keller (1987), *Lectures on Numerical Methods in Bifurcation Problems*, Springer (Berlin, Heidelberg, New York).

R. B. Kellogg, T.-Y. Li and J. A. Yorke (1976), 'A constructive proof of the Brouwer fixed point theorem and computational results', *SIAM J. Numer. Anal.* **13**, 473–483.

L. G. Khachiyan (1979), 'A polynomial algorithm in linear programming', *Sov. Math. Dokl.* **20**, 191–194.

F. Klein (1882–1883), 'Neue Beiträge zur Riemannschen Funktionentheorie', *Math. Ann.*

M. Kojima (1974), 'Computational methods for solving the nonlinear complementarity problem', *Keio Engrg. Rep.* **27**, 1–41.

M. Kojima (1979), 'A complementarity pivoting approach to parametric programming', *Math. Oper. Res.* **4**, 464–477.

M. Kojima and R. Hirabayashi (1984), 'Sensitivity, stability and parametric analysis', Vol. 21 of *Mathematical Programming Study*, North-Holland (Amsterdam), 150–198.

M. Kojima and Y. Yamamoto (1982), 'Variable dimension algorithms: Basic theory, interpretation, and extensions of some existing methods', *Math. Programming* **24**, 177–215.

M. Kojima and Y. Yamamoto (1984), 'A unified approach to the implementation of several restart fixed point algorithms and a new variable dimension algorithm', *Math. Programming* **28**, 288–328.

M. Kojima, N. Megiddo and S. Mizuno (1990a), 'A general framework of continuation methods for complementarity problems', Technical report, IBM Almaden Research Center, San Jose, CA.

M. Kojima, N. Megiddo and S. Mizuno (1991a), 'Theoretical convergence of large-step primal-dual interior point algorithms for linear programming', preprint, Tokyo Inst. of Techn.

M. Kojima, N. Megiddo and T. Noma (1991b), 'Homotopy continuation methods for nonlinear complementarity problems', *Math. Oper. Res.* **16**, 754–774.

M. Kojima, N. Megiddo and Y. Ye (1992), 'An interior point potential reduction algorithm for the linear complementarity problem', *Math. Programming* **54**, 267–279.

M. Kojima, N. Megiddo, T. Noma and A. Yoshise (1991c), *A Unified Approach to Interior Point Algorithms for Linear Complementarity Problems*, Vol. 538 of *Lecture Notes in Computer Science*, Springer (Berlin).

M. Kojima, S. Mizuno and T. Noma (1990b), 'Limiting behavior of trajectories generated by a continuation method for monotone complementarity problems', *Math. Oper. Res.* **15**, 662–675.

M. Kojima, S. Mizuno and A. Yoshise (1988), 'A primal-dual interior point algorithm for linear programming', in *Progress in Mathematical Programming, Interior Point and Related Methods* (N. Megiddo, ed.), Springer (New York), 29–47.

M. Kojima, S. Mizuno and A. Yoshise (1989), 'A polynomial-time algorithm for a class of linear complementarity problems', *Math. Programming* **44**, 1–26.

M. Kojima, S. Mizuno and A. Yoshise (1991d), 'An $O(\sqrt{n}L)$ iteration potential reduction algorithm for linear complementarity problems', *Math. Programming* **50**, 331–342.

M. Kojima, H. Nishino and T. Sekine (1976), 'An extension of Lemke's method to the piecewise linear complementarity problem', *SIAM J. Appl. Math.* **31**, 600–613.

M. A. Krasnosel'skiĭ (1964), *Topological Methods in the Theory of Nonlinear Integral Equations*, Pergamon (New York).

M. Kubíček (1976), 'Algorithm 502. Dependence of solutions of nonlinear systems on a parameter', *ACM Trans. Math. Software* **2**, 98–107.

H. W. Kuhn (1969), 'Approximate search for fixed points', in *Computing Methods in Optimization Problems 2* (L. A. Zadek, L. W. Neustat and A. V. Balakrishnan, eds), Academic (New York), 199–211.

E. Lahaye (1934), 'Une méthode de resolution d'une categorie d'equations transcendantes', *C. R. Acad. Sci. Paris* **198**, 1840–1842.

C. E. Lemke (1965), 'Bimatrix equilibrium points and mathematical programming', *Management Sci.* **11**, 681–689.

C. E. Lemke (1980), 'A survey of complementarity theory', in *Variational Inequalities and Complentarity Problems* (R. W. Cottle, F. Gianessi and J. L. Lions, eds), John Wiley (London).

C. E. Lemke and S. J. Grotzinger (1976), 'On generalizing Shapley's index theory to labelled pseudo manifolds', *Math. Programming* **10**, 245–262.

C. E. Lemke and J. T. Howson (1964), 'Equilibrium points of bimatrix games', *SIAM J. Appl. Math.* **12**, 413–423.

J. Leray and J. Schauder (1934), 'Topologie et Équations fonctionelles', *Ann. Sci. École Norm. Sup.* **51**, 45–78.

K. Li and T.-Y. Li (1992), 'An algorithm for symmetric tridiagonal eigen-problems — divide and conquer with homotopy continuation', Preprint, Michigan State University.

T.-Y. Li and N. H. Rhee (1989), 'Homotopy algorithm for symmetric eigenvalue problems', *Numer. Math.* **55**, 265–280.

T.-Y. Li and X. Wang (1992a), 'Nonlinear homotopies for solving deficient polynomial systems with parameters', *SIAM J. Numer. Anal.* to appear.

T.-Y. Li and X. Wang (1992b), 'Solving real polynomial systems with real homotopies', *Math. Comput.* to appear.

T.-Y. Li and Z. Zeng (1992), 'Homotopy-determinant algorithm for solving nonsymmetric eigenvalue problems', *Math. Comput.* to appear.

T.-Y. Li, T. Sauer and J. A. Yorke (1987), 'Numerical solution of a class of deficient polynomial systems', *SIAM J. Numer. Anal.* **24**, 435–451.

T.-Y. Li, T. Sauer and J. A. Yorke (1989), 'The cheater's homotopy: An efficient procedure for solving systems of polynomial equations', *SIAM J. Numer. Anal.* **26**, 1241–1251.

T.-Y. Li, Z. Zeng and L. Cong (1992), 'Solving eigenvalue problems of real nonsymmetric matrices with real homotopies', *SIAM J. Numer. Anal.* **29**, 229–248.

T.-Y. Li, H. Zhang and X.-H. Sun (1991), 'Parallel homotopy algorithm for the symmetric tridiagonal eigenvalue problem', *SIAM J. Sci. Statist. Comput.* **12**, 469–487.

J. L. Liu and W. C. Rheinboldt (1991), 'A posteriori error estimates for parametrized nonlinear equations', in *Nonlinear Computational Mechanics* (P. Wriggers and W. Wagner, eds), Springer (Heidelberg, Germany), 31–46.

B. N. Lundberg and A. B. Poore (1991), 'Variable order Adams-Bashforth predictors with error-stepsize control for continuation methods', *SIAM J. Sci. Statist. Comput.* **12**, 695–723.

B. N. Lundberg and A. B. Poore (1993), 'Numerical continuation and singularity detection methods for parametric nonlinear programming', *SIAM J. Optim.* to appear.

I. J. Lustig, R. E. Marsten and D. F. Shanno (1991), 'The interaction of algorithms and architectures for interior point methods', Technical Report RRR 36-91, Rutgers Univ., New Brunswick, NJ.

W. Mackens (1989), 'Numerical differentiation of implicitly defined space curves', *Computing* **41**, 237–260.

R. Marsten, R. Subramanian, M. Saltzman, I. J. Lustig and D. F. Shanno (1990), 'Interior point methods for linear programming: Just call Newton, Lagrange and Fiacco and Mccormick!', *Interfaces* **20**, 105–116.

H. Maurer and H. D. Mittelmann (1991), 'The nonlinear beam via optimal control with bounded state variables', *Optimal Control Appl. Methods* **12**, 19–31.

K. A. McShane, C. L. Monma and D. F. Shanno (1989), 'An implementation of a primal-dual interior point method for linear programming', *ORSA J. Comput.* **1**, 70–83.

N. Megiddo (1988), 'Pathways to the optimal set in linear programming', in *Progress in Mathematical Programming, Interior Point and Related Methods* (N. Megiddo, ed.), Springer (New York), 131–158.

N. Megiddo and M. Shub (1989), 'Boundary behavior of interior point algorithms in linear programming', *Math. Oper. Res.* **14**, 97–146.

R. Mejia (1986), 'CONKUB: A conversational path-follower for systems of nonlinear equations', *J. Comput. Phys.* **63**, 67–84.

R. Mejia (1990), 'Interactive program for continuation of solutions of large systems of nonlinear equations', in *Computational Solution of Nonlinear Systems of Equations* (E. L. Allgower and K. Georg, eds), Vol. 26 of *Lectures in Applied Mathematics*, American Mathematical Society (Providence, RI), 429–449.

R. G. Melhem and W. C. Rheinboldt (1982), 'A comparison of methods for determining turning points of nonlinear equations', *Computing* **29**, 201–226.

R. C. Melville, L. Trajkovic, S.-C. Fang and L. T. Watson (1990), 'Globally convergent homotopy methods for the DC operating point problem', Technical Report TR 90-61, Virginia Polytechnic Institute, Blacksburg.

R. Menzel and H. Schwetlick (1978), 'Zur Lösung parameterabhängiger nichtlinearer Gleichungen mit singulären Jacobi-Matrizen', *Numer. Math.* **30**, 65–79.

R. Menzel and H. Schwetlick (1985), 'Parametrization via secant length and application to path following', *Numer. Math.* **47**, 401–412.

O. Merrill (1972), 'Applications and extensions of an algorithm that computes fixed points of a certain upper semi-continuous point to set mapping', PhD thesis, University of Michigan, Ann Arbor, MI.

E. Miersemann and H. D. Mittelmann (1989), 'Continuation for parametrized nonlinear variational inequalities', *J. Comput. Appl. Math.* **26**, 23–34.

E. Miersemann and H. D. Mittelmann (1990a), 'Extension of Beckert's continuation method to variational inequalities', *Math. Nachr.* **148**, 183–195.

E. Miersemann and H. D. Mittelmann (1990b), 'A free boundary problem and stability for the rectangular plate', *Math. Methods Appl. Sci.* **12**, 129–138.

E. Miersemann and H. D. Mittelmann (1990c), 'On the stability in obstacle problems with applications to the beam and plate', *Z. Angew. Math. Mech.* **71**, 311–321.

E. Miersemann and H. D. Mittelmann (1991), 'Stability and continuation of solutions to obstacle problems', *J. Comput. Appl. Math.* **35**, 5–31.

E. Miersemann and H. D. Mittelmann (1992), 'Stability in obstacle problems for the von Karman plate', *SIAM J. Math. Anal.* to appear.

J. W. Milnor (1969), *Topology from the Differentiable Viewpoint*, University Press of Virginia (Charlottesville, VA).

H. D. Mittelmann (1990), 'Nonlinear parametrized equations: New results for variational problems and inequalities', in *Computational Solution of Nonlinear Systems of Equations* (E. L. Allgower and K. Georg, eds), Vol. 26 of *Lectures in Applied Mathematics*, American Mathematical Society (Providence, RI), 451–456.

H. D. Mittelmann and D. Roose, eds (1990), *Continuation Techniques and Bifurcation Problems*, Vol. 92 of *ISNM*, Birkhäuser (Berlin).

S. Mizuno (1992), 'A new polynomial time method for a linear complementarity problem', *Math. Programming* **56**, 31–43.

S. Mizuno, M. J. Todd and Y. Ye (1992), 'On adaptive-step primal-dual interior-point algorithms for linear programming', *Math. Oper. Res.* to appear.

S. Mizuno, A. Yoshise and T. Kikuchi (1989), 'Practical polynomial time algorithms for linear complementarity problems', *J. Oper. Res. Soc. Japan* **32**, 75–92.

R. C. Monteiro and I. Adler (1989), 'Interior path following primal-dual algorithms, I: Linear programming, II: Convex quadratic programming', *Math. Programming* **44**, 27–66.

G. Moore and A. Spence (1980), 'The calculation of turning points of nonlinear equations', *SIAM J. Numer. Anal.* **17**, 567–576.

A. P. Morgan (1986), 'A transformation to avoid solutions at infinity for polynomial systems', *Appl. Math. Comput.* **18**, 77–86.

A. P. Morgan (1987), *Solving Polynomial Systems Using Continuation for Engineering and Scientific Problems*, Prentice-Hall (Englewood Cliffs, NJ).

A. P. Morgan and A. J. Sommese (1987), 'A homotopy for solving general polynomial systems that respects m-homogeneous structures', *Appl. Math. Comput.* **24**, 101–113.

A. P. Morgan and A. J. Sommese (1989), 'Coefficient parameter polynomial continuation', *Appl. Math. Comput.* **29**, 123–160.

A. P. Morgan, A. J. Sommese and C. W. Wampler (1991a), 'Computing singular solutions to nonlinear analytic systems', *Numer. Math.* **58**, 669–684.

A. P. Morgan, A. J. Sommese and C. W. Wampler (1991b), 'Computing singular solutions to polynomial systems', *Adv. Appl. Math.* to appear.

A. P. Morgan, A. J. Sommese and C. W. Wampler (1992), 'A power series method for computing singular solutions to nonlinear analytic systems', *Numer. Math.* to appear.

A. P. Morgan, A. J. Sommese and L. T. Watson (1989), 'Finding all isolated solutions to polynomial systems using HOMPACK', *ACM Trans. Math. Software* **15**, 93–122.

B. A. Murtagh and M. A. Saunders (1987), *MINOS 5.1 Users Guide*, Stanford University, CA.

J. M. Ortega and W. C. Rheinboldt (1970), *Iterative Solution of Nonlinear Equations in Several Variables*, Academic Press (New York, London).

H.-O. Peitgen (1982), 'Topologische Perturbationen beim globalen numerischen Studium nichtlinearer Eigenwert- und Verzweigungsprobleme', *Jahresber. Deutsch. Math.-Verein.* **84**, 107–162.

H.-O. Peitgen and H. W. Siegberg (1981), 'An $\bar{\varepsilon}$-perturbation of Brouwer's definition of degree', in *Fixed Point Theory* (E. Fadell and G. Fournier, eds), Vol. 886 of *Lecture Notes in Mathematics*, Springer (Berlin, Heidelberg, New York), 331–366.

P. Percell (1980), 'Note on a global homotopy', *Numer. Funct. Anal. Optim.* **2**, 99–106.

H. Poincaré (1881–1886), *Sur les Courbes Defi_é par une Équation Differentielle. I–IV*, Oeuvres I. Gauthier-Villars (Paris).

E. Polak, J. E. Higgins and D. Q. Mayne (1992), 'A barrier function method for minimax problems', *Math. Programming* **54**, 155–176.

G. Pönisch and H. Schwetlick (1981), 'Computing turning points of curves implicitly defined by nonlinear equations depending on a parameter', *Computing* **26**, 107–121.

A. B. Poore and C. A. Tiahrt (1987), 'Bifurcation problems in nonlinear parametric programming', *Math. Programming* **39**, 189–205.

A. B. Poore and C. A. Tiahrt (1990), 'A bifurcation analysis of the nonlinear parametric programming problem', *Math. Programming* **47**, 117–141.

P. H. Rabinowitz (1971), 'Some global results for nonlinear eigenvalue problems', *J. Funct. Anal.* **7**, 487–513.

J. Rakowska, R. T. Haftka and L. T. Watson (1991), 'An active set algorithm for tracing parametrized optima', *Structural Optimization* **3**, 29–44.

J. Renegar (1985), 'On the complexity of a piecewise linear algorithm for approximating roots of complex polynomials', *Math. Programming* **32**, 301–318.

J. Renegar (1987), 'On the efficiency of Newton's method in approximating all zeros of systems of complex polynomials', *Math. Oper. Res.* **12**, 121–148.

J. Renegar (1988a), 'A polynomial-time algorithm, based on Newton's method, for linear programming', *Math. Programming* **40**, 59–93.

J. Renegar (1988b), 'Rudiments of an average case complexity theory for piecewise-linear path following algorithms', *Math. Programming* **40**, 113–163.

J. Renegar and M. Shub (1992), 'Unified complexity analysis for Newton LP methods', *Math. Programming* **53**, 1–16.

W. C. Rheinboldt (1978), 'Numerical methods for a class of finite dimensional bifurcation problems', *SIAM J. Numer. Anal.* **15**, 1–11.

W. C. Rheinboldt (1980), 'Solution fields of nonlinear equations and continuation methods', *SIAM J. Numer. Anal.* **17**, 221–237.

W. C. Rheinboldt (1981), 'Numerical analysis of continuation methods for nonlinear structural problems', *Comput. and Structures* **13**, 103–113.

W. C. Rheinboldt (1986), *Numerical Analysis of Parametrized Nonlinear Equations*, John Wiley (New York).

W. C. Rheinboldt (1987), 'On a moving-frame algorithm and the triangulation of equilibrium manifolds', in *Bifurcation: Analysis, Algorithms, Applications* (T. Küpper, R. Seydel and H. Troger, eds), Vol. 79 of *ISNM*, Birkhäuser (Basel), 256–267.

W. C. Rheinboldt (1988a), 'On a theorem of S. Smale about Newton's method for analytic mappings', *Appl. Math. Lett.* 1, 69–72.

W. C. Rheinboldt (1988b), 'On the computation of multi-dimensional solution manifolds of parametrized equations', *Numer. Math.* 53, 165–182.

W. C. Rheinboldt (1991), 'On the sensitivity of solutions of parametrized equations', Technical Report ICMA-91-158, University of Pittsburgh.

W. C. Rheinboldt (1992a), 'On the sensitivity of parametrized equations', *SIAM J. Numer. Anal.* to appear.

W. C. Rheinboldt (1992b), 'On the theory and numerics of differential-algebraic equations', in *Advances in Numerical Analysis* (W. Light, ed.), Oxford University Press (Oxford), 237–175.

W. C. Rheinboldt and J. V. Burkardt (1983a), 'Algorithm 596: A program for a locally-parametrized continuation process', *ACM Trans. Math. Software* 9, 236–241.

W. C. Rheinboldt and J. V. Burkardt (1983b), 'A locally-parametrized continuation process', *ACM Trans. Math. Software* 9, 215–235.

W. C. Rheinboldt, D. Roose and R. Seydel (1990), 'Aspects of continuation software', in *Continuation and Bifurcations: Numerical Techniques and Applications* (D. Roose, B. de Dier and A. Spence, eds), Vol. 313 of *NATO ASI Series C*, Kluwer (Dordrecht), 261–268.

S. M. Robinson (1987), 'Local structure of feasible sets in nonlinear programming, Part III: Stability and sensitivity', *Math. Programming Study* 30, 45–66.

C. Roos and J.-P. Vial, eds (1991), *Interior Point Methods for Linear Programming: Theory and Practice*, Vol. 52 of *Math. Programming, Ser. B*, Mathematical Programming Society, North-Holland (Amsterdam).

D. Roose, B. de Dier and A. Spence, eds (1990), *Continuation and Bifurcations: Numerical Techniques and Applications*, Vol. 313 of *NATO ASI Series C*, Kluwer (Dordrecht).

R. Saigal (1971), 'Lemke's algorithm and a special linear complementarity problem', *Oper. Res.* 8, 201–208.

R. Saigal (1976), 'Extension of the generalized complementarity problem', *Math. Oper. Res.* 1, 260–266.

R. Saigal (1977), 'On the convergence rate of algorithms for solving equations that are based on methods of complementary pivoting', *Math. Oper. Res.* 2, 108–124.

R. Saigal (1984), 'Computational complexity of a piecewise linear homotopy algorithm', *Math. Programming* 28, 164–173.

R. Saigal and M. J. Todd (1978), 'Efficient acceleration techniques for fixed point algorithms', *SIAM J. Numer. Anal.* 15, 997–1007.

D. Saupe (1982), 'On accelerating PL continuation algorithms by predictor-corrector methods', *Math. Programming* 23, 87–110.

H. E. Scarf (1967), 'The approximation of fixed points of a continuous mapping', *SIAM J. Appl. Math.* 15, 1328–1343.

H. Schwetlick (1984a), 'Algorithms for finite-dimensional turning point problems from viewpoint to relationships with constrained optimization methods', in *Numerical methods for bifurcation problems* (T. Küpper, H. Mittelmann and H. Weber, eds), Birkhäuser (Basel), 459–479.

H. Schwetlick (1984b), 'Effective methods for computing turning points of curves implicitly defined by nonlinear equations', in *Computational Mathematics* (A. Wakulicz, ed.), Vol. 13 of *Banach Center Publications*, PWN (Polish Scientific Publ.) (Warsaw), 623–645.

H. Schwetlick and J. Cleve (1987), 'Higher order predictors and adaptive stepsize control in path following algorithms', *SIAM J. Numer. Anal.* **24**, 1382–1393.

R. Seydel (1988), *From Equilibrium to Chaos. Practical Bifurcation and Stability Analysis*, Elsevier (New York).

R. Seydel (1991a), *BIFPACK: A Program Package for Continuation, Bifurcation and Stability Analysis, Version 2.3+*, University of Ulm (Germany).

R. Seydel (1991b), 'On detecting stationary bifurcations', *Int. J. Bifurcation and Chaos* **1**, 335–337.

R. Seydel, F. W. Schneider, T. Küpper and H. Troger, eds (1991), *Bifurcation and Chaos: Analysis, Algorithms, Applications*, Vol. 97 of *ISNM*, Birkhäuser (Basel).

L. F. Shampine and M. K. Gordon (1975), *Computer Solutions of Ordinary Differential Equations. The Initial Value Problem*, W. H. Freeman (San Francisco).

L. S. Shapley (1974), 'A note on the Lemke-Howson algorithm', in *Pivoting and Extensions: In Honor of A. W. Tucker* (M. L. Balinski, ed.), Vol. 1 of *Math. Programming Study*, North-Holland (New York), 175–189.

G. M. Shroff and H. B. Keller (1991), 'Stabilization of unstable procedures: A hybrid algorithm for continuation', *SIAM J. Numer. Anal.* to appear.

M. Shub and S. Smale (1991), 'Complexity of Bezout's theorem, I, geometric aspects', IBM Research Report.

S. Smale (1976), 'A convergent process of price adjustement and global Newton methods', *J. Math. Econom.* **3**, 1–14.

S. Smale (1986), 'Newton's method estimates from data at one point', in *The Merging of Disciplines in Pure, Applied and Computational Mathematics*, Springer (New York), 185–196.

G. Sonnevend (1985), 'An analytical center for polyhedrons and new classes of global algorithms for linear (smooth, convex) programming', Vol. 84 of *Lecture Notes in Control and Information Sciences*, Springer (New York), 866–876.

G. Sonnevend, J. Stoer and G. Zhao (1989), 'On the complexity of following the central path of linear programs by linear extrapolation', *Meth. Oper. Res.* **63**, 19–31.

G. Sonnevend, J. Stoer and G. Zhao (1991), 'On the complexity of following the central path of linear programs by linear extrapolation II', in *Interior Point Methods for Linear Programming: Theory and Practice* (C. Roos and J.-P. Vial, eds), Vol. 52 of *Math. Programming, Ser. B*, Mathematical Programming Society, North-Holland (Amsterdam), 527–553.

A. J. J. Talman and Y. Yamamoto (1989), 'A simplicial algorithm for stationary point problems on polytopes', *Math. Oper. Res.* **14**, 383–399.

M. J. Todd (1976a), *The Computation of Fixed Points and Applications*, Vol. 124 of *Lecture Notes in Economics and Mathematical Systems*, Springer (Berlin, Heidelberg, New York).

M. J. Todd (1976b), 'Extensions of Lemke's algorithm for the linear complementarity problem', *J. Optim. Theory Appl.* **20**, 397–416.

M. J. Todd (1976c), 'Orientation in complementary pivot algorithms', *Math. Oper. Res.* **1**, 54–66.

M. J. Todd (1978), 'Fixed-point algorithms that allow restarting without extra dimension', Technical Report, Cornell University, Ithaca, NY.

M. J. Todd (1981), *PLALGO: A FORTRAN Implementation of a Piecewise-linear Homotopy Algorithm for Solving Systems of Nonlinear Equations*, School of Operations Research and Industrial Engineering, Cornell University (Ithaca, NY).

M. J. Todd (1982), 'On the computational complexity of piecewise-linear homotopy algorithms', *Math. Programming* **24**, 216–224.

M. J. Todd (1986), 'Polynomial expected behavior of a pivoting algorithm for linear complementarity and linear programming problems', *Math. Programming* **35**, 173–192.

M. J. Todd (1989), 'Recent developments and new directions in linear programming', in *Mathematical Programming, Recent Developments and Applications* (N. Iri and K. Tanabe, eds), Kluwer (London), 109–157.

A. Ushida and L. O. Chua (1984), 'Tracing solution curves of nonlinear equations with sharp turning points', *Int. J. Circuit Theory Appl.* **12**, 1–21.

P. M. Vaidya (1990), 'An algorithm for linear programming which requires $O(((m+n)n^2 + (m+n)^{1.5}n)L)$ arithmetic operations', *Math. Programming* **47**, 175–202.

G. van der Laan and A. J. J. Talman (1979), 'A restart algorithm for computing fixed points without an extra dimension', *Math. Programming* **17**, 74–84.

A. Vanderbauwhede (1982), *Local Bifurcation Theory and Symmetry*, Pitman (London).

G. Vasudevan, F. H. Lutze and L. T. Watson (1990), 'A homotopy method for space flight rendezvous problems', in *Astrodynamics 1989* (C. L. Thornton, R. J. Proulx, J. E. Prussing and F. R. Hoots, eds), Vol. 71 of *Advances in the Astronautical Sciences*, 533–548.

C. W. Wampler and A. P. Morgan (1991), 'Solving the 6R inverse position problem using a generic-case solution methodology', *Mech. Mach. Theory* **26**, 91–106.

C. W. Wampler, A. P. Morgan and A. J. Sommese (1990), 'Numerical continuation methods for solving polynomial systems arising in kinematics', *ASME J. on Design* **112**, 59–68.

C. W. Wampler, A. P. Morgan and A. J. Sommese (1992), 'Complete solution of the nine-point path synthesis problem for four-bar linkages', *ASME J. Mech. Des..* to appear.

D. S. Watkins (1984), 'Isospectral flows', *SIAM Rev.* **26**, 379–391.

L. T. Watson (1981), 'Engineering application of the Chow-Yorke algorithm', *Appl. Math. Comput.* **9**, 111–133.

L. T. Watson (1986), 'Numerical linear algebra aspects of globally convergent homotopy methods', *SIAM Rev.* **28**, 529–545.

L. T. Watson and C. Y. Wang (1981), 'A homotopy method applied to elastica problems', *Int. J. Solids Structures* **17**, 29–37.

L. T. Watson and W. H. Yang (1980), 'Optimal design by a homotopy method', *Applicable Anal.* **10**, 275–284.

L. T. Watson, S. C. Billups and A. P. Morgan (1987), 'HOMPACK: A suite of codes for globally convergent homotopy algorithms', *ACM Trans. Math. Software* **13**, 281–310.

L. T. Watson, T.-Y. Li and C. Y. Wang (1978), 'Fluid dynamics of the elliptic porous slider', *J. Appl. Mech.* **45**, 435–436.

B. Werner (1992), 'Test functions for bifurcation points and Hopf points in problems with symmetries', in *Bifurcation and Symmetry* (E. L. Allgower, K. Böhmer and M. Golubitsky, eds), Vol. 104 of *ISNM*, Birkhäuser (Basel), 317–327.

R. Widmann (1990a), 'An efficient algorithm for the triangulation of surfaces in $\mathbf{R}^3$', Preprint, Colorado State University.

R. Widmann (1990b), 'Efficient triangulation of 3-dimensional domains', Preprint, Colorado State University.

A. H. Wright (1981), 'The octahedral algorithm, a new simplicial fixed point algorithm', *Math. Programming* **21**, 47–69.

A. H. Wright (1985), 'Finding all solutions to a system of polynomial equations', *Math. Comput.* **44**, 125–133.

M. H. Wright (1992), 'Interior methods for constrained optimization', *Acta Numerica* **1**, 341–407.

Z.-H. Yang and H. B. Keller (1986), 'A direct method for computing higher order folds', *SIAM J. Sci. Statist. Comput.* **7**, 351–361.

Y. Ye, O. Güler, R. A. Tapia and Y. Zhang (1991), 'A quadratically convergent $O(\sqrt{n}L)$-iteration algorithm for linear programming', Preprint.

Y. Yomdin (1990), 'Sard's theorem and its improved versions in numerical analysis, in *Computational Solution of Nonlinear Systems of Equations* (E. L. Allgower and K. Georg, eds), Vol. 26 of *Lectures in Applied Mathematics*, American Mathematical Society, Providence, RI, pp. 701–706.

Acta Numerica (1993), pp. 65–109

Multivariate piecewise polynomials

C. de Boor*

Center for Mathematical Sciences
University of Wisconsin-Madison
Madison WI 53705 USA
E-mail: deboor@cs.wisc.edu

CONTENTS

1. Introduction

This article was supposed to be on 'multivariate splines'. An informal survey, taken recently by asking various people in Approximation Theory what they consider to be a 'multivariate spline', resulted in the answer that a multivariate spline is a possibly smooth, piecewise polynomial function of several arguments. In particular, the potentially very useful thin-plate spline was thought to belong more to the subject of radial basis functions than in the present article. This is all the more surprising to me since I am convinced that the variational approach to splines will play a much greater role in multivariate spline theory than it did or should have in the univariate theory. Still, as there is more than enough material for a survey of multivariate piecewise polynomials, this article is restricted to this topic, as is indicated by the (changed) title.

The available material concerning the space

$$\Pi_{k,\Delta}^{\rho} = \Pi_{k,\Delta}^{\rho}(\mathbb{R}^d)$$

* Supported by the United States Army, the National Science Foundation, and the Alexander von Humboldt Stiftung.

of all **pp** (:= piecewise polynomial) functions in $C^{(\rho)}(\mathbb{R}^d)$ of degree $\leq k$ with some partition Δ is quite vast, as is evidenced by the bibliography Franke and Schumaker (1987) (which contains over 1100 items, yet, e.g., only skims the available engineering literature on finite elements) and the supplementary bibliographies in Schumaker (1988, 1991). This means that, in an article such as this, it is only possible to sketch some of the ideas underlying some of the recent developments in this area.

After a short section on notation, the major topics addressed here are:

(i)   the BB-form;
(ii) the dimension of $\Pi^{\rho}_{k,\Delta}$;
(iii) polyhedral splines;
(iv) the Strang–Fix condition;
(v)   upper bounds for the approximation power of $\Pi^{\rho}_{k,\Delta}$.

Of these, the **BB-** (:= Bernstein–Bézier-) form is perhaps the most immediately useful. Although approximation theorists became aware of it (through the work of Farin and others in CAGD) in the early 1980s, it should be much better known. For example, people in Finite Elements could benefit greatly from its use. For this reason, I am giving a rather leisurely introduction to it, in the generality of functions of several (rather than just one or two) variables.

The second topic, the dimension of $\Pi^{\rho}_{k,\Delta}$, has been a major topic since Strang published some conjectures concerning the bivariate case. It turned out to be a hard problem, perhaps solvable only for 'generic' partitions if at all. However, it gives me the opportunity to illustrate further the use of the BB-form in the process of indicating the difficulty of the problem.

Much effort has been expended in the last 15 years to understand and make use of polyhedral splines, especially simplex splines and box splines. These are multivariate generalizations of Schoenberg's highly successful univariate B-spline. Although some beautiful mathematics has been, and is still being, generated in pursuit of a better understanding, these multivariate B-splines have not yet become standard tools for approximation. However (or, perhaps, because of this), it is important to be aware of the basic idea underlying them, if only because it is the only general principle available at present for the construction of compactly supported pp functions of two or more arguments of degree $\leq k$ and in $C^{(\rho)}$ for ρ 'near' k. Also, the recent introduction, by Dahmen, Micchelli and Seidel, of what looks in hindsight to be the 'right' construction principle for a basis of simplex splines suitable for a given triangulation, awakens new hope for the ultimate usefulness of polyhedral splines.

The Strang–Fix condition (as it is called in Approximation Theory) relates the approximation power of the space spanned by the integer translates of some compactly supported function φ to the behaviour of its Fourier transform $\widehat{\varphi}$ 'at' the discrete set $2\pi\mathbb{Z}^d\backslash 0$. Since its formulation in the early 1970s as the result of a mathematical analysis of the Finite Element Method, it has been the main tool for the determination of approximation orders for shift-invariant pp spaces (such as those generated from box splines, or those on regular partitions). Recent understanding of the structure of shift-invariant spaces has led to a better understanding of what underlies the Strang–Fix condition.

The last section provides a simple discussion of the basic technique for determining upper bounds for the approximation power of a pp space.

The omission of any discussion of *parametric* pp functions, such as curves and surfaces, is likely to be remedied by an entire article on this topic, perhaps in the next volume of this journal. It is to be hoped that another major omission in the context of splines, the discussion of thin-plate splines and other radial functions, will be similarly remedied. Finally, the discussion of numerical methods for approximation by multivariate pp functions is better postponed to a time when these are better understood.

Incidentally, with the exception of numerical methods and, perhaps, the dimension question, none of the topics mentioned (as being discussed or omitted here) appears in the early survey Birkhoff and de Boor (1965) on piecewise polynomial interpolation and approximation.

Finally, a comment concerning the term 'multivariate'. To the annoyance and confusion of statisticians, the term 'multivariate' has become standard in Approximation Theory for what statisticians (and, perhaps, others) would call 'multivariable'. It is too late to change this.

2. Polynomials

The collection of all polynomials in d arguments is denoted here by

$$\Pi \; = \; \Pi(\mathbb{R}^d).$$

For multivariate polynomials, multi-index notation is standard. A **multi-index** is, by definition, any vector with nonnegative integer entries. The **length** of such a multi-index α is the sum of its entries,

$$|\alpha| := \sum_i \alpha(i).$$

Further, $\alpha \le \beta$ iff $\alpha(i) \le \beta(i)$ for all i, and $\alpha < \beta$ iff $\alpha \le \beta$ yet $\alpha \ne \beta$.

With $x(i)$ the ith component of $x \in \mathbb{R}^d$, one uses the abbreviation

$$x^\alpha := \prod_{i=1}^d x(i)^{\alpha(i)}, \quad x \in \mathbb{R}^d, \alpha \in \mathbb{Z}_+^d.$$

The notation

$$(\cdot)^\alpha : \mathbb{R}^d \to \mathbb{R} : x \mapsto x^\alpha$$

for the **monomial of degree** α is convenient (though nonstandard). With $\alpha \in \mathbb{Z}_+^d$,

$$\Pi_\alpha := \Pi_{\leq \alpha} := \text{span}\{(\cdot)^\beta : \beta \leq \alpha\}$$

is the space of all polynomials of **degree** $\leq \alpha$. For any integer k,

$$\Pi_k := \Pi_{\leq k} := \text{span}\{(\cdot)^\beta : |\beta| \leq k\}$$

is the space of all polynomials of **total** degree $\leq k$. The spaces $\Pi_{<\alpha}$ and $\Pi_{<k}$ are defined analogously.

Many expressions simplify if one uses the **normalized power function**

$$[\cdot]^\alpha : x \mapsto x^\alpha / \alpha!,$$

with

$$\alpha! := \prod_i \alpha(i)!,$$

with the understanding that $[\cdot]^\alpha = 0$ if $\alpha \in \mathbb{Z}^d \setminus \mathbb{Z}_+^d$. For example, with $\alpha, \xi, \upsilon, \zeta \in \mathbb{Z}_+^d$, the **Multinomial Theorem** takes the simple form

$$[x + y + \cdots + z]^\alpha = \sum_{\xi + \upsilon + \cdots + \zeta = \alpha} [x]^\xi [y]^\upsilon \cdots [z]^\zeta. \tag{2.1}$$

The multinomial theorem is immediate (by induction on the number of summands in the sum on the left-hand side) once one knows it for two summands. For two summands, though, it is just the special case $p = [\cdot]^\alpha$ of the *Taylor expansion*

$$p(x + y) = \sum_\xi [x]^\xi D^\xi p(y),$$

in which

$$D^\xi := D_1^{\xi(1)} \cdots D_d^{\xi(d)},$$

(with D_i differentiation with respect to the ith argument), hence

$$D^\xi [\cdot]^\alpha (y) = [y]^{\alpha - \xi}.$$

A more sophisticated example is provided by the **Leibniz–Hörmander formula**

$$p(sD)(fg) = \sum_\beta \left(\left([D]^\beta p \right) (sD)f \right) [sD]^\beta g$$

concerning the differentiation of the product fg of two functions, in which s is an arbitrary scalar, and p an arbitrary polynomial, $p = \sum_\alpha [\cdot]^\alpha c(\alpha)$ say, therefore

$$p(D) := \sum D^\alpha / \alpha! \, c(\alpha)$$

the corresponding constant-coefficient differential operator.

Since $D^\alpha[\![\cdot]\!]^\beta(0) = \delta_{\alpha\beta}$, Π_k has dimension

$$\dim \Pi_k = \#\{\alpha \in \mathbb{Z}_+^d : |\alpha| \le k\} = \binom{k+d}{d};$$

the last equality can be verified, e.g., with the aid of the invertible map

$$\{\alpha \in \mathbb{Z}_+^d : |\alpha| \le k\} \to \binom{\{1,\dots,d+k\}}{d} : \alpha \mapsto \left\{\sum_{i \le j}(\alpha(i)+1) : j = 1, \dots, d\right\},$$

with $\binom{X}{d}$ the collection of all d-sets, i.e., all subsets of cardinality d, in X.

While there are various *uni*variate polynomial forms available, there is, aside from the (possibly shifted and/or normalized) power form, only one *multi*variate polynomial form in general use, namely the BB-form, to be discussed next. In particular, the equivalent of a Chebyshev form (or similar form of good condition with respect to the max-norm on some domain) is, as yet, not readily available. The BB-form illustrates that it is often good to give up on the power form altogether in favour of forms which employ more general homogeneous polynomials of the form $x \mapsto \prod_{y \in Y} y^T x$, with

$$y^T x := \sum_i y(i)x(i)$$

the standard inner product.

3. BB-form

The BB-form is, at present, the most effective polynomial form for work with pp functions on a simplicial partition (or, more generally, a simploidal partition). For, the BB-form of a polynomial, with respect to a given simplex

$$\langle V \rangle := \operatorname{conv}(V)$$

spanned by some $(d+1)$-set $V \subset \mathbb{R}^d$, is symmetric with respect to the vertices of that simplex, and readily provides information about the behaviour of the polynomial on all the faces $\langle W \rangle$, $W \subset V$, of that simplex. This facilitates the smooth matching of two polynomial pieces across the intersection of their respective simplicial cells. For more details than are (or can be) offered here, see Farin (1986) (which concentrates on the bivariate case) as well as de Boor (1987). The presentation here is based on the latter, albeit with certain changes in notation. For the use of the BB-form in the treatment of finite elements, see, e.g., Luscher (1987).

The BB-form can be viewed as a generalization of the standard representation

$$p = \sum_{v \in V} \xi_v p(v)$$

of the linear interpolant to data given at a $(d+1)$-point set $V \subset \mathbb{R}^d$ in general position, with $\xi_v = \xi_{v,V}$ the unique linear polynomial which takes the value 1 at v and vanishes on

$$V \backslash v := \{w \in V : w \neq v\}.$$

In this connection, 'general position' is tautological since it means nothing more than that such a representation exists for every $p \in \Pi_1(\mathbb{R}^d)$, hence is necessarily unique since $\dim \Pi_1(\mathbb{R}^d) = d + 1 = \#V$.

The $(d+1)$-vector

$$\xi_V(x) := (\xi_v(x))_{v \in V}$$

provides the **barycentric coordinates** of x with respect to the point set V. Equivalently, $\xi_V(x)$ is the unique solution of the linear system

$$\sum_{v \in V} \xi_v(x)\,(v,1) \;=\; (x,1) \quad \in \mathbb{R}^{d+1}, \tag{3.1}$$

and this provides the opportunity to write out a formula for its components $\xi_v(x)$ as a ratio of determinants and so explains the alternative name **areal coordinates**.

The BB-form for $p \in \Pi_k$ employs all possible products of k of the linear polynomials ξ_v, $v \in V$, with repetitions permitted, i.e., all the functions

$$\xi_V^\alpha : x \mapsto \xi_V(x)^\alpha$$

with α any multi-index (indexed by V, i.e., in $\mathbb{Z}_+^V$) of length k. However, it turns out to be very convenient to use the particular normalization

$$B_\alpha := B_{\alpha,V} := \binom{|\alpha|}{\alpha} \xi_V^\alpha = |\alpha|!\, [\![\xi_V]\!]^\alpha,$$

which arises when we apply the multinomial theorem (2.1) to obtain

$$1 = k! [\![\sum_{v \in V} \xi_v(x)]\!]^k = \sum_{|\alpha|=k} B_\alpha(x). \tag{3.2}$$

The fact that

$$\#\{\alpha \in \mathbb{Z}_+^V : |\alpha| = k\} = \#\{\beta \in \mathbb{Z}^d : |\beta| \leq k\} = \dim \Pi_k$$

implies that *the collection $(B_\alpha)_{|\alpha|=k}$ is a basis for Π_k* since (i) any $p \in \Pi_k$ can be written as a linear combination of products of k linear polynomials (e.g., the linear polynomials $x \mapsto x(i), i = 1, \ldots, d$ and $x \mapsto 1$); and (ii) any linear polynomial can be written as a linear combination of the ξ_v, $v \in V$, hence $\Pi_k \subseteq \mathrm{span}\{B_\alpha : |\alpha| = k\}$. The resulting representation

$$p = \sum_{|\alpha|=k} B_\alpha\, b_{p,V}(\alpha)$$

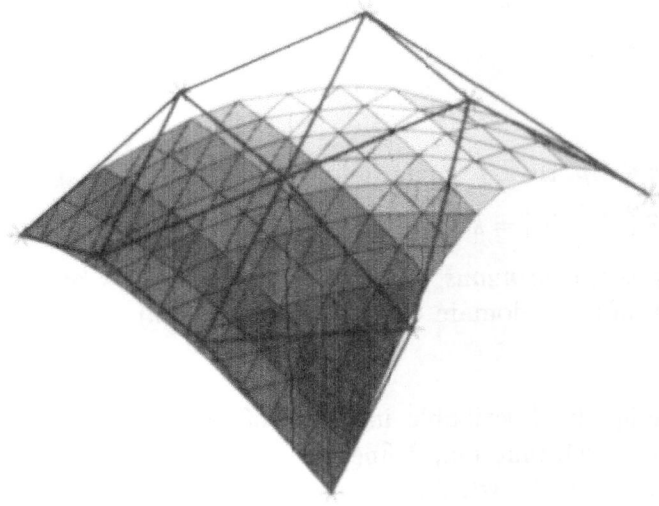

Fig. 3.3. A cubic patch and its control net.

for $p \in \Pi_k$ constitutes the BB-form (a form associated with the names of Bernstein (Lorentz 1953; p. 51), de Casteljau (1963, 1985), Bézier (1970, 1977), Farin (1977, 1986, 1988), and perhaps others).

Since ξ_v vanishes at all the points in $V \backslash v$ and is linear, it vanishes on the simplex $\langle V \backslash v \rangle$ spanned by these points. It follows that, for any subset U of V, the restriction of B_α to $\langle U \rangle$ is not the zero function if and only if $\mathrm{supp}\,\alpha \subseteq U$. In particular, the only B_α not zero on $\{v\} = \langle\{v\}\rangle$ is the one with $\alpha = k\mathbf{i}_v$, where

$$\mathbf{i}_v(u) := \delta_{vu}, \quad v \in V.$$

With (3.2), this implies that $B_{k\mathbf{i}_v}(v) = 1$, hence further that

$$p(v) = b_{p,V}(k\mathbf{i}_v), \quad v \in V.$$

This fact and others have made it customary to associate, more generally, the coefficient $b_{p,V}(\alpha)$ with the corresponding **domain point**

$$V_\alpha := \sum_{v \in V} v\,\alpha(v)/|\alpha|,$$

thereby obtaining the **(Bézier) control net**

$$C_p := C_{p,V,k} := (V_\alpha, b_{p,V}(\alpha))_{|\alpha|=k}$$

for p. Note that $\mathrm{supp}\,\alpha \subseteq U$ for some $U \subseteq V$ if and only if $V_\alpha \in \langle U \rangle$. Hence, on $\langle U \rangle$, p is entirely determined by $b_{p,V}(\alpha)$ with $V_\alpha \in \langle U \rangle$. To put it differently, the restriction of p to $\langle U \rangle$ has the control net

$$C_{p,U,k} = (V_\alpha, b_{p,V}(\alpha))_{|\alpha|=k, V_\alpha \in \langle U \rangle}.$$

In particular, if

$$f = \begin{cases} p & \text{on } \langle V \rangle, \\ q & \text{on } \langle W \rangle, \end{cases} \tag{3.4}$$

for some $p, q \in \Pi_k$, then f is continuous on $\langle V \rangle \cup \langle W \rangle$ if and only if

$$\forall \{\alpha \in \mathbb{Z}_+^{V \cup W} : |\alpha| = k, \operatorname{supp}\alpha \subset V \cap W\} \quad b_{p,V}(\alpha_{|V}) = b_{q,W}(\alpha_{|W}).$$

Thus, if f is a *continuous* pp function of degree $\leq k$ on some **complex** (:= partition of some domain $G \subset \mathbb{R}^d$ into simplices) Δ, in formulæ:

$$f \in \Pi^0_{k,\Delta},$$

then it is uniquely describable in terms of its **BB-net**, b_f. This is, by definition, the mesh-function, defined on the union of all the domain points V_α, $|\alpha| = k$, $\langle V \rangle \in \Delta$, which, for each $\langle V \rangle \in \Delta$, agrees with $b_{p,V}$ on the points in $\langle V \rangle$.

It is well worth stressing that, as d increases, the ratio of domain points in the boundary of a $\langle V \rangle$ over the total number of domain points in $\langle V \rangle$ increases for fixed k, reaching the limiting value 1 as soon as $d > k$. In effect, with increasing d, the polynomial pieces in a pp function of fixed degree $\leq k$ become increasingly 'superficial', with more and more of their degrees of freedom needed just to maintain continuity.

3.1. The BB-form as a k-fold difference

For a discussion of a smoother join as well as for its own sake, we need to know how to differentiate the BB-form. For this, and for various other properties, we observe the following striking

Fact 3.5 For $\omega \in \mathbb{R}^V$, let ωE denote the 'difference operator' which acts on the mesh-function $c : \mathbb{Z}^V \to \mathbb{R}$ by the rule

$$(\omega E)c := \sum_{v \in V} \omega(v)c(\cdot + \mathbf{i}_v).$$

Then

$$\sum_{|\alpha|=k} B_\alpha(x)c(\alpha) = (\xi_V(x)E)^k c(0).$$

Indeed,

$$(\xi_V(x)E)^k c(0) = \sum_{u \in V}\sum_{v \in V} \cdots \sum_{w \in V} \xi_u(x)\xi_v(x) \cdots \xi_w(x) \; c(\mathbf{i}_u + \mathbf{i}_v + \cdots + \mathbf{i}_w)$$

with exactly k summations, hence all summands are of the form $\xi_V(x)^\alpha c(\alpha)$

for some $\alpha \in \mathbb{Z}_+^V$ with $|\alpha| = k$, and this particular summand occurs exactly $\binom{k}{\alpha}$ times. See Figure 3.13 for an illustration.

With this,

$$\sum_{|\alpha|=k} B_\alpha \, b_{p,V}(\alpha) = p = (\xi_V E)^k b_{p,V}(0), \qquad p \in \Pi_k.$$

With this formula in hand, differentiation of the BB-form requires nothing more than the chain rule, as follows. If $y \in \mathbb{R}^d \backslash 0$, then

$$D_y p = D_y(\xi_V E)^k c(0) = k(\xi_V E)^{k-1}(D_y \xi_V E)c(0).$$

We obtain the vector $D_y \xi_V$ by the observation that, by (3.1), $\xi_V(x + ty) - \xi_V(x) = t\eta_V(y)$, with $\eta_V(y) \in \mathbb{R}^V$ the unique solution of

$$\sum_{v \in V} \eta_v(y) \, (v, 1) = (y, 0).$$

Hence, altogether,

$$D_y p = k \sum_{|\alpha|=k-1} B_\alpha(\eta_V(y)E) b_{p,V}(\alpha) \tag{3.6}$$

for $p \in \Pi_k$ and $\sum_{v \in V} \eta_v(y) \, (v,1) := (y,0) \in \mathbb{R}^{d+1} \backslash 0$.

For example, for two distinct points $v, u \in V$, $\eta_V(v - u) = \mathbf{i}_v - \mathbf{i}_u$, hence

$$b_{D_{v-u}p, V}(\alpha) = \frac{b_{p,V}(\alpha + \mathbf{i}_v) - b_{p,V}(\alpha + \mathbf{i}_u)}{1/k}.$$

Repeated application of (3.6) provides the BB-form for any derivative of p of the form $D_Y p$ with Y any finite subset of $\mathbb{R}^d \backslash 0$ and

$$D_Y := \prod_{y \in Y} D_y.$$

3.2. Smooth matching of polynomial pieces

Since we now know how to obtain the BB-form of any derivative of a polynomial p from the BB-form of p, we can describe the matching of derivatives across the common interface $\langle V \cap W \rangle$ of two simplicial cells $\langle V \rangle$ and $\langle W \rangle$. Simply put, the derivative in question of the polynomial p on $\langle V \rangle$ and the polynomial q on $\langle W \rangle$ must agree on $\langle V \cap W \rangle$, i.e., their corresponding control points with domain point in $\langle V \cap W \rangle$ must agree.

It is not hard to write specific smoothness conditions in the form of an equality between the expressions, obtained by application of (3.6), for the relevant control points (see, e.g., Chui and Lai (1987) and Chui (1988, Theorem 5.1) or Farin (1986)) of the relevant derivatives. However, if the goal is a $C^{(\rho)}$-match, i.e., a matching of all derivatives of order $\le \rho$, then the uniformity of the BB-form permits a more unexpected formulation of the corresponding smoothness conditions, as follows.

For $p \in \Pi_k$ and $\beta \in \mathbb{Z}_+^V$ with $|\beta| \le k$, let

$$p_\beta := \sum_{|\gamma|=k-|\beta|} b_{p,V}(\beta+\gamma)B_\gamma.$$

These are the **subpolynomials** introduced in de Boor (1987); see also Farin (1986; (2.5)). For example, if $|\beta| = k$, then p_β is the constant polynomial with value $b_{p,V}(\beta)$. Consequently, (3.4) is continuous if and only if $p_\beta = q_\beta$ for all β with $\mathrm{supp}\,\beta \subset V \cap W$ and $|\beta| = k$. As another example, if $|\beta| = k-1$, then p_β is the linear polynomial whose value at $v \in V$ is $b_{p,V}(\beta+\mathbf{i}_v)$, and, for any y, its derivative $D_y p_\beta$ is the constant $b_{D_y p,V}(\beta)$. Consequently, (3.4) is in $C^{(1)}$ if and only if $p_\beta = q_\beta$ for all β with $\mathrm{supp}\,\beta \subset V \cap W$ and $|\beta| = k-1$.

Here is the general theorem.

Theorem 3.7 *The pp function f, defined in (3.4), is in $C^{(r)}$ for some $r \le k$ if and only if*

$$\forall\{\beta \in \mathbb{Z}_+^V : \mathrm{supp}\,\beta \subset V \cap W, |\beta| = k - r\} \quad p_\beta = q_\beta. \tag{3.8}$$

In particular, since $q_\beta(w) = b_{q,W}(\beta+r\mathbf{i}_w)$ for each such β and each $w \in W$, $C^{(r)}$-continuity requires that

$$\forall\{\beta \in \mathbb{Z}_+^V : \mathrm{supp}\,\beta \subset V \cap W, |\beta| = k-r, w \in W\backslash V\} \quad b_{q,W}(\beta+r\mathbf{i}_w) = p_\beta(w). \tag{3.9}$$

Conversely, if our f is already in $C^{(r-1)}$, hence $p_\beta = q_\beta$ for all β with $\mathrm{supp}\,\beta \subset V \cap W$ and $|\beta| = k - r + 1$, then the conditions (3.8) are equivalent to the conditions (3.9). In particular, (3.9) supplies a complete and independent set of conditions for $C^{(r)}$-continuity across $\langle V \cap W \rangle$ in the presence of $C^{(r-1)}$-continuity. Consequently, the union over $r = 0, \ldots, \rho$ of these conditions constitutes a complete and independent set of conditions for $C^{(\rho)}$-continuity across $\langle V \cap W \rangle$.

Note the remarkable *uniformity* of the conditions (3.9): The weights in the right-hand side $p_\beta(w)$, considered as a linear combination of the BB-coefficients $b_{p,V}(\alpha)$ for p, depend only on w and r (and V) and not on β or k.

Note also that the smoothness conditions of order r, i.e., the conditions (3.9), involve only control points of f in the first r 'layers' along $\langle V \cap W \rangle$. Note finally, that we might have, equally well, used the complementary conditions

$$\forall\{\beta \in \mathbb{Z}_+^V : \mathrm{supp}\,\beta \subset V \cap W, |\beta| = k - r, v \in V\backslash W\} \quad b_{p,V}(\beta + r\mathbf{i}_v) = q_\beta(v). \tag{3.10}$$

In effect, the subpolynomials $p_\beta = q_\beta$ with $\operatorname{supp}\beta \subset V \cap W$ and $|\beta| = k - r$ give a complete description of the behaviour of all derivatives of f of order $\leq r$ on $\langle V \cap W \rangle$, and enforcement of (3.9) and (3.10) makes certain that the corresponding derivatives of p and q agree with these of f on $\langle V \cap W \rangle$.

It is this remarkably explicit geometric connection between the control points and the behaviour 'near' any particular face of $\langle V \rangle$ that makes the BB-form so attractive for work with pp functions.

The simplest nontrivial case, $r = 1$, is of particular practical interest. It requires that, for each $\beta \in \mathbb{Z}_+^V$ with $\operatorname{supp}\beta \subset V \cap W$ and $|\beta| = k - 1$, $q_\beta(w) = p_\beta(w)$, i.e., that the control point $(W_{\beta+i_w}, b_{q,W}(\beta + i_w))$ lie on the (hyper)plane spanned by the control points $(V_{\beta+i_v}, b_{p,V}(\beta + i_v))$, $v \in V$, a particularly nice geometric interpretation rightfully stressed in the CAGD literature (see, e.g., Boehm, Farin and Kahmann (1984), Farin (1988) and Hoschek and Lasser (1989)).

3.3. Simple examples

As an illustration of the strength and efficiency of the BB-form, here is a discussion of three standard topics concerning bivariate pp functions.

Quintic Hermite interpolant In bivariate quintic Hermite interpolation, one matches value and first and second derivatives at three points, thus using up eighteen of the available $21 = \binom{7}{2} = \binom{k+d}{d}$ degrees of freedom, and then uses the remaining three degrees of freedom for a possible $C^{(1)}$-join with neighbouring quintic patches. Here are the details, well known, but particularly evident when discussed in terms of the BB-form.

Let $d = 2$, $V = \{u, v, w\}$, $k = 5$, and $p \in \Pi_k$.

Then $p(u) = b_p(5i_u)$.

Further, for $\nu \in V \backslash u$,

$$D_{\nu-u}p(u) = b_{D_{\nu-u}p}(4i_u) = 5(b_p(4i_u + i_\nu) - b_p(4i_u + i_u)),$$

showing that the coefficients

$$b_p(4i_u + i_\nu), \quad \nu \in V \backslash u,$$

are determined by

$$D_{\nu-u}p(u), \quad \nu \in V \backslash u$$

and *vice versa* (once $p(u) = b_p(5i_u)$ is known). More than that, it shows that *the tangent plane to p at u is the plane spanned by the control points at and next to u.* (This discussion actually applies for arbitrary d and k.)

Finally, with $\mu, \nu \in V \backslash u$, all second derivatives are linear combinations of the second derivatives of the form $D_{\mu,\nu}$, of which there are exactly as many as there are distinct points of the form $3i_u + i_\mu + i_\nu$, i.e., control points in the

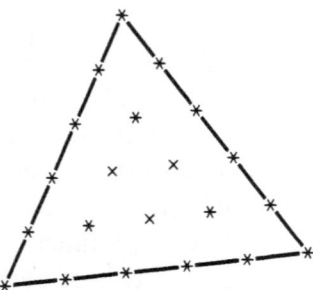

Fig. 3.11. The quintic Hermite interpolant.

second layer of control points near u, and, correspondingly, with the tangent plane at u already determined, the specification of all second derivatives of p at u is equivalent to the specification of all the control points in that second layer. (Again, this discussion applies for arbitrary d and arbitrary k.)

In other words, the behaviour of all derivatives of p at u of order ≤ 2 is determined by the subpolynomial

$$p_{3i_u} = \sum_{|\gamma|=2} b_p(3i_u + \gamma)B_\gamma,$$

and it involves the control points in the zeroth, first and second layer for u. Since $d = 2$ and $k = 5$, this 'triangle' of control points associated with u has no intersection with the corresponding coefficient 'triangles' associated with the other vertices. This implies that one can freely specify value, first and second derivatives of $p \in \Pi_5$ at each of these three vertices, and this specifies the 18 control points in those 'triangles', and leaves free exactly one control point per edge. This control point is in the first layer for that edge, hence determines the middle control point for that edge for any particular first derivative of p. Equivalently, for the control point associated in this way with the edge $\langle u, v \rangle$, it is the only piece of information for the (linear) subpolynomial $p_{2i_u + 2i_v}$ not yet specified (and this is the only linear subpolynomial p_β with supp $\beta \subset \langle \{u, v\} \rangle$ not yet completely specified). Consequently, if the control point is determined in such a way that it equals the corresponding control point of the same derivative of a quintic Hermite interpolant (to the same vertex data) in the triangle sharing this edge, then the two quintic polynomials form a $C^{(1)}$ pp function. This can be achieved, e.g., by specifying the normal derivative at the midpoint of that edge (or any other particular, transversal, derivative).

To re-iterate, the point of this example (and the two to follow) is not to derive a new result, but to show how easily these known results are derivable in the language of the BB-form.

Clough–Tocher　Here, one subdivides a given triangle arbitrarily into

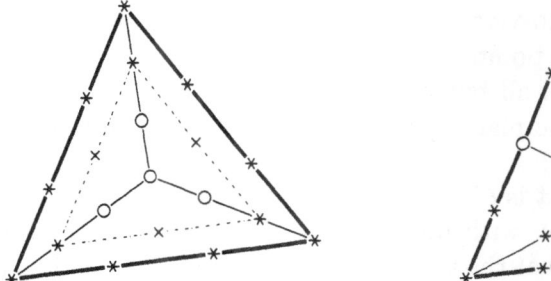

Fig. 3.12. The Clough–Tocher split and the Powell–Sabin split.

three, by connecting its vertices to an arbitrarily chosen point in the interior. Prescribing the tangent plane at each vertex determines the vertex control points and the next-to-vertex control points (see the points marked * in Figure 3.12).

That leaves the points marked x still undetermined, hence allows matching of some transversal derivative at some point. Traditionally, this has been the normal derivative at the midpoint, with the value either given, or else estimated from the vertex information. In this way, value and first derivatives along an edge are entirely determined by information specified on that edge. Hence $C^{(1)}$ matching across that edge is ensured provided the abutting triangle is handled in the same way.

That leaves the control points marked o. These must be determined so that the $C^{(1)}$ conditions hold across the interior edges. At this point, the uniformity of the BB-form comes into play, as follows. One determines the unknown control points to be the control points (with respect to the triangle(s) to which they are assigned) of the unique quadratic polynomial for which the six points on the dot-dashed triangle are the control points (with respect to the triangle to which they are assigned, i.e., the dot-dashed triangle). This can be done by one application of the de Casteljau algorithm to evaluate the given BB-form of this quadratic polynomial at the 'dividing' point chosen in the interior; see the next subsection for details. The resulting control points will satisfy the $C^{(1)}$-conditions since they represent a piecewise quadratic which is even in $C^{(2)}$. In particular, the resulting piecewise cubic is $C^{(2)}$ at the interior vertex (in addition to being $C^{(1)}$ everywhere).

Powell–Sabin There is a corresponding construction of a piecewise quadratic $C^{(1)}$ element, the Powell–Sabin macro-element. Here, one subdivides the triangle into six pieces, starting with some interior point as an additional vertex, but connecting it not only to the vertices, but also to a point on each edge. But, as we shall see, this has to be done just right, to ensure a $C^{(1)}$ match between such macro-elements.

As before, prescription of the tangent plane at each (exterior) vertex pins down vertex and next-to-vertex control points (marked ∗ in Figure 3.12), leaving a 'Y' of control points (marked o). The $C^{(1)}$-conditions across the interior edges determine all but the interior vertex one, and that will necessarily have to lie in the plane spanned by the three control points next to it.

With this, the element is $C^{(1)}$, and any first derivative is piecewise linear along an (exterior) edge, with its extreme values determined explicitly by the given tangent planes at the two vertices of interest. The middle corner of this piecewise linear function is also determined by this information, but in ways that depend strongly on the choice of that the interior vertex and the additional vertex on the edge, as well as on the particular derivative direction. Since only one particular transversal derivative needs to be matched in order to achieve $C^{(1)}$ across the edge, choose a particular direction and then make certain that the interior and the additional edge vertices are so chosen that this particular transversal derivative is actually linear (i.e., has no active interior vertex). Powell and Sabin do this by choosing the midpoint of the edge as the edge vertex and, correspondingly, the interior vertex as the intersection of midpoint normals, i.e., as the centre of the circumscribed circle. This makes the derivative in the direction normal to the edge linear.

More generally, pick, in each macro-triangle to be, the interior vertex in such a way (e.g., as the centre of the inscribed circle) that the line from it to the corresponding point in any neighbouring triangle cuts the common edge at some point strictly between the two common vertices, and use this intersection point as the additional vertex on that edge. Then the three new control points along that midline, as the average of two triples of points with each triple on a straight line, lie themselves on a straight line, thus ensuring $C^{(1)}$.

3.4. Evaluation of the BB-form

As a final advertisement for the BB-form, I discuss the de Casteljau algorithm (de Casteljau (1963)) for its evaluation. This algorithm obtains the value $p(x)$ by carrying out the k-fold application of the difference operator $\xi_V(x)E$ to the mesh-function b_p, as described in Fact 3.5. Since only the value of $(\xi_V(x)E)^k b_p$ at 0 is wanted, we only require $(\xi_V(x)E)^{k-1} b_p$ at α with $|\alpha| = 1$, $(\xi_V(x)E)^{k-2} b_p$ at α with $|\alpha| = 2$, ..., b_p at α with $|\alpha| = k$. It is instructive to visualize the entire discrete $(d+1)$-simplex of mesh points α involved here, as is done in Figure 3.13. For $j = k-1, k-2, \ldots, 0$, the algorithm derives the 'layer' of values associated with $|\alpha| = j$ from the layer associated with $|\alpha| = j+1$, with each value computed as exactly the same averave of the corresponding d-simplex of values in the next layer.

As a remarkable bonus, the calculations provide (Goldman (1983)), simultaneously, the BB-form for p with respect to $W := (V \backslash v) \cup x$ for any

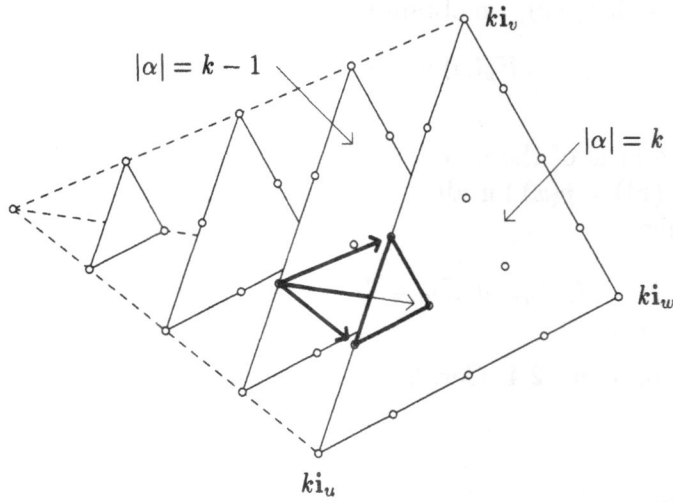

Fig. 3.13. The mesh-point simplex for evaluation.

particular $v \in V$: If we denote by c the mesh-function whose values at α, $|\alpha| \leq k$, are being generated during the algorithm from the numbers $c(\alpha) := b_{p,V}(\alpha)$, $|\alpha| = k$, then

$$b_{p,W}(\alpha + (k - |\alpha|)\mathbf{i}_v) = c(\alpha), \quad \alpha(v) = 0.$$

This is another effect of the uniformity of the BB-form. As we evaluate the BB-form of some polynomial at some point, we are simultaneously evaluating all associated subpolynomials at the same point. On the other hand, the coefficient $b_{p,V}(\alpha)$ is the value at v of the subpolynomial $p_{\alpha - \alpha(v)\mathbf{i}_v}$. See the discussion of the Clough–Tocher element in the preceding section for a ready application of this.

The evaluation at x of a particular derivative, of the form D_Y with the entries of the sequence Y taken from V, proceeds similarly, except that, during the first $\#Y$ steps, one applies the difference operators $\eta_V(y)E$ corresponding to the entries y of Y, and uses the 'evaluation' difference operator $\xi_V(x)E$ only for the remaining $k - \#Y$ steps. Of course, since any two such difference operators commute, one is entitled to apply the relevant difference operators in any order. In particular, it might be most efficient and stable to apply the $k - \#Y$ 'evaluation' operators first, leaving the application of the 'differentiation' operators for the remaining $\#Y$ layers, which are smaller.

Finally, the de Casteljau algorithm in no way relies on the fact (except, perhaps in the argument for its stability) that the weights ω in the difference operator ωE sum to one. If we employ it with some arbitrary weight vector

ω instead of with $\xi_V(x)$, we obtain the number

$$H_p(\omega) := \sum_{|\alpha|=k} b_p(\alpha)k![\![\omega]\!]^\alpha,$$

i.e., the value at ω of the unique *homogeneous* polynomial H_p on $\mathbb{R}^{d+1}$ for which $H_p(\xi_V(x)) = p(x)$ for all $x \in \mathbb{R}^d$. In conjunction with (3.6), this leads to the formulæ

$$\frac{k!}{(k-\rho)!} \sum_{|\alpha|=k-\rho} H_{p_\alpha}(\eta_V(y))B_\alpha = D_y^\rho p = \frac{k!}{(k-\rho)!} \sum_{|\alpha|=k-\rho} p_\alpha H_{B_\alpha}(\eta_V(y))$$

of Farin (1986; Thm. 2.4, Cor. 2.5), sometimes stated with somewhat less care.

4. The space $\Pi_{k,\Delta}^\rho$

While automotive and aerospace engineers have been working with tensor product spline functions since the early 1960s and structural engineers have been working with pp finite elements just as long, mathematicians in Approximation Theory began to study spaces of multivariate pp functions of non-tensor product type seriously only in the 1970s.

The initial focus was the 'spline' space

$$\Pi_{k,\Delta}^\rho$$

(also denoted by $S_k^\rho(\Delta)$) of all pp functions of degree $\leq k$ in $C^{(\rho)}$ with partition Δ. Here, in full generality, Δ is a collection of 'cells', i.e., closed convex sets δ, with pairwise disjoint, nonempty interiors, whose union is some domain $G \subset \mathbb{R}^d$ of interest, and $\Pi_{k,\Delta}^\rho$ consists of exactly all those $f \in C^{(\rho)}(G)$ for which $f_{|\delta} \in \Pi_k(\delta)$ for all $\delta \in \Delta$. Any such space is contained in the space

$$\Pi_{k,\Delta} =: \Pi_{k,\Delta}^{-1}$$

of all pp functions of degree $\leq k$ with partition Δ. However, as soon as we impose some smoothness condition, i.e., as soon as $\rho \geq 0$, the 'cells' of Δ are chosen to be **polytopes**, i.e., the convex hull of a finite set (the **vertex set** for the cell), since the task of matching polynomial pieces across the common boundary of two such cells becomes too difficult otherwise. Further, the partition Δ is taken to be **regular** in the sense that the intersection of two cells is the convex hull of the intersection of their vertex sets. In the simplest case, Δ is a **complex**, i.e., a regular partition consisting of simplices. Such a partition is often called a **triangulation** even when $d > 2$.

Initially, there were high hopes that it would be possible to generate a theory of these spaces to parallel the theory of univariate splines (as recorded, e.g., in Schoenberg (1969), de Boor (1976, 1978), Schumaker (1981) and Powell (1981)). For example, here is a list of desirable goals, from Schumaker (1988, 1991):

1. Explicit formulæ for the dimension of spline spaces;
2. Explicit bases consisting of locally-supported elements;
3. Convenient algorithms for storing and evaluating the splines, their derivatives, and integrals;
4. Estimates of the approximation power of spline spaces;
5. Conditions under which interpolation is well defined;
6. Algorithms for interpolation and approximation.

However, the experience gained so far has led to some doubt as to whether these goals are likely to be achieved fully even in the bivariate case.

It is also not clear whether the restriction to polynomials of total degree $\leq k$ is reasonable *a priori*. On a cell which is the Cartesian product $\delta_1 \times \delta_2$ of lower-dimensional cells δ_1 and δ_2, it seems, offhand, more reasonable to use elements from the tensor product $\Pi_k(\delta_1) \otimes \Pi_k(\delta_2)$ of polynomials of total degree $\leq k$ on those lower-dimensional sets. For example, in a bivariate context, a typical practical partition involves triangles and quadrilaterals, and, in such a setting, the restriction to polynomials of total degree $\leq k$ seems reasonable only if one first refines the partition, by subdividing each quadrilateral into triangles. This does have the advantage of uniformity and, if properly done, may produce partitions which support locally supported smooth pp functions of smaller degree than did the original partition. In fact, for a general partition, this is certain to be so if even the triangles are subdivided appropriately. On the other hand, as of this writing and as a consequence of the early dominance of tensor product methods, most commercially used software packages for surface design and manufacturing can only handle partitions with quadrilateral cells and, correspondingly, bicubic, or biquintic, polynomial pieces.

4.1. The dimension of $\Pi^\rho_{k,\Delta}$

When $\rho = -1$, then $\dim \Pi^\rho_{k,\Delta} = \dim \Pi_k(\mathbb{R}^d) \cdot \#\Delta$. However, already for $\rho = 0$, there is no hope for a formula for $\dim \Pi^\rho_{k,\Delta}$, except in the simplest case, when Δ is a triangulation. In this case, the BB-nets for the polynomial pieces of $f \in \Pi^0_{k,\Delta}$ associated with two neighbouring cells, $\langle V \rangle$ and $\langle W \rangle$, necessarily agree at all domain points in the intersection $\langle V \rangle \cap \langle W \rangle = \langle V \cap W \rangle$. Consequently, the map

$$f \mapsto b_f$$

from f to its BB-net sets up a 1–1 correspondence between $\Pi^0_{k,\Delta}$ and all scalar-valued functions on the mesh

$$A_{k,\Delta} := \{V_\alpha : |\alpha| = k, \langle V \rangle \in \Delta\}.$$

In particular,

$$\dim \Pi^0_{k,\Delta} = \#A_{k,\Delta}.$$

For $\rho > 1$, one would think of $\Pi^\rho_{k,\Delta}$ as the linear subspace of $\Pi^0_{k,\Delta}$ singled out by the $C^{(\rho)}$-conditions across facets, hence could, in principle, determine its dimension as the difference between $\dim \Pi^0_{k,\Delta}$ and the *rank* of the collection of $C^{(\rho)}$-conditions. While, as we have seen, it is easy to specify this rank for the collection of all $C^{(\rho)}$-conditions across *one* facet, it is, in general, very difficult to determine the rank of all conditions, as a simple example below will illustrate. Already for $\rho = 1$, there are real difficulties in ascertaining $\dim \Pi^\rho_{k,\Delta}$. Strang's articles (1973, 1974) called attention to this by providing a conjecture concerning $\dim \Pi^\rho_{k,\Delta}$ in the *bivariate* case, namely that the lower bound in the following theorem, due to Schumaker, is the exact dimension for 'generic' triangulations.

Theorem 4.1 *Let Δ be a finite triangulation in $\mathbb{R}^2$, let V_I, E_I denote the collection of its interior vertices and edges, respectively. Further, for each $v \in V_I$, let E_v denote the collection of all edges having v as an endpoint, and denote by $\tilde{E}_v \subset E_v$ those with different slopes.*
   Then

$$\dim \Pi^\rho_{k,\Delta} - (\dim \Pi_k + \dim \Pi_{k-\rho-1} \cdot \#E_I - (k^2 + 3k - \rho^2 - 3\rho)/2 \cdot \#V_I) \in [\sigma .. \tilde{\sigma}],$$

with

$$\sigma := \sum_{v \in V_I} \sum_{j=1}^{k-\rho} (\rho + j + 1 - j \cdot \#E_v)_+$$

and $\tilde{\sigma}$ defined in the same way, but with E_v replaced by $\tilde{E}_v$.

(Here and elsewhere, $[a .. b]$ specifies the (closed) interval with endpoints a and b, since the more customary notation $[a, b]$ is also used for the divided difference at two points as well as for the matrix with columns a and b.) See Schumaker (1979 (1984)) for a proof of the lower (upper) bound.

Perhaps the simplest example indicating that it is not possible to be more precise than this is provided by consideration of $\dim \Pi^1_{2,\Delta}$, with the partition Δ obtained by connecting the four points of a (convex) quadrilateral with some point in its interior. Assume first that the interior point was chosen 'generically', in which case the four interior edges for Δ have four distinct slopes, as in the left half of Figure 4.2. In search for some $f \in \Pi^1_{2,\Delta} \backslash \Pi_2$, we consider the BB-net for f. We assume without loss that f vanishes on the bottom triangle, and have indicated this in Figure 4.2 by drawing a

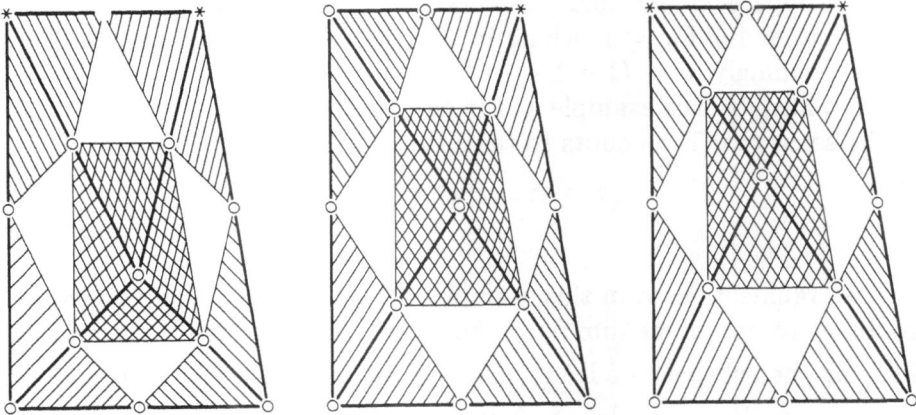

Fig. 4.2. Generic and related nongeneric partitions.

'o' at the six domain points in that triangle for the BB-net for f. Now, as discussed in the last paragraph of subsection 3.2 above, $C^{(1)}$-continuity requires the coplanarity of the four control points associated with each of the shaded quadrilaterals. In particular, this forces all the control points in the first layer outside the edges of the bottom triangle to be zero, and this is also indicated in the figure. Offhand, the control points associated with the two top corners are freely choosable *except* that the control point associated with the midpoint of the top edge (the one left blank) must lie on the plane spanned by the three control points to the left as well as on the plane spanned by the three control points to the right. In the generic case, this imposes one constraint on the two vertex control points, and we conclude that dim $\Pi^1_{2,\Delta} = 7$ in this case.

The same conclusion can be reached when the interior vertex lies on one but not the other of the two diagonals of the quadrilateral, as shown in the middle of Figure 4.2. In terms of that figure, the domain point in the middle of the upper edge lies on the straight line through the domain points of the two zero control points to the right of it, hence the corresponding control point must be zero. Since its domain point does *not* lie on the straight line through the domain points of the two zero control points to the left of it, this implies that also the remaining control point associated with the upper left shaded quadrilateral, the vertex control point, must be zero. The other upper vertex control point, however, is freely choosable.

Finally, if that interior vertex happens to be the intersection of the two diagonals of the quadrilateral (as shown in the right of Figure 4.2), then the argument just given shows that the control point associated with the middle of the upper edge must be zero, and both upper vertex control points are freely choosable. Hence, dim $\Pi^1_{2,\Delta} = 8$ in this case.

For comparison, for this particular example, we have just one interior vertex, v, and $\#E_v = 4$, while, in the three distinct cases, $\#\tilde{E} = 4, 3, 2$. Correspondingly, $\sigma = (1 + 1 + 1 - 4)_+ = 0$, while $\tilde{\sigma} = 0, 0, 1$ in the three cases. Thus, for this example and in these three cases, the theorem is sharp in the sense that it amounts to the assertion that

$$(7, 7, 8) - 7 \in [0 .. (0, 0, 1)].$$

The arguments used in this example illustrate how, in general, one might go about to determine $\dim \Pi_{k,\Delta}^\rho$. As already stressed, one rightly thinks of $\Pi_{k,\Delta}^\rho$ as the subspace of $\Pi_{k,\Delta}^0$ characterized by the $C^{(\rho)}$-conditions. A pp function on the triangulation Δ is in $C^{(\rho)}$ precisely when it is in $C^{(\rho)}$ on any two simplices of Δ which share a whole facet, i.e., whose vertex sets differ only by one point. For this reason, $\Pi_{k,\Delta}^\rho$ is linearly isomorphic to all the mesh-functions b_f on $A_{k,\Delta}$ which, for each such simplex pair, satisfy the corresponding conditions (3.9) across their common facet for $r = 1, \ldots, \rho$. Moreover, for each such facet, this provides a maximally linearly independent set of $C^{(\rho)}$-conditions imposed across *one* such facet. However, conditions across different (but neighbouring) facets may well be linearly dependent. For example, Figure 4.2 shows four $C^{(1)}$-conditions involving the control point at the interior vertex. Yet, since they all require that their respective control points lie on a certain plane, it takes just two such conditions to ensure that all five control points involved lie on the same plane, hence the other two conditions must be dependent on them. Unfortunately, it is in general impossible to provide a basis for the collection of *all* smoothness conditions imposed. This has made it a challenge (unsolved so far and not likely to be solved in any generality) to determine the dimension of $\Pi_{k,\Delta}^\rho$ when $\rho > 0$.

As the example shows, there is no hope to express $\dim \Pi_{k,\Delta}^\rho$ entirely in such combinatorial terms as the number of (interior or boundary) vertices, edges, triangles. However, even the hope that, as in this case, the counting of such things as nonparallel edges incident to a vertex might suffice is dashed by a more subtle example due to Morgan and Scott in 1977 (Morgan and Scott (1990)), which uses the partition Δ obtained by placing a scaled and reflected copy of an equilateral triangle concentrically inside that triangle and connecting each vertex of the inner triangle to the two closer vertices of the outer triangle. As Morgan and Scott show (and use of the BB-net would show more readily), for this Δ, $\dim \Pi_{2,\Delta}^1 = 7$ while, for any generic perturbation Δ' of Δ, $\dim \Pi_{2,\Delta'}^1 = \dim \Pi_2 = 6$.

Since the arguments for Theorem 4.1 make essential use of the fact that one knows how to construct bases for arbitrary univariate spline spaces, while we do not know how to do this in general for bivariate spline spaces, it is unlikely that one can obtain even the trivariate analogon of Theorem 4.1. An observation of Alfeld (in Alfeld, Schumaker and Sirvent (1992), see Schumaker (1991)) makes this precise. The latter reference gives a very good summary of what is presently known about $\dim \Pi_{k,\Delta}^\rho$. In particular, the recent paper Alfeld, Whiteley and Schumaker (199x) gives first specific results concerning the dimension of *trivariate* spline spaces. In addition, Billera and his colleagues initiated and pursued an investigation of $\dim \Pi_{k,\Delta}^\rho$ for arbitrary d with tools from Homological Algebra, which, however, forces them to consider only the case of a 'generic' Δ (which is difficult enough); see Billera (1988, 1989), Billera and Haas (1987) and Billera and Rose (1989, 1991). For example, Billera (1988) shows Strang's conjecture for $\rho = 1$ to be correct 'generically', using a specific construction of Whiteley (1991) to make certain that a certain determinant is not identically zero, hence must be generically nonzero.

Those with an urge to get a feeling for the difficulties one might encounter in considering arbitrary partitions should try the still unsolved problem of providing a formula for $\dim \Pi_{3,\Delta}^1(\mathbb{R}^2)$ for arbitrary Δ.

4.2. Subspaces of $\Pi_{k,\Delta}^\rho$

It is not only the difficulty of determining $\dim \Pi_{k,\Delta}^\rho$, hence of constructing bases for $\Pi_{k,\Delta}^\rho$, that makes the full space more of a challenge than of real interest. For certain partitions, $\Pi_{k,\Delta}^\rho$ contains elements of no use for approximation (such as the **half-space** spline $\mathbb{R}^d \to \mathbb{R} : x \mapsto (\langle y, x \rangle - c)_+^k$, with y a certain element of $\mathbb{R}^d$ and c some constant). Also, if k is large enough compared with ρ, then there are often subspaces of $\Pi_{k,\Delta}^\rho$ with the same 'approximation power' as $\Pi_{k,\Delta}^\rho$ itself.

For example, in the Finite Element method, bivariate pp spaces studied by Ženíšek (1970, 1973, 1974) and recently termed super-spline spaces in Chui and Lai (1987) consist of all elements of $\Pi_{k,\Delta}^\rho$ which, at each vertex, are in $C^{(2\rho)}$. In terms of the BB-net, the motivation (as explained, e.g., in Farin (1986)) for consideration of such subspaces is simple: if, for some $\delta \in \Delta$, we want to determine the polynomial piece $p = f_{|\delta}$ on δ so as to have a $C^{(\rho)}$-join with its neighbouring pieces, then its first ρ layers of control points along each edge of δ are determined by the polynomial piece adjoining that edge. However, certain of these control points are in the first ρ layers of two edges, hence in danger of being overdetermined. For any two edges, these endangered control points are contained in the first 2ρ layers for the vertex common to those two edges (and in no smaller set of layers). Hence, the enforcement of $C^{(2\rho)}$-continuity at the vertices ensures consistency for the competing smoothness conditions.

There are certain questions to be raised here. First, it has become popular, because of the success of the multigrid method, to work with a sequence of spaces, each obtained from the previous one by *refinement*, typically looking at the space of the same type on a refinement of the triangulation of the preceding one. If the spaces involved are super-spline spaces, then, because of the higher smoothness requirement at the vertices, the finer space will fail to contain the rougher space. Also, the degree k must be large enough so that the only questions of consistency of the smoothness conditions are of the kind described. For $d = 2$, this means that $k \geq 4\rho + 1$. Analogous considerations for arbitrary d (though not using BB-nets) led Le Méhauté (1990) to the conclusion that $k \geq 2^d\rho + 1$ was necessary (and sufficient) to provide such a super-spline space, in which an approximation can be constructed in a totally local way, with the approximant f on the simplex δ depending only on data on δ.

Such degrees are daunting. One response is to give up on using arbitrary triangulations, but use instead triangulations Δ obtained, e.g., by proper refinement of a given triangulation. The standard example is the Clough–Tocher element (although, because of its greater smoothness at its interior vertex, the space spanned by it does not properly refine, either). The extreme case of partitions (in general, they are not even triangulations) which will support compactly supported pp functions of low degree compared with the required smoothness are those provided by the multivariate B-spline construct to be discussed next.

5. Multivariate B-splines

The central role ultimately played by the univariate B-splines of (Curry and) Schoenberg (1946, 1966) in univariate spline theory (as illustrated, e.g., in Schoenberg (1969), de Boor (1976), or Schumaker (1981)) provided the impetus for the study of a certain multivariate generalization. Offhand, this generalization is based on preserving the somewhat obscure property of the univariate B-spline illustrated in Figure 5.1 and originally proved, in Curry and Schoenberg (1966), for the purpose of showing that the univariate B-spline is log-concave. Here are the details.

The **univariate B-spline $M(\cdot|\Theta)$ with knot sequence $\Theta = (\theta_0, \ldots, \theta_s)$** is, by one of its definitions, the Peano kernel for the divided difference (functional) $[\theta_0, \ldots, \theta_s]$, i.e., it is the unique function for which

$$[\theta_0, \ldots, \theta_s]f = \int_{\mathbb{R}} M(t|\Theta)\, D^s f(t)\, dt/s!$$

for all sufficiently smooth functions f. On combining this with the Hermite-Genocchi formula (Nörlund, 1924) for the divided difference, Schoenberg

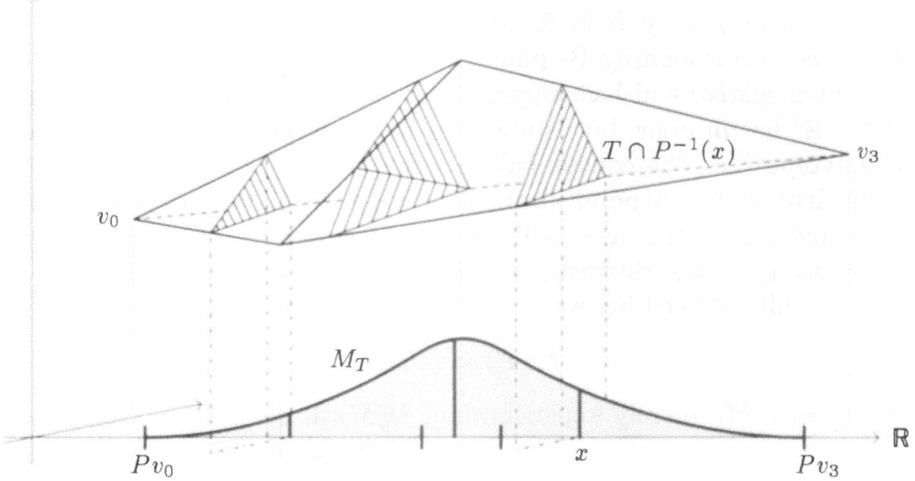

Fig. 5.1. A quadratic B-spline as the shadow of a tetrahedron.

obtains the equation

$$\int_{\mathbb{R}} M(t|\Theta)D^s f(t)\mathrm{d}t/s! =$$

$$\int_{\Theta} D^s f := \int_0^1 \int_0^{\tau_1} \cdots \int_0^{\tau_{s-1}} D^s f(\theta_0 + \tau_1\nabla\theta_1 + \cdots + \tau_s\nabla\theta_s)\mathrm{d}\tau_s \cdots \mathrm{d}\tau_2\,\mathrm{d}\tau_1$$

(with $\nabla\theta_j := \theta_j - \theta_{j-1}$, as usual). This equation implies that $M(t|\Theta)$ is the $(s-1)$-dimensional volume of the set

$$\{\tau \in T_s : \theta_0 + \tau_1\nabla\theta_1 + \cdots + \tau_s\nabla\theta_s = t\},$$

with T_s the **standard s-simplex**

$$T_s := \{\tau \in \mathbb{R}^s : 1 \geq \tau_1 \geq \tau_2 \geq \cdots \geq \tau_s \geq 0\}.$$

This simplex has vertices $v_j := \sum_{i=1}^j i_i$, $j = 0, \ldots, s$. Hence,

$$P : \mathbb{R}^s \to \mathbb{R} : \tau \mapsto \theta_0 + \tau_1\nabla\theta_1 + \cdots + \tau_s\nabla\theta_s$$

is the affine map which carries v_j to θ_j, all j. Consequently, $M(\cdot|\Theta)$ represents the distribution (aka continuous linear functional on $C(\mathbb{R})$)

$$f \mapsto \int_{T_s} f \circ P$$

which carries f to the sum over T_s of its extension $f \circ P$ to a function on $\mathbb{R}^s$. This is illustrated in Figure 5.1 for $s = 3$.

Once this is recognized, there is much scope for generalization (initiated in Schoenberg (1965) and followed up in de Boor (1976), Micchelli (1980), de Boor and DeVore (1983) and de Boor and Höllig (1982)), as follows. For a given (convex) body B in $\mathbb{R}^s$ and a given affine map $P : \mathbb{R}^s \to \mathbb{R}^d$, one defines the corresponding B-spline M_B as the distribution $f \mapsto \int_B f \circ P$. M_B is nonnegative and has support $P(B)$. M_B is a function exactly when $P(B) \subset \mathbb{R}^d$ has interior, but is always a function on affine$(P(B))$. When B is a polytope (i.e., the convex hull of some *finite* set), then M_B is called a **polyhedral** spline. A polyhedral spline is pp, with the junction places the images under P of the $(d-1)$-dimensional faces of B. This is most readily seen by using Stokes' theorem, as follows.

After a shift, if need be, we can assume that P is a linear map. Then

$$D_z(f \circ P) = (D_{Pz}f) \circ P.$$

Further, with M_B merely a distribution, $D_y M_B$ is defined by integration by parts,

$$D_y M_B f = -M_B(D_y f).$$

Therefore, for arbitrary $y \in \mathbb{R}^d$ and for any $z \in P^{-1}\{y\}$,

$$
\begin{aligned}
(D_{Pz} M_B) f &= -\int_B (D_{Pz} f) \circ P = -\int_B D_z(f \circ P) \\
&= -\int_{\partial B} z^T n \, (f \circ P) = - \sum_{F \in B^{(s-1)}} z^T n_F \, M_F f.
\end{aligned}
\tag{5.2}
$$

Here, ∂B is the (oriented) boundary of B. Since B is a polytope, ∂B is the essentially disjoint union of the collection $B^{(s-1)}$ of **facets** (i.e., $(s-1)$-dimensional faces) of B. Further, n is the outward unit normal, and n_F is its constant value on the facet F.

Iteration of this recurrence relation shows that any derivative of M_B of order $> s - d$ is a linear combination of distributions of the form M_F with F itself less than d-dimensional. Hence, on any connected component of the complement of the set

$$\bigcup_{F \in B^{(d-1)}} P(F),$$

(with $B^{(d-1)}$ the collection of all $(d-1)$-dimensional faces of B), M_B is a polynomial of degree $\leq k := s - d$. Further, if the polytope B is in general position and P is onto $\mathbb{R}^d$, then any d-face of B is mapped by P to a set with interior, hence all derivatives of M_B of order $\leq s - d$ are L_∞ functions. This means that, generically, M_B is pp of degree $\leq s - d$ and in $C^{(s-d-1)}$. However, in the interest of obtaining a relatively simple partition (or a partition which is not too different from a given one), one may have to choose B in a special way, and then M_B may not be maximally smooth. For, as the argument shows, M_B is in $C^{(s-m-1)}$, with m the smallest integer for which P maps every $F \in B^{(m)}$ to a set with interior.

For example, if $B = [0..1]^s$ is the s-dimensional unit cube, and $\theta_j := Pi_j$, $j = 1, \ldots, s$, and $\theta_0 := P0 = 0$, then the *bivariate* B-spline M_B may have discontinuities in some derivative across any image under P of an edge of B, i.e., across any segment of the form $[\sum_{\theta \in U} \theta .. \sum_{\theta \in W} \theta]$, with U, W arbitrary subsequences of the sequence $(\theta_0, \ldots, \theta_s)$. If each of these segments is also required to be part of the so-called square mesh (or, **two-direction** mesh) (formed by all the lines of the form $\{x \in \mathbb{R}^2 : x(j) = h\}$ with $j \in \{1,2\}$ and $h \in \mathbb{Z}$), then, up to scaling and certain translations, each θ_j is necessarily one of the two unit vectors i_1, i_2. This implies that some face of B of dimension $\lceil s/2 \rceil$ is mapped by P to a set without (two-dimensional) interior, hence M_B is, at best, in $C^{(s/2-2)}$ if s is even. The situation is slightly better for the **three-direction** mesh (formed by all lines of the form $\{x \in \mathbb{R}^2 : x(j) = h\}$ with $j \in \{1,2,3\}$ and $h \in \mathbb{Z}$, and $x(3) := x(1) - x(2)$). Now, θ_j may, in addition to i_1 and i_2, also take on the value $i_3 := i_1 + i_2$. In fact, if $s = 3$ and $\theta_j = i_j$, $j = 1, 2, 3$, then the resulting M_B is the hat function, the standard linear finite element at times associated with Courant because of Courant (1943).

Of course, one uses not just one polyhedral spline but linear combinations of sufficiently many to effect good approximation. At a minimum, this means that, after normalization if need be, such a collection $(M_B)_{B \in \mathcal{B}}$ of polyhedral splines should form a **partition of unity**, i.e., satisfy

$$\sum_{B \in \mathcal{B}} M_B = 1.$$

This is quite easy to achieve, as follows. Simply choose the collection $\mathcal{B}$ so that its elements are pairwise essentially disjoint, and their union is a set of the form $\mathbb{R}^d \times C$ for some suitable (convex) $(s-d)$-dimensional set C. For, in that case,

$$\sum_{B \in \mathcal{B}} M_B(x) = \underset{s-d}{\mathrm{vol}}(C),$$

while $M_B \geq 0$ in any case. If $B = [0..1]^s$ (hence M_B is a 'box spline') and P is given by an integer matrix, then the collection $M_B(\cdot - j)$, $j \in \mathbb{Z}^d$, of all integer shifts can be shown to be a partition of unity. Standard arguments concerning approximation order (see the next section) require, more generally, that it be possible to write every $p \in \Pi_{<r}$ as a linear combination of the M_B, $B \in \mathcal{B}$, and this is clearly satisfied for $r = 1$ in case $(M_B)_{B \in \mathcal{B}}$ forms a partition of unity. Much work has gone into constructing $\mathcal{B}$ for which r is large, preferably as large as $s - d + 1$ (it could be no larger), or, alternatively, into determining the largest possible such r for a given $\mathcal{B}$.

It is also important to have the means for reliable evaluation of such a polyhedral spline. It was only after the discovery of stable recurrence relations that univariate B-splines became an effective computational tool. In the same way, work on polyhedral splines only flourished after Micchelli (1980) established stable recurrence relations for simplex splines. The following generalization, to arbitrary polyhedral splines, was given in de Boor and Höllig (1982); it connects M_B to the M_F with F a facet of B:

$$(s - d)M_B(Pz) = \sum_{F \in B^{(s-1)}} (z - a_F)^T n_F M_F(Pz), \qquad (5.3)$$

with a_F an arbitrary point in affine(F). But there are only very few bodies B for which such a facet F is again a body of the same kind: the simplex, the cube or 'box', and the (polyhedral) cone. The corresponding B-splines are called, correspondingly, **simplex** spline, **box** spline, and **cone** spline (the last introduced in Dahmen (1979)). Each of these can be described entirely in terms of $P(B)$. In other words, any such B-spline is (a shift of) M_B with B a standard simplex, e.g., $\langle 0, i_1, \ldots, i_s \rangle$, a standard box $\square := [0 .. 1]^s$, or a standard cone $\mathbb{R}_+^s$, and P a suitable linear map (which is specified as soon as we know $P i_j$ for all j).

A first survey of multivariate B-splines is given in Dahmen and Micchelli (1983), an introduction to both simplex splines and box splines is given in Höllig (1986). The only book so far devoted entirely to multivariate B-splines is de Boor, Höllig and Riemenschneider (1992), a book on box splines. Box splines also figure prominently in the survey Chui (1988).

The first multivariate B-spline (and for some still the only one worthy of this appellation) was the simplex spline. If $v_0, \ldots, v_s$ is the sequence of vertices of the underlying simplex, then $M_{\langle v_0, \ldots, v_s \rangle}$ is, up to a scale factor, uniquely determined by the sequence $\Theta := (P v_j)_j$. For this reason, it has become standard to denote the typical simplex spline by

$$M(\cdot | \Theta),$$

with Θ some finite sequence in $\mathbb{R}^d$ (the images under P of the vertices of the underlying simplex) and to choose the underlying simplex to have unit volume, whence $\int_{\mathbb{R}^d} M(\cdot | \Theta) = 1$. This is entirely consistent with the notation $M(\cdot | \Theta)$ used earlier for the univariate B-spline.

The relative neglect simplex splines have experienced in spite of the fact that they were the first multivariate B-splines to be considered may have several reasons.

Box splines, like their univariate antecedents, the cardinal B-splines (see Schoenberg's monograph (1969)), lead very quickly to a rich mathematical theory, as exemplified by the beautiful results of Dahmen and Micchelli (announced in Dahmen and Micchelli (1984)). This theory concerns mainly the shift-invariant space spanned by the integer translates of one box spline, and these are pp spaces with a regular partition Δ, and this regularity makes them amenable to Fourier transform techniques.

In contrast, the simplex splines were expected to be the multivariate equivalent of the general univariate B-spline, of use in the understanding and handling of *arbitrary* multivariate spline spaces. Since any polytope is the essentially disjoint union of simplices, any multivariate B-spline is a linear combination of simplex splines. However, use of the recurrence relations for the evaluation of simplex splines turned out to be much more expensive than had been hoped, for the simple reason (Grandine (1986)) that the recurrrence relation connects a d-variate simplex spline to at least $d + 1$ simplex splines of one order less, while it connects it to at most two simplex splines of one order higher. Further, as already pointed out, for an arbitrary partition Δ and positive ρ, $\Pi^\rho_{k,\Delta}$ may not contain any compactly supported element unless k is very much larger than ρ. This means that, for k 'close' to ρ, only some suitably chosen refinement Δ' of Δ may support enough simplex splines so that their span has some approximation power. Unfortunately, the first scheme proposed for this (in Goodman and Lee (1981), Dahmen and Micchelli (1982) and Höllig (1982)) did not lead to a spline space with easily constructed quasi-interpolant schemes. However, very recently, a scheme has become available, in Dahmen, Micchelli and Seidel (1992), that, in hindsight, appears to be the 'right' one. It is based on the multivariate 'B-patch' of Seidel (1991). Given a triangulation Δ, it provides a suitable basis of simplex splines for the space $\Pi^{k-1}_{k,\Delta'}$, with Δ' obtained, in effect, as the roughest partition that contains all the cells for the simplex splines employed, thus known, at least in principle, once these simplex splines are in hand. These simplex splines are all possible ones of the form

$$M(\cdot|V^\beta),$$

where

(i) V is a $(d+1)$-set with $\langle V \rangle \in \Delta$;
(ii) $\beta \in \mathbb{Z}^V_+$ with $|\beta| = k$;
(iii) $V^\beta := \{v_j : 0 \le j \le \beta(v); v \in V\}$;
(iv) the points v_j are obtained, by choosing, for each v in the vertex set $V(\Delta) := \cup_{\langle V \rangle \in \Delta} V$ for Δ, k additional points $v_1, \ldots, v_k \in \mathbb{R}^d$, and setting $v_0 := v$.

The only condition imposed upon the choice of these additional points v_j, $j = 1, \ldots, k$, $v \in V(\Delta)$, is the following. For any $(d+1)$-set V with $\langle V \rangle \in \Delta$,

$$\Omega_{V,k} := \bigcap \{\langle (v_{\beta(v)})_{v \in V} \rangle : \beta \in \mathbb{Z}^V_+; |\beta| \le k\} \neq \emptyset.$$

Under these assumptions, Seidel (1992) proves that, for any $f \in \Pi^{k-1}_{k,\Delta}$,

$$f = \sum_{V,\beta} M(\cdot|V^\beta) w(V,\beta) F_V(V^{\beta-i}), \qquad (5.5)$$

with $w(V, \beta)$ certain explicitly known normalizing factors, with

$$\beta - \mathbf{i} : v \mapsto \beta(v) - 1,$$

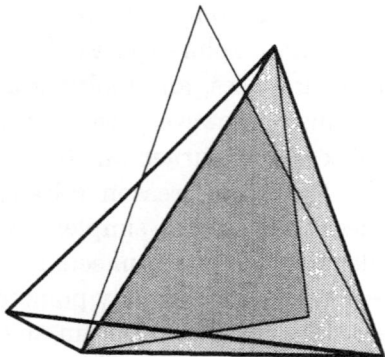

Fig. 5.4. With $k = 1$, the triangle $\langle V \rangle$ (lightly shaded and partially covered by) the set $\Omega_{V,k}$ (strongly shaded), and the meshlines (heavy) for one of the three related simplex splines.

hence $\#V^{\beta-\mathbf{i}} = |\beta| = k$, and with F_V the **blossom** of the polynomial which agrees with f on the cell $\langle V \rangle \in \Delta$. This means that F_V is the unique symmetric multi-linear form with k arguments for which

$$f(x) = F_V(x, x, \ldots, x), \quad \forall x \in \langle V \rangle.$$

The proof uses the validity of this result for any $f \in \Pi_k$, as established in Dahmen *et al.* (1992).

This is a most surprising and unexpected result. It captures completely the now standard formula for the coefficients in the B-spline expansion of an arbitrary *univariate* spline as stated in de Casteljau (1963) and beautifully explained in Ramshaw (1987, 1989). It is to be hoped that the computational aspects of this formulation are equally favourable.

6. Approximation order

The treatment of approximation order given here follows in part the survey article by de Boor (1992). The approximation power of a subspace S of $\Pi_{k,\Delta}$ is, typically, measured in terms of the **mesh(size)**

$$|\Delta| := \sup_{\delta \in \Delta} \operatorname{diam} \delta$$

of the partition Δ and the smoothness of the function f being approximated. The typical result is a statement of the following sort:

$$\operatorname{dist}(f, S) \leq \operatorname{const}|\Delta|^r \|D^r f\|,$$

in which $\|D^r f\|$ is some appropriate measure of the derivatives of order r of f, and const is independent of f and Δ, provided Δ is chosen from some appropriate class of partitions. For example, the constant may, offhand, depend on the **uniformity measure**

$$R_\Delta := \sup_{\delta \in \Delta} \inf\{M/m : B_m(x) \subset \delta \subset B_M(y)\}$$

(with $B_m(x)$ the open ball with centre x and radius m), hence be independent of Δ only if Δ is restricted to have $R_\Delta \leq R$ for some finite R.

A particularly simple version of the approximation order of S is the following. One considers not just S, but the entire **scale** $(\sigma_h S)_h$ with

$$\sigma_h S := \{f(\cdot/h) : f \in S\},$$

and says that S **has (exact) approximation order** r and writes

$$\mathbf{ao}(S) = r,$$

provided

(i)   for all 'smooth' f, $\mathrm{dist}(f, \sigma_h S) = O(h^r)$;
(ii) for some 'smooth' f, $\mathrm{dist}(f, \sigma_h S) \neq o(h^r)$.

By itself, (i) provides a *lower* bound for $\mathbf{ao}(S)$, and such lower bounds are usually established by exhibiting a particular approximation scheme, Q_h say, for which $\mathrm{ran}\, Q_h$ (= the range of Q_h) lies in $\subset \sigma_h S$, and $\|f - Q_h f\| \leq \mathrm{const} h^r \|D^r f\|$. So-called quasi-interpolants are a favourite choice for the Q_h, of which more below.

By itself, (ii) provides an *upper* bound on $\mathbf{ao}(S)$, and there seems to be only duality (as made clear below) to establish such upper bounds.

Of course, for completeness, this definition requires specification of the norm in which the distance is to be measured, i.e., the normed linear space X in which the approximation is to take place. Typically, it is $L_p(G)$, with G some suitable subset of $\mathbb{R}^d$, and $p = 1, 2$ or ∞. It also requires a definition of 'smooth'. Often, it is sufficient to mean 'polynomial' or 'complex exponential'. However, it usually means that some norm involving certain derivatives is finite.

Somewhat more generally, one considers an indexed family $(S_h)_h$ of spaces, and denotes its approximation order, correspondingly, by $\mathbf{ao}((S_h)_h)$ to stress the fact that it is not (necessarily) obtained by scaling. In the latter situation, it turns out to be helpful to consider S_h to be of the form

$$S_h =: \sigma_h S^h.$$

If S^h is independent of h, we are back to the scaling case which, therefore, is also referred to as the **stationary** case, to distinguish it from the more general **nonstationary** case.

Questions of approximation order, particularly from (multivariate) pp spaces, have been dominated by what in Approximation Theory is called the Strang–Fix theory, which, on careless reading, seems to imply that $\mathbf{ao}((S_h)_h)$

cannot be $\geq r$ unless $\Pi_{<r} \subset \cap_h S^h$. In fact, such a conclusion can only be reached in the stationary case, and even there only for very special situations. See Example 6.4 below for a simple counterexample; Ron (1991, 1992) and Beatson and Light (1992) treat approximation order specifically in the absence of polynomial reproduction. A similarly careless reading has also led to the *wrong* conclusion that, if all of $\Pi_{<r}$ is contained in each S^h locally, uniformly in h, then $\mathbf{ao}((S_h)_h) \geq r$. Even in the stationary case, the situation is more subtle, as is indicated in the subsections to follow. A first counter-example to that careless reading was given in de Boor and Höllig (1983).

In any event, the Strang–Fix theory applies only to the stationary case $S_h = \sigma_h S$, with S a *shift-invariant* space.

6.1. Shift-invariance

A collection S of functions on $\mathbb{R}^d$ is called **shift-invariant** if it is invariant under any translation by an integer, i.e., if

$$g \in S \quad \Longrightarrow \quad g(\cdot + \alpha) \in S \text{ for all } \alpha \in \mathbb{Z}^d.$$

For example, the space $\Pi_{k,\Delta}^\rho$ is shift-invariant in case Δ is shift-invariant in the sense that

$$\Delta + \alpha = \Delta \text{ for all } \alpha \in \mathbb{Z}^d.$$

Examples of interest include the three- and four-direction mesh popular in the bivariate box spline literature.

With $\ell_0(\mathbb{Z}^d)$ the collection of all *finitely* supported sequences $c : \mathbb{Z}^d \mapsto \mathbb{R}$, the simplest (nontrivial) example of a shift-invariant space is the space

$$S_0(\varphi) := \left\{ \sum_{\alpha \in \mathbb{Z}^d} \varphi(\cdot - \alpha)\, c(\alpha) : c \in \ell_0(\mathbb{Z}^d) \right\}$$

of all finite linear combinations of the shifts of one (nontrivial) function, φ. This is the **shift-invariant space generated by** φ since it is the smallest shift-invariant space containing φ. Following de Boor, DeVore and Ron (1991), its closure, in whatever norm the context suggests, is denoted by

$$S(\varphi) := \overline{S_0(\varphi)}$$

and called the **principal shift-invariant**, or **PSI**, space generated by φ. For example, approximation by box splines has been discussed almost entirely in terms of the scale $(\sigma_h S(\varphi))_h$ with φ a box spline.

More generally, if Φ is a finite collection of functions on $\mathbb{R}^d$, then one defines

$$S_0(\Phi) := \sum_{\varphi \in \Phi} S_0(\varphi)$$

and calls

$$S(\Phi) := \overline{S_0(\Phi)}$$

the **finitely generated shift-invariant**, or **FSI**, space, and calls Φ its set of **generators**. The structure of PSI and FSI spaces in $L_2(\mathbb{R}^d)$ is detailed in de Boor *et al.* (1991, 1992a), with particular emphasis on the construction of generating sets for a given FSI space having good properties (such as 'stability' or 'linear independence').

It is natural to consider approximations from $S(\varphi)$ in the form

$$\varphi * c := \sum_{\alpha \in \mathbb{Z}^d} \varphi(\cdot - \alpha)\, c(\alpha) \qquad (6.1)$$

for a suitable coefficient sequence c. However, offhand, such a sum makes sense only for finitely supported c, and one of the technical difficulties in ascertaining the approximation order of $S(\varphi)$ derives from the fact that, in general, $S(\varphi)$ may contain elements which cannot be represented in the form $\varphi * c$ for some sequence c, with the series $\varphi * c$ converging in norm. This is a problem even in the present context, where φ is, typically, some pp finite element and, in particular, compactly supported, hence the sum (6.1) converges pointwise (and even uniformly on compact sets) for arbitrary c. To give a simple example, from de Boor, DeVore and Ron (1992a), take for φ the Haar function, specifically $\varphi := \chi_{[-1..0)} - \chi_{[0..1)}$, with χ_I the characteristic function of the set I. Then $S(\varphi) = \Pi_{0,z} \cap L_2(\mathbb{R})$ and, in particular, $\chi_{[0..1)} \in S(\varphi)$. However, if the equation $\chi_{[0..1)} = \varphi * c$ is to hold even only in some weak sense, e.g., in the sense of pointwise convergence, then necessarily $c(\alpha) = c(0) + (\alpha - .5)_+^0$, all $\alpha \in \mathbb{Z}$, and $\varphi * c$ fails to converge in norm.

6.2. *Quasi-interpolants*

In the spline and finite-element literature, lower bounds for $\mathbf{ao}((S_h)_h)$ are usually obtained with the aid of a corresponding sequence $(Q_h)_h$ of linear maps, with $\operatorname{ran} Q_h \subseteq S_h$, which is a 'good quasi-interpolant sequence of order r' in the sense of the following definition.

Definition 6.2 $(Q_h)_h$ *is a* **good quasi-interpolant sequence of order** r *if it satisfies the following two conditions:*

(i) **uniformly local:** *For some h-independent finite ball B and all $x \in G$,*
$$|(Q_h f)(x)| \le \operatorname{const} \|f_{|x+hB}\|;$$
(ii) **polynomial reproduction:** $Q_h f = f$ *for all $f \in \Pi_{<r}$.*

For example, if $(\varphi)_{\varphi \in \Phi}$ is a **stable and local partition of unity**, i.e.,

$$\Big\| \sum_{\varphi \in \Phi} |\varphi| \Big\|_\infty < \infty, \qquad \sup_{\varphi \in \Phi} \operatorname{diam} \operatorname{supp} \varphi < \infty, \qquad \sum_{\varphi \in \Phi} \varphi = 1,$$

then $(\sigma_h Q \sigma_{1/h})_h$ with

$$Q : f \mapsto \sum_{\varphi \in \Phi} \varphi f(\tau_\varphi)$$

and $\tau_\varphi \in \operatorname{supp} \varphi$, all $\varphi \in \Phi$, is a good quasi-interpolant sequence of order 1.

As a more substantial example, it is part of the attraction of (5.5) that it provides an expansion of any $f \in \Pi_k$ in the form

$$f = \sum_{V, \beta} M(\cdot|V^\beta) w(V, \beta) \lambda_{V,\beta}(f), \tag{6.3}$$

with each $\lambda_{V,\beta}$ an explicitly known linear functional on Π_k. In particular (see Dahmen et al. (1992)) it is possible, as in the univariate case, to specify points $\tau_{V,\beta}$ so that the **Schoenberg operator**

$$Qf := \sum_{V, \beta} M(\cdot|V^\beta) w(V, \beta) f(\tau_{V,\beta})$$

reproduces every $f \in \Pi_1$. Since $\tau_{V,\beta}$ necessarily lies in the support of $M(\cdot|V^\beta)$ and this support is compact (and of the size of $\langle V \rangle$), it follows that $(\sigma_h Q \sigma_{1/h})_h$ is a good quasi-interpolant sequence of order 2. In fact, Dahmen et al. (1992) are able to lift the entire univariate quasi-interpolation argument (see, e.g., de Boor (1976)) to their multivariate setting, by showing the uniform linear independence of the functions $M(\cdot|V^\beta) w(V, \beta)$ which, in conjunction with (5.5), implies that any norm-preserving extension of $\lambda_{V,\beta}$ from $\Pi_k(\langle V \rangle)$ to some linear functional $\mu_{V,\beta}$, all V and β, provides a bounded linear projector

$$P : f \mapsto \sum_{V, \beta} M(\cdot|V^\beta) w(V, \beta) \mu_{V,\beta}(f)$$

onto the span of the simplex splines involved, and now, $(\sigma_h P \sigma_{1/h})_h$ is a good quasi-interpolant sequence of order $k + 1$.

The term 'quasi-interpolant' is used in the finite element literature (see, e.g., Strang and Fix (1973)) to stress the fact that $Q_h f$ does not necessarily match function values at all the nodes of the finite elements used, but 'merely' reproduces certain polynomials. For a recent survey of the use of quasi-interpolants in spline theory, see de Boor (1990).

To recall, the standard use made of such a good quasi-interpolant sequence is to observe that, for arbitrary f and arbitrary $g \in \Pi_{<r}$,

$$|f(x) - Q_h f(x)| = |(1 - Q_h)(f - g)(x)| \leq \operatorname{const} \|(f - g)_{|x+hB}\|,$$

which provides a bound on $\|f - Q_h f\|$ in terms of how well f can be approximated from $\Pi_{<r}$ on a set of the form $x + hB$, giving the error bound $\operatorname{const}_B h^r \|D^r f\|$ in which $\|D^r f\|$ measures the 'size' of the rth derivatives of f and which provides the desired $O(h^r)$. If our space X is L_p for some

$p < \infty$, then this argument has to be fleshed out a bit (see, e.g., Jia and Lei (1991)).

There are certain costs associated with the quasi-interpolant approach, even when one only considers shift-invariant spaces with compactly supported generators. For example, it works, offhand, only with integer values of r. Also, offhand, it requires that $\cap_h S_h$ contain some nontrivial polynomial space. The artificiality of this last restriction is nicely illustrated by the following simple example, from Dyn and Ron (1990):

Example 6.4. Let $d = 1$, $p = \infty$, and let S_h be the span of the $h\mathbb{Z}$-translates of the piecewise linear function

$$\varphi_h : x \mapsto \begin{cases} x + 1, & 0 \leq x < h; \\ 0, & \text{otherwise.} \end{cases}$$

Thus S_h consists of certain piecewise linear functions, with breakpoint sequence $h\mathbb{Z}$, but the only polynomial (hence the only analytic function) it contains is the zero polynomial. In particular, it is not possible to construct a quasi-interpolant of positive order for it. Nevertheless, the approximation

$$Q_h f := \sum_{j \in h\mathbb{Z}} \varphi_h(\cdot - j) f(j)$$

has the error

$$f - Q_h f = f - \sum_{j \in h\mathbb{Z}} \chi_h(\cdot - j) f(j) + \sum_{j \in h\mathbb{Z}} (\chi_h - \varphi_h)(\cdot - j) f(j),$$

with χ_h the characteristic function of the interval $[0 .. h)$. Since $\|\chi_h - \varphi_h\|_\infty = h$,

$$\|f - Q_h f\|_\infty \leq \omega_f(h) + \|f\|_\infty h,$$

where ω_f is the modulus of continuity of f. It follows that $Q_h f$ converges to f uniformly in case f is uniformly continuous and bounded. More than that, if f has a bounded first derivative, then $\|f - Q_h f\|_\infty \leq (\|Df\|_\infty + \|f\|_\infty)h$, giving approximation order 1 in the uniform norm.

This example could still be treated by an appropriate generalization of the notion of quasi-interpolant. Specifically, one could consider a good quasi-interpolant sequence (Q_h) of positive **local** order r, meaning that (Q_h) is uniformly local and that

$$Q_h f = f + O(\|f_{|B}\| \, |h|^r)$$

on hB for any $f \in \Pi_{<r}$. However, the point is made that a sequence $(S_h)_h$ of spaces does not need to contain a nontrivial polynomial space in order to have positive approximation order.

Finally, the quasi-interpolant approach is of no help with upper bounds.

6.3. The Strang–Fix condition

The literature on $\mathbf{ao}(\mathcal{S}(\varphi))$ for a compactly supported φ has been dominated by the Strang–Fix condition. It concerns the behaviour of the Fourier transform

$$\widehat{\varphi} : \xi \mapsto \int_{\mathbb{R}^d} \varphi\, e_{-\xi}$$

of φ at the points of $2\pi\mathbb{Z}^d$. Here and below,

$$e_\theta : \mathbb{R}^d \to \mathbb{C} : x \mapsto \exp(i\theta^T x)$$

denotes the exponential function (with purely imaginary frequency $i\theta$). In one of its many versions, the Strang–Fix condition reads as follows.

Definition 6.5 *We say that φ satisfies* SF$_r$ *in case*

(i) $\widehat{\varphi}(0) = 1$;
(ii) *For all multi-indices α satisfying $|\alpha| < r$ we have $D^\alpha\widehat{\varphi} = 0$ on $2\pi\mathbb{Z}^d\backslash 0$.*

Its importance derives from the following theorem (see Schoenberg (1946) for $d = 1$ and Strang and Fix (1973) for the general case), in which we use the convenient notation

$$\varphi*'f := \sum_{j\in\mathbb{Z}^d} \varphi(\cdot - j)f(j)$$

for the **semidiscrete convolution** of the two functions φ and f even if it requires further discussion of just what exactly is meant by it when the sum is not (locally) finite. Also, for any set X of functions on $\mathbb{R}^d$, we denote by

$$X_{\mathbf{c}}$$

the compactly supported functions in X.

Theorem 6.6 *For $\varphi \in L_1(\mathbb{R}^d)_{\mathbf{c}}$, the following are equivalent:*

(a) *$\varphi*'$ is* **degree-preserving** *on $\Pi_{<r}$, i.e., $\varphi*'p \in p+\Pi_{<\deg p}$, for all p in $\Pi_{<r}$;*
(b) *φ satisfies* SF$_r$.

The proof is via the Poisson summation formula (for which see, e.g., Stein and Weiss (1971; p. 252)). Starting with Strang and Fix (1973), the theorem is used to construct a good quasi-interpolant sequence (Q_h) of order r with $\operatorname{ran} Q_h \subseteq \sigma_h\mathcal{S}(\varphi)$. More than that, it forms part of an argument that seems to show that $\mathbf{ao}(\mathcal{S}(\varphi)) \geq r$ if and only if $\varphi/\widehat{\varphi}(0)$ satisfies SF$_r$. The precise statement of this equivalence for $X = L_2(\mathbb{R}^d)$ (see Strang and Fix (1973)) involves, unfortunately, a restricted notion of approximation order called 'controlled' approximation.

For $X = L_2(\mathbb{R}^d)$, the recent paper de Boor *et al.* (1991) contains a complete characterization of the approximation order of a not necessarily stationary scale of closed shift-invariant spaces. A crucial ingredient is the following theorem from the same reference, in which $P_S f$ denotes the orthogonal projection of f onto S, hence $\text{dist}(f, S) = \|f - P_S f\|$.

Theorem 6.7 *Let S be a closed shift-invariant subspace of $L_2(\mathbb{R}^d)$, and let $f, g \in L_2(\mathbb{R}^d)$. Then*

$$\text{dist}(f, S) \leq \text{dist}(f, S(P_S g)) \leq \text{dist}(f, S) + 2\,\text{dist}(f, S(g)).$$

This theorem shows that the approximation power of a general shift-invariant subspace of L_2 is already attained by one of its PSI subspaces, provided one can, for given r, supply an element $g \in L_2(\mathbb{R}^d)$ for which **ao**$(S(g)) > r$. But that is easy to do, as follows.

Lemma 6.8 *There are simple functions g (e.g., the inverse Fourier transform of the characteristic function of some small neighbourhood of the origin) for which, for any r,*

$$\text{dist}(f, \sigma_h S(g)) = o(h^r \|f\|_{W_2^r(\mathbb{R}^d)}).$$

Here,

$$\|f\|_{W_2^r(\mathbb{R}^d)} := \|(1 + |\cdot|)^r \hat{f}\|_2.$$

For a directed family $(\sigma_h S^h)_h$ with each S^h a PSI space, de Boor *et al.* (1991) provide the following characterization of the approximation order, in which

$$\Lambda_\varphi := 1 - \frac{|\hat{\varphi}|^2}{[\hat{\varphi}, \hat{\varphi}]} = \frac{\sum_{\alpha \in \mathbb{Z}^d \backslash 0} |\hat{\varphi}(\cdot + 2\pi\alpha)|^2}{\sum_{\alpha \in \mathbb{Z}^d} |\hat{\varphi}(\cdot + 2\pi\alpha)|^2},$$

$\mathbb{T}^d$ is the d-dimensional torus, i.e.,

$$\mathbb{T}^d := [-\pi \mathrel{..} \pi]^d$$

with the appropriate identification of boundary points, and

$$[f, g] : \mathbb{T}^d \to \mathbb{C} : x \mapsto \sum_{\alpha \in \mathbb{Z}^d} f(x + 2\pi\alpha)\overline{g}(x + 2\pi\alpha)$$

is the very convenient **bracket product** of $f, g \in L_2(\mathbb{R}^d)$.

Theorem 6.9 *For any $(\varphi_h)_h$ in $X := L_2(\mathbb{R}^d)$,*

$$\textbf{ao}((\sigma_h S(\varphi_h))_h) \geq r \iff \sup\nolimits_h \left\| \frac{\Lambda_{\varphi_h}}{(h + |\cdot|)^{2r}} \right\|_{L_\infty(\mathbb{T}^d)} < \infty.$$

This result focuses attention on the behaviour of Λ_{φ_h} near 0, hence, if $\hat{\varphi}_h$ is bounded away from zero near 0 (uniformly in h), it focuses attention on the ratios

$$\widehat{\varphi}_h(\cdot + 2\pi\alpha)/\widehat{\varphi}_h \,, \qquad \alpha \in \mathbb{Z}^d\backslash 0. \tag{6.10}$$

Here is a typical corollary (from the same reference) which shows the relationship of this characterization to the Strang–Fix condition.

Corollary 6.11 *If $\varphi \in L_2(\mathbb{R}^d)$, and $1/\widehat{\varphi}$ is essentially bounded near 0, and $\widehat{\varphi} \in W_2^\rho(U)$ for some $\rho > r + d/2$ and some neighbourhood U of $2\pi\mathbb{Z}^d\backslash 0$, and if φ satisfies SF_r, then $\mathbf{ao}(\mathcal{S}(\varphi)) \geq r$.*

Finally, as a consequence of Theorem 6.7 (and a good understanding of the structure of FSI spaces), de Boor *et al.* (1992a) obtains the following result which finishes a job left undone in Strang and Fix (1973) (see de Boor *et al.* (1992a) for historical commentary).

Theorem 6.12 *The approximation order in $L_2(\mathbb{R}^d)$ of the FSI space $\mathcal{S}(\Phi)$ with $\Phi \subset L_2(\mathbb{R}^d)$ is already attained by some PSI space $\mathcal{S}(\varphi)$ with $\varphi \in \mathcal{S}_0(\Phi)$.*

In particular, if Φ consists of compactly supported functions, then the 'super element' φ of the theorem is also compactly supported. This follows, more explicitly, from a representation of the Fourier transform of $P_S g$ as a sum of the form $\sum_{\varphi\in\Phi} \tau_\varphi \widehat{\varphi}$, in which the τ_φ are ratios of 2π-periodic functions, each a linear combination of products of functions of the form $[\widehat{\phi}, \widehat{\psi}]$ with $\phi, \psi \in \Phi \cup \{g\}$. Now, for any particular r, it is possible to choose g compactly supported and such that $\mathbf{ao}((\,)\mathcal{S}(g)) \geq r$, while all the elements of Φ are compactly supported by assumption. This means that, with such a choice for g, each τ_φ is the ratio of two trigonometric polynomials, hence, there are trigonometric polynomials T_g, T_φ, $\varphi \in \Phi$, so that $T_g \widehat{P_S g} = \sum_{\varphi\in\Phi} T_\varphi \widehat{\varphi}$. This implies that the inverse Fourier transform of $T_g \widehat{P_S g}$ is in $\mathcal{S}_0(\Phi)$ and generates the same shift-invariant space as does $P_S g$, hence may be taken as the desired 'super-element'.

The paper by de Boor and Ron (1991) deals with approximation from PSI spaces in $L_\infty(\mathbb{R}^d)$. The results are surprisingly similar in form, even if, due to the greater difficulties expected in this norm, there is a gap between lower and upper bounds for the approximation order obtained.

The main tool is Ron's (1991) surprisingly simple observation that, since

$$\varphi*'f = f*'\varphi \qquad \forall\, f \in \mathcal{S}(\varphi) \tag{6.13}$$

(as hinted at in Chui, Jetter and Ward (1987)), therefore

$$\varphi*'e_\theta - e_\theta*'\varphi = \varphi*'(e_\theta - f) - (e_\theta - f)*'\varphi, \qquad \forall\, f \in \mathcal{S}(\varphi)$$

(recall that $e_\theta : x \mapsto \exp(i\theta^T x)$), and this leads to the conclusion that

$$\|\varphi*'e_\theta - e_\theta*'\varphi\|_\infty \leq 2\|\varphi*'\|_\infty \operatorname{dist}_\infty(e_\theta, \mathcal{S}(\varphi)), \tag{6.14}$$

with

$$\|\varphi*'\|_\infty := \|\sum_{\alpha \in \mathbb{Z}^d} |\varphi(\cdot - \alpha)|\,\|_\infty.$$

Since (as pointed out by A. Ron)

$$\frac{\varphi*'e_\theta - e_\theta*'\varphi}{e_\theta} \sim c + \sum_{\alpha \in \mathbb{Z}^d \backslash 0} \widehat{\varphi}(\theta + 2\pi\alpha)\,e_\alpha,$$

and the left-hand side has the same norm as $\|\varphi*'e_\theta - e_\theta*'\varphi\|_\infty$, this throws new light on the connection between $\mathbf{ao}(S(\varphi))$ in L_∞ and the behaviour of $\widehat{\varphi}$ 'at' $2\pi\mathbb{Z}^d \backslash 0$, and provides both upper and lower bounds for $\mathbf{ao}((S(\varphi_h))_h)$.

As to lower bounds, these are obtained (in de Boor and Ron (1991)) by the approximation

$$f(\theta) = \int_{\mathbb{R}^d} e_\theta \widehat{f}/(2\pi)^d \sim \int_{\mathbb{R}^d} \varepsilon_\theta \widehat{f}/(2\pi)^d$$

(and a related one), with

$$\varepsilon_\theta := \varphi*'e_\theta / \sum_{\alpha \in \mathbb{Z}^d} \varphi(\alpha)e_{-\alpha}$$

an approximation from $S(\varphi)$ to e_θ suggested by (6.14). In particular, the following theorem is proved there, in which $S(\varphi)$ is not the norm-closure of $S_0(\varphi)$ in $L_\infty(\mathbb{R}^d)$ but, in effect, the largest shift-invariant space containing $S_0(\varphi)$ and satisfying (6.13). Also, the 'size' of the rth derivatives of f is measured in terms of its Fourier transform, as follows. It is assumed that f is 'smooth' in the sense that its Fourier transform is a Radon measure for which

$$\|f\|_{(r)} := \|(1 + |\cdot|^r)\widehat{f}\|_1 < \infty,$$

with the suffix '1' intended to indicate that the total variation of the measure in question is meant.

Theorem 6.15 *Assume that $\|\varphi_h*'\| < \infty$ for every h. Then, for any positive η,*

$$\operatorname{dist}(f, \sigma_h S(\varphi_h)) \leq h^r (2\pi)^{-d} \|f\|_{(r)} A + o(h^r)$$

with

$$A := \sup_h \sum_{\alpha \in \mathbb{Z}^d \backslash 0} \left\| \frac{1}{(h^r + |\cdot|^r)} \frac{\widehat{\varphi}_h(\cdot + 2\pi\alpha)}{\widehat{\varphi}_h} \right\|_{L_\infty(B_\eta)}.$$

Since this theorem gives $\mathbf{ao}((\sigma_h S(\varphi_h))_h) \geq r$ only if $A < \infty$, this focuses, once again, attention on the behaviour near zero of each of the ratios

$$\widehat{\varphi}_h(\cdot + 2\pi\alpha)/\widehat{\varphi}_h, \qquad \alpha \in \mathbb{Z}^d \backslash 0$$

mentioned already in (6.10). Specifically, in the stationary case, if this ratio is a smooth function in a neighbourhood of 0, then the finiteness of A would require the ratio to have a zero of order r at 0, and conversely, provided $\hat{\varphi}$ has some decay. From this vantage point, the Strang–Fix condition SF_r is seen to be neither necessary nor sufficient for $\mathbf{ao}(S(\varphi)) \geq r$, but to come close to being necessary and sufficient for appropriately restricted φ.

The striking observation (6.14) actually provides more immediately an *upper* bound on the approximation order (see Ron (1991)). The main result of de Boor and Ron (1991) concerning this is the following.

Theorem 6.16 *Let* (φ_h) *be an indexed collection of elements of* $X :=$ $L_\infty(\mathbb{R}^d)$. *Assume that* $\sup_h \|\varphi_h*'\| < \infty$, *and that* $\theta \in \mathbb{R}^d$.
If $\mathrm{dist}(e_\theta, \sigma_h S(\varphi_h)) = O(h^r)$, *then*

$$\sum_{\alpha \in \mathbb{Z}^d \backslash 0} |\hat{\varphi}_h(h\theta + 2\pi\alpha)|^2 \leq \mathrm{const}_\theta\, h^{2r}.$$

In particular, then

$$|\hat{\varphi}_h(h\theta + 2\pi\alpha)| \leq \mathrm{const}_\theta h^r \quad \text{for all nonzero } \alpha \text{ in } \mathbb{Z}^d.$$

Note that nothing is said here about $\hat{\varphi}_h(0)$ (which is particularly important if $\hat{\varphi}_h(0)$ is zero). On the other hand, it is easy to recover from this the rest of SF_r in the stationary case, i.e., in case $\varphi_h = \varphi$, for all h.

6.4. Upper bounds

Upper bounds for $\mathbf{ao}((S_h)_h)$ have to be fashioned separately for each case. However, one always employs duality, which provides the following well-known observation.

If Y is a linear subspace of the normed linear space X, and $\lambda \in X^*$ with $\lambda \perp Y$ (i.e., λ is a continuous linear functional on X which vanishes on all of Y), then, for any $x \in X$ and any $y \in Y$, $\lambda x = \lambda(x - y) \leq \|\lambda\|\|x - y\|$, hence $|\lambda x| \leq \|\lambda\| \mathrm{dist}(x, Y)$. In other words,

$$\lambda \perp Y \quad \Longrightarrow \quad \mathrm{dist}(x, Y) \geq \frac{|\lambda x|}{\|\lambda\|}.$$

For example, Ron's upper-bound argument mentioned in the preceding subsection is based on the linear map $f \mapsto \varphi*'f - f*'\varphi$ which vanishes on all of $S(\varphi)$.

As a more direct example, consider $\mathbf{ao}(S)$ for

$$X = L_\infty(G), \quad S = \Pi^\rho_{k,\Delta}.$$

Assume without loss of generality that G is the d-dimensional cube,

$$G = C := [-1 .. 1]^d,$$

let δ be any cell in the partition Δ, and let g be any nontrivial homogeneous polynomial of degree $k+1$. If γ is the error in the best $L_2(\delta)$-approximation to g from Π_k, then the mapping

$$\lambda : L_\infty \to \mathbb{R} : f \mapsto \int_\delta \gamma f$$

(i)   is a bounded linear functional;
(ii)   is orthogonal to S, since all λ sees of $f \in S$ is its restriction to δ, and on δ each $f \in S$ is just a polynomial of degree $\leq k$;
(iii) satisfies $\lambda g = \int_\delta \gamma\gamma > 0$.

Now consider $\lambda_h f := \int_\delta \gamma f(h\cdot)$. Then

(i) λ_h is a bounded linear functional, with h-independent norm

$$\|\lambda_h\| = \int_\delta |\gamma| = \lambda\,\text{signum}(\gamma),$$

where $\text{signum}(\gamma) : x \mapsto \text{signum}(\gamma(x))$.
(ii) $\lambda_h \perp S_h := \sigma_h S$, since $g \in S_h$ is of the form $f(\cdot/h)$ for some $f \in S$.
(iii) Using the homogeneity of g, one computes that

$$\lambda_h g = \int_\delta \gamma g(h\cdot) = h^{k+1} \int_\delta \gamma g = h^{k+1}\lambda g$$

with $\lambda g > 0$.

So, altogether,

$$\text{dist}(g, S_h) \geq h^{k+1}(\lambda g / \lambda\,\text{signum}(\gamma)),$$

showing that $\mathbf{ao}(\Pi^\rho_{k,\Delta}) \leq k+1$.

If we try the same argument for $p < \infty$, we hit a little snag. Take, in fact, p at the other extreme, $p = 1$. There is no difficulty with (ii) or (iii), but the conclusion is weakened because (i) now reads

(i)' $\|\lambda_h\| = \sup_{f \in L_1} |\int_\delta \gamma f(h\cdot)| / \|f\|_1 \leq \|\gamma_{|\delta}\|_\infty \sup_{f \in L_1(\delta)} \int_\delta |f(h\cdot)| / \|f\|_1,$

and the best we can say about that last supremum is that it is at most h^{-d} since $\int_\delta f(h\cdot) = \int_{h\delta} f/h^d$. Hence, altogether, $\|\lambda_h\| \leq \text{const}/h^d$. Thus, now our bound reads

$$\text{dist}_1(g, S_h) \geq h^{k+1}\text{const}/(\text{const}/h^d) \neq o(h^{k+1+d})$$

which is surely correct, but not very helpful.

What we are witnessing here is the fact that the error in a max-norm approximation is indeed localized, i.e., it occurs at a point, while, for $p < \infty$, the error 'at a point' is less relevant; the error is more global; one needs to consider the error over a good part of G. Further, in the argument below, I need some kind of uniformity of the partition Δ, of the following (very weak) sort (in which $|A|$ denotes the d-dimensional volume of the set A, and C continues to denote the cube $[-1..1]^d$):

Assumption 6.17 *There exists an open set b and a locally finite set $I \subset \mathbb{R}^d$ (meaning that I meets any bounded set only in finitely many points) so that*

(α) $b + I$ is the disjoint union of $b + i$, $i \in I$, with each $b + i$ lying in some $\delta \in \Delta$ (the possibility of several lying in the same δ is not excluded);
(β) for some const > 0 and all n, $|(b + I) \cap nC| \geq \text{const}|nC|$.

For example, any uniform partition of $\mathbb{R}$ satisfies this condition. As another example, if $d = 2$ and Δ is the three-direction mesh, then Δ consists of triangles of two kinds, and taking b to be the interior of one of these triangles and $I = \mathbb{Z}^2$ guarantees (α), while (β) holds with const $= 1/2$. On the other hand, Shayne Waldron (a student at Madison) has constructed a neat example to show that the Assumption 6.17 is, in general, necessary for the conclusion that $\mathbf{ao}(\Pi_{k,\Delta}^\rho) \leq k + 1$. The example uses $\rho = -1$ and arbitrary k, $d = 1$, $G = [-1 .. 1]$, $p = 1$, and Δ obtained from $\mathbb{Z}$ by subdividing $[j .. j + 1]$ into $2^{|j|}$ equal pieces, $j \in \mathbb{Z}$.

With Assumption 6.17 holding, define λ as before, but with b replacing the element δ of Δ. Further, assume without loss that $C \subseteq G$, and define

$$\lambda_h f := \int_b \gamma \sum_{i \in I_h} f(h \cdot + i),$$

where

$$I_h := \{i \in I : b + i \subseteq C/h\}.$$

This gives

(i)$_1$

$$\|\lambda_h\| \leq \sup_{f \in L_1} \frac{\sum_{i \in I_h} \int_{b+i} |\gamma| |f(h\cdot)|}{\sum_{i \in I_h} \int_{h(b+i)} |f|} = \|\gamma_{|b}\|_\infty / h^d,$$

using the fact that the union $b + I_h$ is disjoint.

Hence, we have not worsened our situation here. Neither have we sacrificed (ii) because, by assumption, each $b + i$ lies in the interior of some $\delta \in \Delta$, and therefore $\int_b \gamma f(h \cdot + i) = 0$ for every $f \in S_h$. But we have materially improved the situation as regards (iii), for we now obtain

(iii)$_1$

$$\lambda_h g = \int_b \gamma \sum_{i \in I_h} g(h \cdot + i) = h^{k+1} \int_b \gamma \sum_{i \in I_h} g = h^{k+1} \text{const} \, \#I_h$$

with

$$\#I_h = |b + I_h|/|b| \geq \text{const}|C/h|/|b| = \text{const}/h^d.$$

With this, our conclusion is back to what we want:

$$\text{dist}_1(g, S_h) \geq (h^{k+1} \text{const}/h^d)/(\text{const}/h^d) \neq o(h^{k+1}).$$

Note that this lower bound on the distance only sees S as a space of pp's of degree $\leq k$, hence is valid even when we take the biggest such space, i.e., the space $\Pi_{k,\Delta}$ of *all* pp functions of degree $\leq k$ on the partition Δ. For this space, it is not hard to show that the approximation order is at least $k+1$, since approximations can be constructed entirely locally. Thus,

$$\mathbf{ao}(\Pi_{k,\Delta}) = k+1.$$

For this reason, this is called the **optimal** approximation order for a pp space of degree $\leq k$.

Such a local construction of approximations is still possible for $\Pi^0_{k,\Delta}$, hence,

$$\mathbf{ao}(\Pi^\rho_{k,\Delta}) = k+1 \quad \text{for} \quad \rho \leq 0.$$

However, for $\rho > 0$, the story is largely unknown. Here are some working conjectures.

Conjecture (Ming-Yun Lai) *If* $\mathbf{ao}(\Pi^\rho_{k,\Delta}) = k+1$, *then* $\mathbf{ao}(\Pi^\rho_{k',\Delta}) = k'+1$ *for all* $k' \geq k$.

Conjecture $\mathbf{ao}(\Pi^\rho_{k,\Delta}) > 0 \implies \Pi^\rho_{k,\Delta}$ *contains elements with compact support.*

Conjecture $\mathbf{ao}(\Pi^\rho_{k,\Delta}) > 0 \implies \Pi^\rho_{k,\Delta}$ *contains a local partition of unity.*

First results (and more conjectures) can be found in de Boor and DeVore (1985) and Jia (1989).

Further illustrations of the use of duality in the derivation of upper bounds on $\mathbf{ao}(S)$ (albeit only for bivariate pp S) can be found in de Boor and Jia (199x) and its references. In particular, in conjunction with de Boor and Höllig (1988), it is proved there that, with Δ the three-direction mesh, the approximation order of $\Pi^\rho_{k,\Delta}$ (in the uniform norm) is $k+1$ (i.e., optimal) if and only if $k > 3\rho + 1$.

Acknowledgements

It is a pleasure to thank Martin Buhmann for his many constructive comments on a draft of this paper.

REFERENCES

P. Alfeld, L.L. Schumaker and Maritza Sirvent (1992), 'On dimension and existence of local bases for multivariate spline spaces', *J. Approx. Theory* **70**, 243–264.

P. Alfeld, W. Whiteley and L.L. Schumaker (199x), 'The generic dimension of the space of C^1 splines of degree $d \geq 8$ on tetrahedral decompositions', *SIAM J. Numer. Anal.* **xx**, xxx–xxx.

R. K. Beatson and W. A. Light (1992), 'Quasi-interpolation in the absence of polynomial reproduction', Univ. of Leicester Mathematics Techn. Report 1992/15, May.

P.E. Bézier (1970), *Emploi des Machines à Commande Numérique*, Masson et Cie (Paris).

P.E. Bézier (1977), 'Essai de définition numérique des courbes et des surfaces expérimentale', dissertation, Univ. Paris.

L.J. Billera (1988), 'Homology of smooth splines: generic triangulations and a conjecture of Strang', *Trans. Amer. Math. Soc.* **310**, 325–340.

L.J. Billera (1989), 'The algebra of continuous piecewise polynomials', *Advances of Math.* **76(2)**, 170–183.

L.J. Billera and R. Haas (1987), 'Homology of divergence-free splines', Cornell.

L.J. Billera and L.L. Rose (1989), 'Gröbner basis methods for multivariate splines', in *Mathematical Methods in Computer Aided Geometric Design* (T. Lyche and L. Schumaker, eds), Academic Press (New York), 93–104.

Louis J. Billera and Lauren L. Rose (1991), 'A dimension series for multivariate splines', *Discrete & Computational Geometry* **6**, 107–128.

G. Birkhoff and C.R. de Boor (1965), 'Piecewise polynomial interpolation and approximation', in *Approximation of Functions* (H.L. Garabedian, ed.), Elsevier (Amsterdam), 164–190.

W. Boehm, G. Farin and J. Kahmann (1984), 'A survey of curve and surface methods in CAGD', *Comput. Aided Geom. Design* **1**, 1–60.

C. de Boor (1976), 'Splines as linear combinations of B–splines, a survey', in *Approximation Theory II* (G. G. Lorentz, C. K. Chui and L. L. Schumaker, eds), Academic Press (New York), 1–47.

C. de Boor (1978), *A Practical Guide to Splines*, Springer Verlag (New York).

C. de Boor (1987), 'B–form basics', in *Geometric Modeling: Algorithms and New Trends* (G. E. Farin, ed.), SIAM Publications (Philadelphia), 131–148.

C. de Boor (1990), 'Quasiinterpolants and approximation power of multivariate splines', in *Computation of Curves and Surfaces* (W. Dahmen, M. Gasca and C. Micchelli, eds), Kluwer, 313–345.

C. de Boor (1992), 'Approximation order without quasi-interpolants', in *Approximation Theory VII* (E. W. Cheney, C. Chui and L. Schumaker, eds), Academic Press (New York), xxx–xxx.

C. de Boor and R. DeVore (1983), 'Approximation by smooth multivariate splines', *Trans. Amer. Math. Soc.* **276**, 775–788.

C. de Boor and R. DeVore (1985), 'Partitions of unity and approximation', *Proc. Amer. Math. Soc.* **93**, 705–709.

C. de Boor and K. Höllig (1982), 'Recurrence relations for multivariate B-splines', *Proc. Amer. Math. Soc.* **85**, 397–400.

C. de Boor and K. Höllig (1983), 'Approximation order from bivariate C^1-cubics: a counterexample', *Proc. Amer. Math. Soc.* **87**, 649–655.

C. de Boor and K. Höllig (1988), 'Approximation power of smooth bivariate pp functions', *Math. Z.* **197**, 343–363.

C. de Boor and Rong-Qing Jia (199x), 'A sharp upper bound on the approximation order of smooth bivariate pp functions', *J. Approx. Theory* **xx**, xxx–xxx.

C. de Boor and A. Ron (1992), 'Fourier analysis of the approximation power of principal shift-invariant spaces', *Constr. Approx.* **8**, 427–462.

C. de Boor, R. DeVore and A. Ron (1991), 'Approximation from shift-invariant subspaces of $L_2(\mathbb{R}^d)$', CMS-TSR **92-2**, University of Wisconsin (Madison).

C. de Boor, R. DeVore and A. Ron (1992a), 'The structure of finitely generated shift-invariant spaces in $L_2(\mathbb{R}^d)$', CMS-TSR **92-8**, University of Wisconsin (Madison).

C. de Boor, K. Höllig and S.D. Riemenschneider (1992), *Box Splines*, Springer-Verlag (Berlin).

P. de Casteljau (1963), 'Courbes et Surfaces à Pôles', André Citroën Automobiles SA (Paris).

P. de Casteljau (1985), *Formes à Pôles*, Hermes (Paris).

C.K. Chui (1988), *Multivariate Splines*, CBMS-NSF Reg. Conf. Series in Appl. Math., vol. 54, SIAM (Philadelphia).

C.K. Chui and M.J. Lai (1987), 'On multivariate vertex splines and applications', in *Topics in Multivariate Approximation* (C. K. Chui, L. L. Schumaker and F. Utreras, eds), Academic Press (New York), 19–36.

C.K. Chui, K. Jetter and J.D. Ward (1987), 'Cardinal interpolation by multivariate splines', *Math. Comput.* **48**, 711–724.

R. Courant (1943), 'Variational methods for the solution of problems in equilibrium and vibrations', *Bull. Amer. Math. Soc.* **49**, 1–23.

H.B. Curry and I.J. Schoenberg (1966), 'On the Pólya frequency functions IV: the fundamental spline functions and their limits', *J. Analyse Math.* **17**, 71–107.

W. Dahmen (1979), 'Multivariate B-splines—Recurrence relations and linear combinations of truncated powers', in *Multivariate Approximation Theory* (W. Schempp and K. Zeller, eds), Birkhäuser (Basel), 64–82.

W. Dahmen and C.A. Micchelli (1982), 'On the linear independence of multivariate B-splines I. Triangulations of simploids', *SIAM J. Numer. Anal.* **19**, 992–1012.

W. Dahmen and C.A. Micchelli (1983), 'Recent progress in multivariate splines', in *Approximation Theory IV* (C. Chui, L. Schumaker and J. Ward, eds), Academic Press (New York), 27–121.

W. Dahmen and C.A. Micchelli (1984), 'Some results on box splines', *Bull. Amer. Math. Soc.* **11**, 147–150.

W. Dahmen, C.A. Micchelli and H.-P. Seidel (1992), 'Blossoming begets B-spline bases built better by B-patches', *Math. Comput.* **59**, 97–115.

N. Dyn and A. Ron (1990), 'Local approximation by certain spaces of multivariate exponential-polynomials, approximation order of exponential box splines and related interpolation problems', *Trans. Amer. Math. Soc.* **319**, 381–404.

G. Farin (1977), 'Konstruktion und Eigenschaften von Bézier-Kurven und Bézier Flächen', Diplom-Arbeit, Univ. Braunschweig.

G. Farin (1986), 'Triangular Bernstein–Bézier patches', *Comput. Aided Geom. Design* **3**, 83–127.

G. Farin (1988), *Curves and Surfaces for Computer Aided Geometric Design*, Academic Press (New York).

Ivor D. Faux and Michael J. Pratt (1979), *Computational Geometry for Design and Manufacture*, Ellis Horwood (Chichester).

R. Franke and L.L. Schumaker (1987), 'A bibliography of multivariate approximation', in *Topics in Multivariate Approximation* (C. K. Chui, L. L. Schumaker
   and F. Utreras, eds), Academic Press (New York), 275–335.

R.N. Goldman (1983), 'Subdivision algorithms for Bézier triangles', *Computer-
   Aided Design* **15**, 159–166.

T.N. Goodman and S.L. Lee (1981), 'Spline approximation operators of Bernstein–
   Schoenberg type in one and two variables', *J. Approx. Theory* **33**, 248–263.

T.A. Grandine (1986), 'The computational cost of simplex spline functions', MRC
   Rpt. 2926, Univ. Wis. (Madison WI).

K. Höllig (1982), 'Multivariate splines', *SIAM J. Numer. Anal.* **19**, 1013–1031.

K. Höllig (1986), 'Multivariate splines', in *Approximation Theory, Proc. Symp.
   Appl. Math.* **36** (C. de Boor, ed.), Amer. Math. Soc. (Providence, RI), 103–
   127.

J. Hoschek and D. Lasser (1989), *Grundlagen der geometrischen Datenverarbeitung*,
   B.G. Teubner (Stuttgart).

Rong-Qing Jia (1989), 'Approximation order of translation invariant subspaces of
   functions', in *Approximation Theory VI* (C. Chui, L. Schumaker and J. Ward,
   eds), Academic Press (New York), 349–352.

Rong-Qing Jia (1992), 'Approximation by multivariate splines: an application of
   Boolean methods', ms.

Rong-Qing Jia and Junjiang Lei (1991), 'Approximation by multiinteger translates
   of functions having global support', *J. Approx. Theory* **x**, xxx–xxx.

A. Le Méhauté (1990), 'A finite element approach to surface reconstruction', in
   Computation of Curves and Surfaces (W. Dahmen, M. Gasca and C. Micchelli,
   eds), Kluwer, 237–274.

G.G. Lorentz (1953), *Bernstein Polynomials*, Toronto Press (Toronto).

N. Luscher (1987), 'Die Bernstein–Bézier Technik in der Methode der finiten Ele-
   mente', dissertation, Univ. Braunschweig.

C.A. Micchelli (1980), 'A constructive approach to Kergin interpolation in $\mathbb{R}^k$:
   multivariate B-splines and Lagrange interpolation', *Rocky Mountain J. Math.*
   10, 485–497.

J. Morgan and R. Scott (1990), 'The dimension of the space of C^1 piecewise polynomials', Research Report UH/MD-78, University of Houston.

N.E. Nörlund (1924), *Vorlesungen über Differenzenrechnung*, Grundlehren Vol.
   XIII, Springer (Berlin).

M.J.D. Powell (1981), *Approximation Theory and Methods*, Cambridge University
   Press (Cambridge).

L. Ramshaw (1987), 'Blossoming: a connect-the-dots approach to splines', Techn.
   Rep., Digital Systems Research Center (Palo Alto).

L. Ramshaw (1989), 'Blossoms are polar forms', *Comput. Aided Geom. Design* **6**,
   23–32.

A. Ron (1991), 'A characterization of the approximation order of multivariate spline
   spaces', *Studia Math.* **98**, 73–90.

A. Ron (1992), 'The L_2-approximation orders of principal shift-invariant spaces
   generated by a radial basis function', in *Numerical Methods in Approximation
   Theory, ISNM 105* (D. Braess, L.L. Schumaker, eds), Birkhäuser (Basel),
   245–268.

I. J. Schoenberg (1946), 'Contributions to the problem of approximation of equidistant data by analytic functions, Part A: On the problem of smoothing or graduation, a first class of analytic approximation formulas', *Quart. Appl. Math.* **4**, 45–99.

I.J. Schoenberg (1965), 'Letter to Philip J. Davis', dated May 31.

I.J. Schoenberg (1969), 'Cardinal interpolation and spline functions', *J. Approx. Theory* **2**, 167–206.

L.L. Schumaker (1979), 'On the dimension of spaces of piecewise polynomials in two variables', in *Multivariate Approximation Theory* (W. Schempp and K. Zeller, eds), Birkhäuser (Basel), 396–412.

L.L. Schumaker (1981), *Spline Functions: Basic Theory*, Wiley (New York).

L.L. Schumaker (1984), 'Bounds on the dimension of spaces of multivariate piecewise polynomials', *Rocky Mountain J. Math.* **14**, 251–264.

L.L. Schumaker (1988), 'Constructive aspects of bivariate piecewise polynomials', in *The Mathematics of Finite Elements and Applications VI* (J.R. Whiteman, ed.), Academic Press (London), 513–520.

L.L. Schumaker (1991), 'Recent progress on multivariate splines', in *Mathematics of Finite Elements and Applications VII* (J. Whiteman, ed.), Academic Press (London), 535–562.

H.-P. Seidel (1991), 'Symmetric recursive algorithms for surfaces: B-patches and the de Boor algorithm for polynomials over triangles', *Constr. Approx.* **7**, 257–279.

H.-P. Seidel (1991), 'Representing piecewise polynomials as linear combinations of multivariate B-splines', ms.

E. Stein and G. Weiss (1971), *Introduction to Fourier analysis on Euclidean spaces*, Princeton University Press (Princeton).

G. Strang (1973), 'Piecewise polynomials and the finite element method', *Bull. Amer. Math. Soc.* **79**, 1128–1137.

G. Strang (1974), 'The dimension of piecewise polynomials, and one-sided approximation', in *Numerical Solution of Differential Equations* (G. A. Watson, ed.), Springer (Berlin), 144–152.

G. Strang and G. Fix (1973), 'A Fourier analysis of the finite element variational method', in *Constructive Aspects of Functional Analysis* (G. Geymonat, ed.), C.I.M.E. II Ciclo 1971, 793–840.

W. Whiteley (1991), 'A matrix for splines', in *Progress in Approximation Theory* (P. Nevai and A. Pinkus, eds.), Academic Press (Boston), 821–828.

A. Ženišek (1970), 'Interpolation polynomials on the triangle', *Numer. Math.* **15**, 283–296.

A. Ženišek (1973), 'Polynomial approximation on tetrahedrons in the finite element method', *J. Approx. Theory* **7**, 334–351.

A. Ženišek (1974), 'A general theorem on triangular C^m elements', *Rev. Française Automat. Informat. Rech. Opér., Anal. Numer.* **22**, 119–127.

Acta Numerica (1993), *pp.* 111–197

Parallel numerical linear algebra

James W. Demmel*

Computer Science Division and Mathematics Department
University of California at Berkeley
Berkeley, CA 94720 USA
E-mail: demmel@cs.berkeley.edu

Michael T. Heath[†]

Department of Computer Science
and
National Center for Supercomputing Applications
University of Illinois
Urbana, IL 61801 USA
E-mail: heath@ncsa.uiuc.edu

Henk A. van der Vorst[‡]

Mathematical Institute
Utrecht University
Utrecht, The Netherlands
E-mail: vorst@math.ruu.nl

We survey general techniques and open problems in numerical linear algebra
on parallel architectures. We first discuss basic principles of parallel processing, describing the costs of basic operations on parallel machines, including
general principles for constructing efficient algorithms. We illustrate these
principles using current architectures and software systems, and by showing
how one would implement matrix multiplication. Then, we present direct and
iterative algorithms for solving linear systems of equations, linear least squares
problems, the symmetric eigenvalue problem, the nonsymmetric eigenvalue
problem, and the singular value decomposition. We consider dense, band and
sparse matrices.

* The author was supported by NSF grant ASC-9005933, DARPA contract DAAL03-
91-C-0047 via a subcontract from the University of Tennessee administered by ARO,
and DARPA grant DM28E04120 via a subcontract from Argonne National Laboratory.
This work was partially performed during a visit to the Institute for Mathematics and
its Applications at the University of Minnesota.
† The author was supported by DARPA contract DAAL03-91-C-0047 via a subcontract
from the University of Tennessee administered by ARO.
‡ This work was supported in part by NCF/Cray Research University Grant CRG 92.03.

CONTENTS

1. Introduction

Accurate and efficient algorithms for many problems in numerical linear algebra have existed for years on conventional serial machines, and there are many portable software libraries that implement them efficiently (Dongarra, Bunch, Moler and Stewart, 1979; Dongarra and Grosse, 1987; Garbow, Boyle, Dongarra and Moler, 1977; Smith, Boyle, Dongarra, Garbow, Ikebe, Klema and Moler, 1976). One reason for this profusion of successful software is the simplicity of the cost model: the execution time of an algorithm is roughly proportional to the number of floating point operations it performs. This simple fact makes it relatively easy to design efficient and portable algorithms. In particular, one need not worry excessively about the location of the operands in memory, nor the order in which the operations are performed. That we can use this approximation is a consequence of the progress from drum memories to semiconductor cache memories, software to hardware floating point, assembly language to optimizing compilers, and so on. Programmers of current serial machines can ignore many details earlier programmers could ignore only at the risk of significantly slower programs.

With modern parallel computers we have come full circle and again need to worry about details of data transfer time between memory and processors, and which numerical operations are most efficient. Innovation is very rapid, with new hardware architectures and software models being proposed and implemented constantly. Currently one must immerse oneself in the multitudinous and often ephemeral details of these systems in order to write reasonably efficient programs. Perhaps not surprisingly, a number of techniques for dealing with data transfer in blocked fashion in the 1960s are being rediscovered and reused (Bell, Hatlestad, Hansteen and Araldsen, 1973).

Our first goal is to enunciate two simple principles for identifying the important strengths and weaknesses of parallel programming systems (both hardware and software): locality and regularity of operation. We do this in Section 2. Only by understanding how a particular parallel system embodies these principles can one design a good parallel algorithm for it; we illustrate this in Section 3 using matrix multiplication.*

Besides matrix multiplication, we discuss parallel numerical algorithms for linear equation solving, least squares problems, symmetric and nonsymmetric eigenvalue problems, and the singular value decomposition. We organize this material with dense and banded linear equation solving in Section 4, least squares problems in Section 5, eigenvalue and singular value problems in Section 6, direct methods for sparse linear systems in Section 7, iterative methods for linear systems in Section 8, and iterative methods for eigenproblems in Section 9. We restrict ourselves to general techniques, rather than techniques like multigrid and domain decomposition that are specialized for particular application areas.

We emphasize algorithms that are *scalable*, i.e. remain efficient as they are run on larger problems and larger machines. As problems and machines grow, it is desirable to avoid algorithm redesign. As we will see, we will sometimes pay a price for this scalability. For example, though many parallel algorithms are parallel versions of their serial counterparts with nearly identical roundoff and stability properties, others are rather less stable, and would not be the algorithm of choice on a serial machine.

Any survey of such a busy field is necessarily a snapshot reflecting some of the authors' biases. Other recent surveys include Dongarra, Duff, Sorensen and van der Vorst (1991) and Gallivan, Heath, Ng, Ortega, Peyton, Plemmons, Romine, Sameh and Voigt (1990a), the latter of which includes a bibliography of over 2000 entries.

2. Features of parallel systems

2.1. General principles

A large number of different parallel computers (Gottlieb and Almasi, 1989), languages (see Zima and Chapman (1991) and the references therein), and software tools have recently been built or proposed. Though the details of these systems vary widely, there are two basic issues they must deal with, and these will guide us in understanding how to design and analyse parallel algorithms. These issues are *locality* and *regularity* of computation.

Locality refers to the proximity of the arithmetic and storage components of computers. Computers store data in memories, which are physically separated from the computational units that perform useful arithmetic or logical

---

* This discussion will not entirely prepare the reader to write good programs on any particular machine, since many machine-specific details will remain.

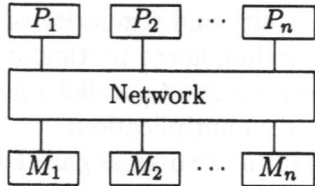

Fig. 1. Diagram of a parallel computer (P = processor, M = memory).

functions. The amount of time it takes to move the data from the memory to the arithmetic unit can far exceed the time required to perform the arithmetic unless the memory is immediately proximate to the arithmetic unit; such memory is usually called the *register file* or *cache*. There are good electrical and economic reasons that not all memory can consist of registers or cache. Therefore all machines, even the simplest PCs, have *memory hierarchies* of fast, small, expensive memory-like registers, then slower, larger and cheaper main memory, and finally down to disk or other peripheral storage. Parallel computers have even more levels, possibly including local memory as well as remote memory, which may serve as the local memory for other processors (see Figure 1). Useful arithmetic or logical work can occur only on data stored at the top of the memory hierarchy, and data must be moved from the lower, slower levels in the hierarchy to the top level to participate in computation. Therefore, much of algorithm design involves deciding where and when to store or fetch data in order to minimize this movement. The action of processor i storing or fetching data in memory j as shown in Figure 1 is called *communication*. Depending on the machine, this may be done automatically by the hardware whenever the program refers to nonlocal data, or it may require the explicit sending and/or receiving of messages on the part of the programmer. Communication among processors occurs over a *network*.

A special kind of communication worth distinguishing is *synchronization*, where two or more processors attempt to have their processing reach a commonly agreed upon stage. This requires an exchange of messages as well, perhaps quite short ones, and so qualifies as communication.

A very simple model for the time it takes to move n data items from one location to another is $\alpha + \beta \cdot n$, where $0 \leq \beta, \alpha$. One way to describe α is the *start up time* of the operation; another term for this is *latency*. The incremental time per data item moved is β; its reciprocal is called *bandwidth*. Typically $0 < \beta \ll \alpha$, i.e. it takes a relatively long time to start up an operation, after which data items arrive at a higher rate of speed. This cost model, which we will see later, reflects the pipeline implementation of the hardware: the pipeline takes a while to fill up, after which data arrive at a high rate.

The constants α and β depend on the parts of the memory between which transfer occurs. Transfer between higher levels in the hierarchy may be orders of magnitude faster than those between lower levels (for example, cache–memory *versus* memory–disk transfer). Since individual memory levels are themselves built of smaller pieces, and may be shared among different parts of the machine, the values of α and β may strongly depend on the location of the data being moved.

Regularity of computation means that the operations parallel machines perform fastest tend to have simple, regular patterns, and efficiency demands that computations be decomposed into repeated applications of these patterns. These regular operations include not only arithmetic and logical operations but communication as well. Designing algorithms that use a very high fraction of these regular operations is, in addition to maintaining locality, one of the major challenges of parallel algorithm design. The simplest and most widely applicable cost models for these regular operations is again $\alpha + \beta \cdot n$, and for the same reason as before: pipelines are ubiquitous.

Amdahl's Law quantifies the importance of using the most efficient parallel operations of the machine. Suppose a computation has a fraction $0 < p < 1$ of its operations which can be effectively parallellized, while the remaining fraction $s = 1 - p$ cannot be. Then with n processors, the most we can decrease the run time is from $s + p = 1$ to $s + p/n$, for a speed up of $1/(s + p/n) \leq 1/s$; thus the serial fraction s limits the speed up, no matter how many parallel processors n we have. Amdahl's Law suggests that only large problems can be effectively parallellized, since for the problems we consider p grows and s shrinks as the problem size grows.

2.2. Examples

We illustrate the principles of regularity and locality with examples of current machines and software systems.

A sequence of machine instructions without a branch instruction is called a *basic block*. Many processors have pipelined execution units that are optimized to execute basic blocks; since there are no branches, the machine can have several instructions in partial stages of completion without worrying that a branch will require 'backing out' and restoring an earlier state. So in this case, regularity of computation means code without branches. An algorithmic implication of this is *loop unrolling*, where the body of a loop like

$$\text{for } i = 1 : n$$
$$a_i = a_i + b * c_i$$

is replicated four times (say) yielding

$$\text{for } i = 1 : n \text{ step } 4$$

$$a_i = a_i + b * c_i$$
$$a_{i+1} = a_{i+1} + b * c_{i+1}$$
$$a_{i+2} = a_{i+2} + b * c_{i+2}$$
$$a_{i+3} = a_{i+3} + b * c_{i+3}$$

In this case the basic block is the loop body, since the end of the loop is a conditional branch back to the beginning. Unrolling makes the basic block four times longer.

One might expect compilers to perform simple optimizations like this automatically, but many do not, and seemingly small changes in loop bodies can make this difficult to automate (imagine adding the line 'if $i > 1$, $d_i = e_i$' to the loop body, which could instead be done in a separate loop from $i = 2$ to n without the 'if'). For a survey of such compiler optimization techniques see Zima and Chapman (1991). A hardware approach to this problem is *optimistic execution*, where the hardware guesses which way the branch will go and computes ahead under that assumption. The hardware retains enough information to undo what it did a few steps later if it finds out it decided incorrectly. But in the case of branches leading back to the beginning of loops, it will almost always make the right decision. This technique could make unrolling and similar low-level optimizations unnecessary in the future.

A similar example of regularity is *vector pipelining*, where a single instruction initiates a pipelined execution of a single operation on a sequence of data items; componentwise addition or multiplications of two arrays or 'vectors' is the most common example, and is available on machines from Convex, Cray, Fujitsu, Hitachi, NEC, and others. Programmers of such machines prefer the unrolled version of the above loop, and expect the compiler to convert it into, say, a single machine instruction to multiply the vector c by the scalar b, and then add it to vector a.

An even higher level of such regularity is so-called *SIMD parallellism*, which stands for Single Instruction Multiple Data, where each processor in Figure 1 performs the same operation in lockstep on data in its local memory. (SIMD stands in contrast to *MIMD* or Multiple Instruction Multiple Data, where each processor in Figure 1 works independently.) The CM-2 and MasPar depend on this type of operation for their speed. A sample loop easily handled by this paradigm is

```
for i = 1 : n
    if c_i > 0 then
        a_i = b_i + √c_i
    else
        a_i = b_i - d_i
    endif
```

A hidden requirement for these examples to be truly regular is that no exceptions arise during execution. An exception might be floating point overflow or address out of range. The latter error necessitates an interruption of execution; there is no reasonable way to proceed. On the other hand, there are reasonable ways to continue computing past floating point exceptions, such as *infinity arithmetic* as defined by the IEEE floating point arithmetic standard (ANSI/IEEE, 1985). This increases the regularity of computations by eliminating branches. IEEE arithmetic is implemented on almost all microprocessors, which are often building blocks for larger parallel machines. Whether or not we can make sense out of results that have overflowed or undergone other exceptions depends on the application; it is true often enough to be quite useful.

Now we give some examples of regularity in communication. The CM-2 (Thinking Machines Corporation, 1987) may be thought of in different ways; for us it is convenient to think of it as 2048 processors connected in an *11-dimensional hypercube*, with one processor and its memory at each of the 2048 corners of the cube, and a physical connection along each edge connecting each corner to its 11 nearest neighbours. All 11×2048 such connections may be used simultaneously, provided only one message is sent on each connection.

We illustrate such a regular communication by showing how to compute

$$f_i = \sum_{j=1}^{N} F(x_i, x_j),$$

i.e. an N-body interaction where f_i is the force on body i, $F(x_i, x_j)$ is the force on body i due to body j, and x_i and x_j are the positions of bodies i and j respectively (Brunet, Edelman and Mesirov, 1990). Consider implementing this on a d-dimensional hypercube, and suppose $N = d2^d$ for simplicity. We need to define the *Gray code*

$$G(d) \equiv (G_{d,0}, ..., G_{d,2^d-1}),$$

which is a permutation of the d-bit integers from 0 to $2^d - 1$, ordered so that adjacent codes $G_{d,k}$ and $G_{d,k+1}$ differ only in one bit. $G(d)$ may be defined recursively by taking the $(d-1)$-bit numbers $G(d-1)$, followed by the same numbers in reverse order and incremented by 2^{d-1}. For example, $G(2) = \{00, 01, 11, 10\}$ and $G(3) = \{000, 001, 011, 010, 110, 111, 101, 100\}$. Now imagining our hypercube as a unit hypercube in d-space with one corner at the origin and lying in the positive orthant, number each processor by the d-bit string whose bits are the coordinates of its position in d-space. Since the physically nearest neighbours of a processor lie one edge away, their coordinates or processor numbers differ in only one bit. Since $G_{d,k}$ and $G_{d,k+1}$ differ in only one bit, the Gray code sequence describes a path among the processors in a minimal number of steps visiting each one only

| $x_{0,0},x_{1,0}$ | $x_{0,2},x_{1,2}$ |
| $x_{0,1},x_{1,1}$ | $x_{0,3},x_{1,3}$ |

| $x_{0,1},x_{1,2}$ | $x_{0,3},x_{1,0}$ |
| $x_{0,0},x_{1,3}$ | $x_{0,2},x_{1,1}$ |

| $x_{0,3},x_{1,3}$ | $x_{0,1},x_{1,1}$ |
| $x_{0,2},x_{1,2}$ | $x_{0,0},x_{1,0}$ |

| $x_{0,2},x_{1,1}$ | $x_{0,0},x_{1,3}$ |
| $x_{0,3},x_{1,0}$ | $x_{0,1},x_{1,2}$ |

Initial data   After $m = 0$ After $m = 1$ After $m = 2$

Fig. 2. Force computation on two-dimensional hypercube.

once; such a path is called *Hamiltonian*. Now define the *shifted Gray code*

$$G^{(s)}(d) = \{G_{d,0}^{(s)}, ..., G_{d,2^d-1}^{(s)}\}$$

where $G_{d,k}^{(s)}$ is obtained by left-circular shifting $G_{d,k}$ by s bits. Each $G^{(s)}(d)$ also defines a Hamiltonian path, and all may be traversed simultaneously without using any edges simultaneously. Let $g_{d,k}^{(s)}$ denote the bit position in which $G_{d,k}^{(s)}$ and $G_{d,k+1}^{(s)}$ differ.

Now we define the program each processor will execute in order to compute f_i for the bodies it owns. Number the bodies $x_{k,l}$, where $0 \leq l \leq 2^d - 1$ is the processor number and $0 \leq k \leq d - 1$; so processor l owns $x_{0,l}$ through $x_{d-1,l}$. Then processor l executes the following code, where 'forall' means each iteration may be done in parallel (a sample execution for $d = 2$ is shown in Figure 2).

Algorithm 1 N-body force computation on a hypercube

> for $k = 0 : d - 1$, $tmp_k = x_{k,l}$
> for $k = 0 : d - 1$, $f_{k,l} = 0$ /* $f_{k,l}$ will accumulate force on $x_{k,l}$ */
> for $m = 0 : 2^d - 1$
> > forall $k = 0 : d - 1$, swap tmp_k with processor in direction $g_{d,m}^{(k)}$
> > for $k = 0 : d - 1$
> > > for $k' = 0 : d - 1$
> > > > $f_{k,l} = f_{k,l} + F(x_{k,l}, tmp_{k'})$

In Section 3 we will show how to use Gray codes to implement matrix multiplication efficiently. Each processor of the CM-2 can also send data to any other processor, not just its immediate neighbours, with the network of physical connections forwarding a message along to the intended receiver like a network of post-offices. Depending on the communication pattern this may lead to congestion along certain connections and so be much slower than the special communication pattern discussed earlier.

Here are some other useful regular communication patterns. A *broadcast* sends data from a single source to all other processors. A *spread* may be described as partitioned broadcast, where the processors are partitioned and a separate broadcast done within each partition. For example, in a square

0	1	2	3	4	5	6	7	8	9	a	b	c	d	e	f
	0:1		2:3		4:5		6:7		8:9		a:b		c:d		e:f
			0:3				4:7				8:b				c:f
							0:7								8:f
															0:f
											0:b				
					0:5				0:9				0:d		
		0:2		0:4		0:6		0:8		0:a		0:c		0:e	

Fig. 3. Parallel prefix on 16 data items.

array of processors we might want to broadcast a data item in the first column to all other processors in its row; thus we partition the processor array into rows and do a broadcast to all the others in the partition from the first column. This operation might be useful in Gaussian elimination, where we need to subtract multiples of one matrix column from the other matrix columns. Another operation is a *reduction*, where data distributed over the machine are reduced to a single datum by applying an associative operation like addition, multiplication, maximum, logical or, and so on; this operation is naturally supported by processors connected in a tree, with information being reduced as it passes from the leaves to the root of the tree.

A more general operation than reduction is the *scan* or *parallel prefix* operation. Let $x_0, ...x_n$ be data items, and $\cdot$ any associative operation. Then the scan of these n data items yields another n data items defined by $y_0 = x_0$, $y_1 = x_0 \cdot x_1, ... , y_i = x_0 \cdot x_1 \cdots x_i$; thus y_i is the reduction of x_0 through x_i. An attraction of this operation is its ease of implementation using a simple tree of processors. We illustrate in Figure 3 for $n = 15$, or f in hexadecimal notation; in the figure we abbreviate x_i by i and $x_i \cdots x_j$ by $i : j$. Each row indicates the values held by the processors; after the first row only the data that change are indicated. Each updated entry combines its current value with one a fixed distance to its left.

Parallel prefix may be used, for example, to solve linear recurrence relations $z_{i+1} = \sum_{j=0}^{n} a_{i,j} z_{i-j} + b_i$; this can be converted into simple parallel operations on vectors plus parallel prefix operations where the associative operators are n by n matrix multiplication and addition. For example, to evaluate $z_{i+1} = a_i z_i + b_i$, $i \geq 0$, $z_0 = 0$, we do the following operations:

Algorithm 2 Linear recurrence evaluation using parallel prefix

Compute $p_i = a_0 \cdots a_i$ using parallel prefix multiplication
Compute $\beta_i = b_i / p_i$ in parallel
Compute $s_i = \beta_0 + \cdots + \beta_{i-1}$ using parallel prefix addition
Compute $z_i = s_i \cdot p_{i-1}$ in parallel

Similarly, we can use parallel prefix to evaluate certain rational recurrences $z_{i+1} = (a_i z_i + b_i)/(c_i z_i + d_i)$ by writing $z_i = u_i/v_i$ and reducing to a linear recurrence for u_i and v_i:

$$\begin{bmatrix} u_{i+1} \\ v_{i+1} \end{bmatrix} = \begin{bmatrix} a_i & b_i \\ c_i & d_i \end{bmatrix} \cdot \begin{bmatrix} u_i \\ v_i \end{bmatrix}. \tag{2.1}$$

We may ask more generally about evaluating the scalar rational recurrence $z_{i+1} = f_i(z_i)$ in parallel. Let d be the maximum of the degrees of the numerators and denominators of the rational functions f_i. Then Kung (1974) has shown that z_i can be evaluated faster than linear time (i.e. z_i can be evaluated in $o(i)$ steps) if and only if $d \leq 1$; in this case the problem reduces to 2×2 matrix multiplication parallel prefix in (2.1). Interesting linear algebra problems that can be cast in this way include tridiagonal Gaussian elimination, solving bidiagonal linear systems of equations, Sturm sequence evaluation for the symmetric tridiagonal eigenproblem, and the bidiagonal dqds algorithm for singular values (Parlett and Fernando, 1992); we discuss some of these later. The numerical stability of these procedures remains open, although it is often good in practice (Swarztrauber, 1992).

We now turn to the principle of locality. Since this is an issue many algorithms do not take into account, a number of so-called *shared memory machines* have been designed in which the hardware attempts to make all memory locations look equidistant from every processor, so that old algorithms will continue to work well. Examples include machines from Convex, Cray, Fujitsu, Hitachi, NEC, and others (Gottlieb and Almasi, 1989). The memories of these machines are organized into some number, say b, of *memory banks*, so that memory address m resides in memory bank m mod b. A memory bank is designed so that it takes b time steps to read/write a data item after it is asked to do so; until then it is busy and cannot do anything else. Suppose one wished to read or write a sequence of $n + 1$ memory locations i, $i + s$, $i + 2s, \ldots$, $i + ns$; these will then refer to memory banks i mod b, $i + s$ mod b, $\ldots$, $i + ns$ mod b. If $s = 1$, so that we refer to consecutive memory locations, or if s and b are relatively prime, b consecutive memory references will refer to b different memory banks, and so after a wait of b steps the memory will deliver a result once per time step; this is the fastest it can operate. If instead $gcd(s, b) = g > 1$, then only b/g memory banks will be referenced, and speed of access will slow down by a factor of g. For example, suppose we store a matrix by columns, and the number of rows is s. Then reading a column of the matrix will be $gcd(s, b)$ times faster than reading a row of the matrix, since consecutive row elements have memory addresses differing by s; this clearly affects the design of matrix algorithms. Sometimes these machines also support *indirect addressing* or *gather/scatter*, where the addresses can be arbitrary rather than forming an arithmetic sequence, although it may be significantly slower.

Another hardware approach to making memory access appear regular are *virtual shared memory machines* like the Kendall Square Research machine and Stanford's Dash. Once the memory becomes large enough, it will necessarily be implemented as a large number of separate banks. These machines have a hierarchy of caches and directories of pointers to caches to enable the hardware to locate quickly and fetch or store a nonlocal piece of data requested by the user; the hope is that the cache will successfully anticipate enough of the user's needs to keep them local. To the extent that these machines fulfil their promise, they will make parallel programming much easier; as of this writing it is too early to judge their success.[†]

For machines on which the programmer must explicitly send or receive messages to move data, there are two issues to consider in designing efficient algorithms. The first issue is the relative cost of communication and computation. Recall that a simple model of communicating n data items is $\alpha + n\beta$; let γ be the average cost of a floating point operation. If $\alpha \gg \beta$, which is not uncommon, then sending n small messages will cost $n(\alpha + \beta)$, which can exceed by nearly a factor n the cost of a single message $\alpha + n\beta$. This forces us to design algorithms that do infrequent communications of large messages, which is not always convenient. If $\alpha \gg \gamma$ or $\beta \gg \gamma$, which are both common, then we will also be motivated to design algorithms that communicate as infrequently as possible. An algorithm which communicates infrequently is said to exhibit *coarse-grained parallellism*, and otherwise *fine-grained parallellism*. Again this is sometimes an inconvenient constraint, and makes it hard to write programs that run efficiently on more than one machine.

The second issue to consider when sending messages is the semantic power of the messages (Wen and Yelick, 1992). The most restrictive possibility is that the processor executing 'send' and the processor executing 'receive' must synchronize, and so block until the transaction is completed. So for example, if one processor sends long before the other receives, it must wait, even if it could have continued to do useful work. At the least restrictive the

---

[†] There is a good reason to hope for the success of these machines: parallel machines will not be widely used if they are hard to program, and maintaining locality explicitly is harder for the programmer than having the hardware do it automatically.

sending processor effectively interrupts the receiving processors and executes an arbitrary subroutine on the contents of the message, without any action by the receiving program; this minimizes time wasted waiting, but places a burden on the user program to do its own synchronization.

To illustrate these points, imagine an algorithm that recursively subdivides problems into smaller ones, but where the subproblems can be of widely varying complexity that cannot be predicted ahead of time. Even if we divide the initial set of problems evenly among our processors, the subproblems generated by each processor may be very different. A simple example is the use of Sturm sequences to compute the eigenvalues of a symmetric tridiagonal matrix. Here the problem is to find the eigenvalues in a given interval, and the subproblems correspond to subintervals. The time to solve a subproblem depends not only on the number but also on the distribution of eigenvalues in the subinterval, which is not known until the problem is solved. In the worst case, all processors but one finish quickly and remain idle while the other one does most of the work. Here it makes sense to do *dynamic load balancing*, which means redistributing to idle processors those subproblems needing further processing. This clearly requires communication, and may or may not be effective if communication is too expensive.

2.3. Important tradeoffs

We are accustomed to certain tradeoffs in algorithm design, such as time *versus* space: an algorithm that is constrained to use less space may have to go more slowly than one not so constrained. There are certain other tradeoffs that arise in parallel programming. They arise because of the constraints of regularity of computation and locality to which we should adhere. For example, load balancing to increase parallellism requires communication, which may be expensive. Limiting oneself to the regular operations the hardware performs efficiently may result in wasted effort or use of less sophisticated algorithms; we will illustrate this later in the case of the nonsymmetric eigenvalue problem.

Another interesting tradeoff is parallellism *versus* numerical stability. For some problems the most highly parallel algorithms known are less numerically stable than the conventional sequential algorithms. This is true for various kinds of linear systems and eigenvalue problems, which we will point out as they arise. Some of these tradeoffs can be mitigated by better floating point arithmetic (Demmel, 1992b). Others can be dealt with by using the following simple paradigm:

1   Solve the problem using a fast method, provided it is rarely unstable.
2   Quickly and reliably confirm or deny the accuracy of the computed

solution. With high probability, the answer just (quickly) computed is accurate enough.

3   Otherwise, fall back on a slower but more reliable algorithm.

For example, the most reliable algorithm for the dense nonsymmetric eigenvalue problem is Hessenberg reduction and QR iteration, but this is hard to parallellize. Other routines are faster but occasionally unreliable. These routines can be combined according to the paradigm to yield a guaranteed stable algorithm which is fast with high probability (see Section 6.5).

3. Matrix multiplication

Matrix multiplication is a very regular computation that is basic to linear algebra and lends itself well to parallel implementation. Indeed, since it is the easiest nontrivial matrix operation to implement efficiently, an effective approach to designing other parallel matrix algorithms is to decompose them into a sequence of matrix multiplications; we discuss this in detail in later sections.

One might well ask why matrix multiplication is more basic than matrix-vector multiplication or adding a scalar times one vector to another vector. Matrix multiplication can obviously be decomposed into these simpler operations, and they also seem to offer a great deal of parallelism. The reason is that matrix multiplication offers much more opportunity to exploit locality than these simpler operations. An informal justification for this is given in the following.

Table 1 gives the number of floating point operations (flops), the minimum number of memory references, and their ratio q for the three Basic Linear Algebra Subroutines, or BLAS: scalar-times-vector-plus-vector (or **saxpy** for short, for $\alpha x + y$), matrix–vector multiplication, and matrix–matrix multiplication (for simplicity only the highest order term in n is given for q). When the data are too large to fit in the top of the memory hierarchy, we wish to perform the most flops per memory reference to minimize data movement; q gives an upper bound on this ratio for any implementation. We see that only matrix multiplication offers us an opportunity to make this ratio large.

This table reflects a hierarchy of operations. Operations like **saxpy** operate on vectors and offer the worst q values; these are called Level 1 BLAS (Lawson, Hanson, Kincaid and Krogh, 1979) and include inner products and other simple operations. Operations like matrix–vector multiplication operate on matrices and vectors, and offer slightly better q values; these are called Level 2 BLAS (Dongarra, Du Croz, Hammarling and Richard Hanson, 1988), and include solving triangular systems of equations and rank-1 updates of matrices ($A + xy^T$, x and y column vectors). Operations like matrix–matrix multiplication operate on pairs of matrices, and offer the best q values; these

Table 1. *Memory references and operation counts for the BLAS.*

Operation	Definition	Floating point operations	Memory references	q
saxpy	$y_i = \alpha x_i + y_i,\ i = 1, ..., n$	$2n$	$3n+1$	$2/3$
Matrix–vector mult	$y_i = \sum_{j=1}^{n} A_{ij} x_j + y_i$	$2n^2$	$n^2 + 3n$	3
Matrix–matrix mult	$C_{ij} = \sum_{k=1}^{n} A_{ik} B_{kj} + C_{ij}$	$2n^3$	$4n^2$	$n/2$

are called Level 3 BLAS (Dongarra, Du Croz, Duff and Hammarling, 1990), and include solving triangular systems of equations with many right-hand sides. These operations have been standardized, and many high performance computers have highly optimized implementations of these that are useful for building more complicated algorithms (Anderson, Bai, Bischof, Demmel, Dongarra, Du Croz, Greenbaum, Hammarling, McKenney, Ostrouchov and Sorensen, 1992); this is the subject of several succeeding sections.

3.1. Matrix multiplication on a shared memory machine

Suppose we have two levels of memory hierarchy, fast and slow, where the slow memory is large enough to contain the $n \times n$ matrices A, B and C, but the fast memory contains only M words where $n < M \ll n^2$. Further assume the data are reused optimally (which may be optimistic if the decisions are made automatically by hardware).

The simplest algorithm one might try consists of three nested loops:

Algorithm 3 Unblocked matrix multiplication

> for $i = 1 : n$
>> for $j = 1 : n$
>>> for $k = 1 : n$
>>>> $C_{ij} = C_{ij} + A_{ik} \cdot B_{kj}$

We count the number of references to the slow memory as follows: n^3 for reading B n times, n^2 for reading A one row at a time and keeping it in fast memory until it is no longer needed, and $2n^2$ for reading one entry of C at a time, keeping it in fast memory until it is completely computed. This comes to $n^3 + 3n^2$ for a q of about 2, which is no better than the Level 2 BLAS and far from the maximum possible $n/2$. If $M \ll n$, so that we cannot keep a full row of A in fast memory, q further decreases to 1, since the algorithm reduces to a sequence of inner products, which are Level 1 BLAS. For every permutation of the three loops on i, j and k, one gets another algorithm with q about the same.

The next possibility is dividing B and C into *column blocks*, and computing C block by block. We use the notation $B(i : j, k : l)$ to mean the submatrix in rows i through j and columns k through l. We partition $B = [B^{(1)}, B^{(2)}, ..., B^{(N)}]$ where each $B^{(i)}$ is $n \times n/N$, and similarly for C. Our column block algorithm is then

Algorithm 4 Column-blocked matrix multiplication

    for $j = 1 : N$
        for $k = 1 : n$
            $C^{(j)} = C^{(j)} + A(1 : n, k) \cdot B^{(j)}(k, 1 : n/N)$

Assuming $M \geq 2n^2/N + n$, so that fast memory can accommodate $B^{(j)}$, $C^{(j)}$ and one column of A simultaneously, our memory reference count is as follows: $2n^2$ for reading and writing each block of C once, n^2 for reading each block of B once, and Nn^2 for reading A N times. This yields $q \approx M/n$, so that M needs to grow with n to keep q large.

Finally, we consider *rectangular blocking*, where A is broken into an $N \times N$ block matrix with $n/N \times n/N$ blocks $A^{(ij)}$, and B and C are similarly partitioned. The algorithm becomes

Algorithm 5 Rectangular-blocked matrix multiplication

    for $i = 1 : N$
        for $j = 1 : N$
            for $k = 1, N$
                $C^{(ij)} = C^{(ij)} + A^{(ik)} \cdot B^{(kj)}$

Assuming $M \geq 3(n/N)^2$ so that one block each from A, B and C fit in memory simultaneously, our memory reference count is as follows: $2n^2$ for reading and writing each block of C once, Nn^2 for reading A N times, and Nn^2 for reading B N times. This yields $q \approx \sqrt{M/3}$, which is much better than the previous algorithms.

In Hong and Kung (1981) an analysis of this problem leading to an upper bound near $\sqrt{M}$ is given, so we cannot expect to improve much on this algorithm for square matrices. On the other hand, this brief analysis ignores a number of practical issues:

1   high level language constructs do not yet support block layout of matrices as described here (but see the discussion in Section 3.3);

2   if the fast memory consists of vector registers and has vector operations supporting **saxpy** but not inner products, a column blocked code may be superior;

3   a real code will have to deal with nonsquare matrices, for which the optimal block sizes may not be square (Gallivan *et al.*, 1990).

Another possibility is Strassen's method (Aho, Hopcroft and Ullman,

1974), which multiplies matrices recursively by dividing them into 2×2 block matrices, and multiplying the subblocks using 7 matrix multiplications (recursively) and 15 matrix additions of half the size; this leads to an asymptotic complexity of $n^{\log_2 7} \approx n^{2.81}$ instead of n^3. The value of this algorithm is not just this asymptotic complexity but its reduction of the problem to smaller subproblems which eventually fit in fast memory; once the subproblems fit in fast memory standard matrix multiplication may be used. This approach has led to speedups on relatively large matrices on some machines (Bailey, Lee and Simon, 1991). A drawback is the need for significant workspace, and somewhat lower numerical stability, although it is adequate for many purposes (Demmel and Higham, 1992; Higham, 1990).

Given the complexity of optimizing the implementation of matrix multiplication, we cannot expect all other matrix algorithms to be equally optimized on all machines, at least not in a time users are willing to wait. Indeed, since architectures change rather quickly, we prefer to do as little machine-specific optimization as possible. Therefore, our shared memory algorithms in later sections assume only that highly optimized BLAS are available and build on top of them.

3.2. Matrix multiplication on a distributed memory machine

In this section it will be convenient to number matrix entries (or subblocks) and processors from 0 to $n - 1$ instead of 1 to n.

A dominant issue is *data layout*, or how the matrices are partitioned across the machine. This will determine both the amount of parallellism and the cost of communication. We begin by showing how best to implement matrix multiplication without regard to the layout's suitability for other matrix operations, and return to the question of layouts in the next section.

The first algorithm is due to Cannon (1969) and is well suited for computers laid out in a square $N \times N$ mesh, i.e. where each processor communicates most efficiently with the four other processors immediately north, east, south and west of itself. We also assume the processors at the edges of the grid are directly connected to the processors on the opposite edge; this makes the topology that of a two-dimensional torus. Let A be partitioned into square subblocks $A^{(ij)}$ as before, with $A^{(ij)}$ stored on processor (i, j). Let B and C be partitioned similarly. The algorithm is given below. It is easily seen that whenever $A^{(ik)}$ and $B^{(kj)}$ 'meet' in processor i, j, they are multiplied and accumulated in $C^{(ij)}$; the products for the different $C^{(ij)}$ are accumulated in different orders.

Algorithm 6 Cannon's matrix multiplication algorithm

forall $i = 0 : N - 1$
    Left circular shift row i by i, so that $A^{(i,j)}$

$A^{(00)}$	$A^{(01)}$	$A^{(02)}$
$A^{(11)}$	$A^{(12)}$	$A^{(10)}$
$A^{(22)}$	$A^{(20)}$	$A^{(21)}$

$A^{(01)}$	$A^{(02)}$	$A^{(00)}$
$A^{(12)}$	$A^{(10)}$	$A^{(11)}$
$A^{(20)}$	$A^{(21)}$	$A^{(22)}$

$A^{(02)}$	$A^{(00)}$	$A^{(01)}$
$A^{(10)}$	$A^{(11)}$	$A^{(12)}$
$A^{(21)}$	$A^{(22)}$	$A^{(20)}$

$B^{(00)}$	$B^{(11)}$	$B^{(22)}$
$B^{(10)}$	$B^{(21)}$	$B^{(02)}$
$B^{(20)}$	$B^{(01)}$	$B^{(12)}$

$B^{(10)}$	$B^{(21)}$	$B^{(02)}$
$B^{(20)}$	$B^{(01)}$	$B^{(12)}$
$B^{(00)}$	$B^{(11)}$	$B^{(22)}$

$B^{(20)}$	$B^{(01)}$	$B^{(12)}$
$B^{(00)}$	$B^{(11)}$	$B^{(22)}$
$B^{(10)}$	$B^{(21)}$	$B^{(02)}$

A, B after skewing $\qquad$ A, B after shift $k = 1$ $\qquad$ A, B after shift $k = 2$

Fig. 4. Cannon's algorithm for $N = 3$.

$\qquad$ is assigned $A^{(i,(j+i)\bmod N)}$.
forall $j = 0 : N - 1$
$\qquad$ Upward circular shift column j by j, so that $B^{(i,j)}$
$\qquad\qquad$ is assigned $B^{((j+i)\bmod N,j)}$.
for $k = 1 : N$
$\qquad$ forall $i = 0 : N - 1$, forall $j = 0 : N - 1$
$\qquad$ $C^{(ij)} = C^{(ij)} + A^{(ij)} \cdot B^{(ij)}$
$\qquad$ Left circular shift each row of A by 1, so $A^{(i,j)}$
$\qquad\qquad$ is assigned $A^{(i,(j+1)\bmod N)}$.
$\qquad$ Upward circular shift each column of B by 1, so $B^{(i,j)}$
$\qquad\qquad$ is assigned $B^{((i+1)\bmod N,j)}$.

Figure 4 illustrates the functioning of this algorithm for $N = 3$. A variation of this algorithm suitable for machines that are efficient at *spreading* sub-blocks across rows (or down columns) is to do this instead of the preshifting and rotation of A (or B) (Fox, Johnson, Lyzenga, Otto, Salmon and Walker, 1988).

This algorithm is easily adapted to a hypercube. The simplest way is to embed a grid (or two-dimensional torus) in a hypercube, i.e. map the processors in a grid to the processors in a hypercube, and the connections in a grid to a subset of the connections in a hypercube (Ho, 1990; Johnsson, 1987). Suppose the hypercube is d-dimensional, so the 2^d processors are labelled by d bit numbers. We embed a $2^n \times 2^m$ grid in this hypercube (where $m + n = d$) by mapping processor (i_1, i_2) in the grid to processor $G_{n,i_1} 2^m + G_{m,i_2}$ in the hypercube; i.e. we just concatenate the n bits of G_{n,i_1} and m bits of G_{m,i_2}. Each row (column) of the grid thus occupies an m- (n-) dimensional subcube of the original hypercube, with nearest neighbours in the grid mapped to nearest neighbours in the hypercube (Ho, Johnsson and

0000	0001	0011	0010
0100	0101	0111	0110
1100	1101	1111	1110
1000	1001	1011	1010

Fig. 5. Embedding a 4×4 grid in a four-dimensional hypercube (numbers are processor numbers in hypercube).

Edelman, 1991). We illustrate for a 4×4 grid in Figure 5. This approach easily extends to multi-dimensional arrays of size $2^{m_1} \times \cdots \times 2^{m_r}$, where $\sum_{i=1}^{r} m_i$ is at most the dimension of the hypercube.

This approach (which is useful for more than matrix multiplication) uses only a subset of the connections in a hypercube, which makes the initial skewing operations slower than they need be: if we can move only to nearest neighbours, each skewing operation takes $N - 1$ communication steps, as many as in the computational loop. We may use all the wires of the hypercube to reduce the skewing to $\log_2 N$ operations. In the following algorithm, $\otimes$ denotes the bitwise exclusive-or operator. We assume the $2^n \times 2^n$ grid of data is embedded in the hypercube so that $A^{(i,j)}$ is stored in processor $i \cdot 2^n + j$ (Dekel, Nassimi and Sahni, 1981):

Algorithm 7 Dekel's matrix multiplication algorithm

> for $k = 1 : n$
> > Let $j_k = (k$th bit of $j) \cdot 2^k$
> > Let $i_k = (k$th bit of $i) \cdot 2^k$
> > forall $i = 0 : 2^n - 1$, forall $j = 0 : 2^n - 1$
> > > Swap $A^{(i,j \otimes i_k)}$ and $A^{(i,j)}$
> > > Swap $B^{(j_k \otimes i,j)}$ to $B^{(i,j)}$
> for $k = 1 : 2^n$
> > forall $i = 0 : 2^n - 1$, forall $j = 0 : 2^n - 1$
> > > $C^{(ij)} = C^{(ij)} + A^{(ij)} \cdot B^{(ij)}$
> > > Swap $A^{(i,j \otimes g_{d,k})}$ and $A^{(i,j)}$
> > > Swap $B^{(i \otimes g_{d,k},j)}$ and $B^{(i,j)}$

Finally, we may speed this up further (Ho *et al.*, 1991; Johnsson and Ho, 1989) provided the $A^{(i,j)}$ blocks are large enough, by using the same algorithm as for force calculations in Section 2. If the blocks are n by n (so A and B are $n2^n \times n2^n$), then the algorithm becomes

Algorithm 8 Ho, Johnsson and Edelman's matrix multiplication algorithm

> for $k = 1 : n$
> > Let $j_k = (k$th bit of $j) \cdot 2^k$

Table 2. *Cost of matrix multiplication on a hypercube.*

Algorithm	Message startups	Data sending steps	Floating point steps
Cannon (6)	$2(2^n - 1)$	$2n^2(2^n - 1)$	$2n^3 2^n$
Dekel (7)	$n + 2^n - 1$	$n^3 + n^2(2^n - 1)$	$2n^3 2^n$
Ho *et al.* (8)	$n + 2^n - 1$	$n^3 + n(2^n - 1)$	$2n^3 2^n$

Let $i_k = (k\text{th bit of } i) \cdot 2^k$
forall $i = 0 : 2^n - 1$, forall $j = 0 : 2^n - 1$
    Swap $A^{(i, j \otimes i_k)}$ and $A^{(i,j)}$
    Swap $B^{(j_k \otimes i, j)}$ to $B^{(i,j)}$
for $k = 1 : 2^n$
    forall $i = 0 : 2^n - 1$, forall $j = 0 : 2^n - 1$
    $C^{(ij)} = C^{(ij)} + A^{(ij)} \cdot B^{(ij)}$
        forall $l = 0 : n - 1$
            Swap $A_l^{(i, j \otimes g_{d,k}^{(l)})}$ and $A_l^{(i,j)}$ ($A_l^{(ij)}$ is the lth row of $A^{(ij)}$)
            Swap $B_l^{(i \otimes g_{d,k}^{(l)}, j)}$ and $B_l^{(i,j)}$ ($B_l^{(ij)}$ is the lth column of $B^{(ij)}$)

Algorithms 6–8 all perform the same number of floating point operations in parallel. Table 2 compares the number of communication steps, assuming matrices are $n2^n \times n2^n$, swapping a datum along a single wire is one step, and the motions of A and B that can occur in parallel do occur in parallel. Note that for large enough n the number of floating point steps overwhelms the number of communication steps, so the efficiency gets better.

In this section we have shown how to optimize matrix multiplication in a series of steps tuning it ever more highly for a particular computer architecture, until essentially every communication link and floating point unit is utilized. Our algorithms are scalable, in that they continue to run efficiently on larger machines and larger problems, with communication costs becoming ever smaller with respect to computation. If the architecture permitted us to overlap communication and computation, we could pipeline the algorithm to mask communication cost further.

On the other hand, let us ask what we lose by optimizing so heavily for one architecture. Our high performance depends on the matrices having just the right dimensions, being laid out just right in memory, and leaving them in a scrambled final position (although a modest amount of extra communication could repair this). It is unreasonable to expect users, who want to do several computations of which this is but one, to satisfy all these requirements. Therefore a practical algorithm will have to deal with many irregularities, and be quite complicated. Our ability to do this extreme optimization is limited to a few simple and regular problems like matrix multiplication on a hypercube, as well as other heavily used kernels like the BLAS, which have indeed been highly optimized for many architectures. We do not expect equal success for more complicated algorithms on all architectures of interest, at least within a reasonable amount of time.[‡] Also, the algorithm is highly tuned to a particular interconnection network topology, which may require redesign for another machine (in view of this, a number of recent machines try to make communication time appear as independent of topology as possible, so the user sees essentially a completely connected topology).

3.3. Data layouts on distributed memory machines

Choosing a data layout may be described as choosing a mapping $f(i, j)$ from location (i, j) in a matrix to the processor on which it is stored. As discussed previously, we hope to design f so that it permits highly parallel implementation of a variety of matrix algorithms, limits communication cost as much as possible, and retains these attractive properties as we scale to larger matrices and larger machines. For example, the algorithms of the previous section use the map $f(i, j) = (\lfloor i/r \rfloor, \lfloor j/r \rfloor)$, where we subscript matrices starting at 0, number processors by their coordinates in a grid (also starting at $(0,0)$), and store an $r \times r$ matrix on each processor.

There is an emerging consensus about data layouts for distributed memory machines. This is being implemented in several programming languages (Fox, Hiranandani, Kennedy, Koelbel, Kremer, Tseng and Wu, 1990; High Peformance Fortran, 1991), that will be available to programmers in the near future. We describe these layouts here.

High Performance Fortran (HPF) (High Peformance Fortran, 1991) permits the user to define a virtual array of processors, align actual data structures like matrices and arrays with this virtual array (and so with respect to each other), and then to layout the virtual processor array on an actual machine. We describe the layout functions f offered for this last step. The range of f is a rectangular array of processors numbered from $(0, 0)$ up to $(p_1 - 1, p_2 - 1)$. Then all f can be parameterized by two integer parameters b_1 and b_2:

$$f_{b_1, b_2}(i, j) = \left(\left\lfloor \frac{i}{b_1} \right\rfloor \bmod p_1, \left\lfloor \frac{j}{b_2} \right\rfloor \bmod p_2 \right).$$

[‡] The matrix multiplication subroutine in the CM-2 Scientific Subroutine Library took approximately 10 person-years of effort (Johnsson, 1990).

0,0	0,0	0,0	0,0	0,1	0,1	0,1	0,1	0,2	0,2	0,2	0,2	0,3	0,3	0,3	0,3
0,0	0,0	0,0	0,0	0,1	0,1	0,1	0,1	0,2	0,2	0,2	0,2	0,3	0,3	0,3	0,3
0,0	0,0	0,0	0,0	0,1	0,1	0,1	0,1	0,2	0,2	0,2	0,2	0,3	0,3	0,3	0,3
0,0	0,0	0,0	0,0	0,1	0,1	0,1	0,1	0,2	0,2	0,2	0,2	0,3	0,3	0,3	0,3
1,0	1,0	1,0	1,0	1,1	1,1	1,1	1,1	1,2	1,2	1,2	1,2	1,3	1,3	1,3	1,3
1,0	1,0	1,0	1,0	1,1	1,1	1,1	1,1	1,2	1,2	1,2	1,2	1,3	1,3	1,3	1,3
1,0	1,0	1,0	1,0	1,1	1,1	1,1	1,1	1,2	1,2	1,2	1,2	1,3	1,3	1,3	1,3
1,0	1,0	1,0	1,0	1,1	1,1	1,1	1,1	1,2	1,2	1,2	1,2	1,3	1,3	1,3	1,3
2,0	2,0	2,0	2,0	2,1	2,1	2,1	2,1	2,2	2,2	2,2	2,2	2,3	2,3	2,3	2,3
2,0	2,0	2,0	2,0	2,1	2,1	2,1	2,1	2,2	2,2	2,2	2,2	2,3	2,3	2,3	2,3
2,0	2,0	2,0	2,0	2,1	2,1	2,1	2,1	2,2	2,2	2,2	2,2	2,3	2,3	2,3	2,3
2,0	2,0	2,0	2,0	2,1	2,1	2,1	2,1	2,2	2,2	2,2	2,2	2,3	2,3	2,3	2,3
3,0	3,0	3,0	3,0	3,1	3,1	3,1	3,1	3,2	3,2	3,2	3,2	3,3	3,3	3,3	3,3
3,0	3,0	3,0	3,0	3,1	3,1	3,1	3,1	3,2	3,2	3,2	3,2	3,3	3,3	3,3	3,3
3,0	3,0	3,0	3,0	3,1	3,1	3,1	3,1	3,2	3,2	3,2	3,2	3,3	3,3	3,3	3,3
3,0	3,0	3,0	3,0	3,1	3,1	3,1	3,1	3,2	3,2	3,2	3,2	3,3	3,3	3,3	3,3

Fig. 6. Block layout of a 16×16 matrix on a 4×4 processor grid.

Suppose the matrix A (or virtual processor array) is $m \times n$. Then choosing $b_2 = n$ yields a column of processors, each containing some number of complete rows of A. Choosing $b_1 = m$ yields a row of processors. Choosing $b_1 = m/p_1$ and $b_2 = n/p_2$ yields a *blocked layout*, where A is broken into $b_1 \times b_2$ subblocks, each of which resides on a single processor. This is the simplest two-dimensional layout one could imagine (we used it in the previous section), and by having large subblocks stored on each processor it makes using the BLAS on each processor attractive. However, for straightforward matrix algorithms that process the matrix from left to right (including Gaussian elimination, QR decomposition, reduction to tridiagonal form, and so on), the leftmost processors will become idle early in the computation and make load balance poor. Choosing $b_1 = b_2 = 1$ is called *scatter mapping* (or *wrapped* or *cyclic* or *interleaved mapping*), and optimizes load balance, since the matrix entries stored on a single processor are as nearly as possible uniformly distributed throughout the matrix. On the other hand, this appears to inhibit the use of the BLAS locally in each processor, since the data owned by a processor are not contiguous from the point of view of the matrix. Finally, by choosing $1 < b_1 < m/p_1$ and $1 < b_2 < n/p_2$, we get a *block-scatter mapping* which trades off load balance and applicability of the BLAS. These layouts are shown in Figures 6 through 8 for a 16×16 matrix laid out on a 4×4 processor grid; each array entry is labelled by the number of the processor that stores it.

By being a little more flexible about the algorithms we implement, we can mitigate the apparent tradeoff between the load balance and the applicability

0,0	0,1	0,2	0,3	0,0	0,1	0,2	0,3	0,0	0,1	0,2	0,3	0,0	0,1	0,2	0,3
1,0	1,1	1,2	1,3	1,0	1,1	1,2	1,3	1,0	1,1	1,2	1,3	1,0	1,1	1,2	1,3
2,0	2,1	2,2	2,3	2,0	2,1	2,2	2,3	2,0	2,1	2,2	2,3	2,0	2,1	2,2	2,3
3,0	3,1	3,2	3,3	3,0	3,1	3,2	3,3	3,0	3,1	3,2	3,3	3,0	3,1	3,2	3,3
0,0	0,1	0,2	0,3	0,0	0,1	0,2	0,3	0,0	0,1	0,2	0,3	0,0	0,1	0,2	0,3
1,0	1,1	1,2	1,3	1,0	1,1	1,2	1,3	1,0	1,1	1,2	1,3	1,0	1,1	1,2	1,3
2,0	2,1	2,2	2,3	2,0	2,1	2,2	2,3	2,0	2,1	2,2	2,3	2,0	2,1	2,2	2,3
3,0	3,1	3,2	3,3	3,0	3,1	3,2	3,3	3,0	3,1	3,2	3,3	3,0	3,1	3,2	3,3
0,0	0,1	0,2	0,3	0,0	0,1	0,2	0,3	0,0	0,1	0,2	0,3	0,0	0,1	0,2	0,3
1,0	1,1	1,2	1,3	1,0	1,1	1,2	1,3	1,0	1,1	1,2	1,3	1,0	1,1	1,2	1,3
2,0	2,1	2,2	2,3	2,0	2,1	2,2	2,3	2,0	2,1	2,2	2,3	2,0	2,1	2,2	2,3
3,0	3,1	3,2	3,3	3,0	3,1	3,2	3,3	3,0	3,1	3,2	3,3	3,0	3,1	3,2	3,3
0,0	0,1	0,2	0,3	0,0	0,1	0,2	0,3	0,0	0,1	0,2	0,3	0,0	0,1	0,2	0,3
1,0	1,1	1,2	1,3	1,0	1,1	1,2	1,3	1,0	1,1	1,2	1,3	1,0	1,1	1,2	1,3
2,0	2,1	2,2	2,3	2,0	2,1	2,2	2,3	2,0	2,1	2,2	2,3	2,0	2,1	2,2	2,3
3,0	3,1	3,2	3,3	3,0	3,1	3,2	3,3	3,0	3,1	3,2	3,3	3,0	3,1	3,2	3,3

Fig. 7. Scatter layout of a 16×16 matrix on a 4×4 processor grid.

0,0	0,0	0,1	0,1	0,2	0,2	0,3	0,3	0,0	0,0	0,1	0,1	0,2	0,2	0,3	0,3
0,0	0,0	0,1	0,1	0,2	0,2	0,3	0,3	0,0	0,0	0,1	0,1	0,2	0,2	0,3	0,3
1,0	1,0	1,1	1,1	1,2	1,2	1,3	1,3	1,0	1,0	1,1	1,1	1,2	1,2	1,3	1,3
1,0	1,0	1,1	1,1	1,2	1,2	1,3	1,3	1,0	1,0	1,1	1,1	1,2	1,2	1,3	1,3
2,0	2,0	2,1	2,1	2,2	2,2	2,3	2,3	2,0	2,0	2,1	2,1	2,2	2,2	2,3	2,3
2,0	2,0	2,1	2,1	2,2	2,2	2,3	2,3	2,0	2,0	2,1	2,1	2,2	2,2	2,3	2,3
3,0	3,0	3,1	3,1	3,2	3,2	3,3	3,3	3,0	3,0	3,1	3,1	3,2	3,2	3,3	3,3
3,0	3,0	3,1	3,1	3,2	3,2	3,3	3,3	3,0	3,0	3,1	3,1	3,2	3,2	3,3	3,3
0,0	0,0	0,1	0,1	0,2	0,2	0,3	0,3	0,0	0,0	0,1	0,1	0,2	0,2	0,3	0,3
0,0	0,0	0,1	0,1	0,2	0,2	0,3	0,3	0,0	0,0	0,1	0,1	0,2	0,2	0,3	0,3
1,0	1,0	1,1	1,1	1,2	1,2	1,3	1,3	1,0	1,0	1,1	1,1	1,2	1,2	1,3	1,3
1,0	1,0	1,1	1,1	1,2	1,2	1,3	1,3	1,0	1,0	1,1	1,1	1,2	1,2	1,3	1,3
2,0	2,0	2,1	2,1	2,2	2,2	2,3	2,3	2,0	2,0	2,1	2,1	2,2	2,2	2,3	2,3
2,0	2,0	2,1	2,1	2,2	2,2	2,3	2,3	2,0	2,0	2,1	2,1	2,2	2,2	2,3	2,3
3,0	3,0	3,1	3,1	3,2	3,2	3,3	3,3	3,0	3,0	3,1	3,1	3,2	3,2	3,3	3,3
3,0	3,0	3,1	3,1	3,2	3,2	3,3	3,3	3,0	3,0	3,1	3,1	3,2	3,2	3,3	3,3

Fig. 8. Block scatter layout of a 16×16 matrix on a 4×4 processor grid with 2×2 blocks.

of BLAS. For example, the layout of A shown in Figure 7 is identical to the layout shown in Figure 6 of $P^T A P$, where P is a permutation matrix. This shows that running the algorithms of the previous section to multiply A times B in scatter layout is the same as multiplying $P A P^T$ and $P B P^T$ to get $P A B P^T$, which is the desired product. Indeed, as long as (1) A and B are both distributed over a square array of processors; (2) the permutations of the columns of A and rows of B are identical; and (3) for all i the number of columns of A stored by processor column i is the same as the number of rows of B stored by processor row i, the algorithms of the previous section will correctly multiply A and B. The distribution of the product will be determined by the distribution of the rows of A and columns of B. We will see a similar phenomenon for other distributed memory algorithms later.

A different approach is to write algorithms that work independently of the location of the data, and rely on the underlying language or run-time system to optimize the necessary communications. This makes code easier to write, but puts a large burden on compiler and run-time system writers (Van de Velde, 1992).

4. Systems of linear equations

We discuss both dense and band matrices, on shared and distributed memory machines. We begin with dense matrices and shared memory, showing how the standard algorithm can be reformulated as a block algorithm, calling the Level 2 and 3 BLAS in its innermost loops. The distributed memory versions will be similar, with the main issue being laying out the data to maximize load balance and minimize communication. We also present some highly parallel, but numerically unstable, algorithms to illustrate the trade-off between stability and parallellism. We conclude with some algorithms for band matrices.

4.1. Gaussian elimination on a shared memory machine

To solve $Ax = b$, we first use Gaussian elimination to factor the nonsingular matrix A as $PA = LU$, where L is lower triangular, U is upper triangular, and P is a permutation matrix. Then we solve the triangular systems $Ly = Pb$ and $Ux = y$ for the solution x. In this section we will concentrate on factoring $PA = LU$, which has the dominant number of floating point operations, $2n^3/3 + \mathcal{O}(n^2)$. Pivoting is required for numerical stability, and we use the standard partial pivoting scheme (Golub and Van Loan, 1989); this means that L has unit diagonal and other entries bounded in magnitude by one. The simplest version of the algorithm involves adding multiples of one row of A to others in order to zero out subdiagonal entries, and overwriting A with L and U:

Algorithm 9 Row oriented Gaussian elimination (kij-LU decomposition)

> for $k = 1 : n - 1$
>   { choose l so $|A_{lk}| = \max_{k \leq i \leq n} |A_{ik}|$, swap rows l and k of A }
>   for $i = k + 1 : n$
> $A_{ik} = A_{ik}/A_{kk}$
>   for $j = k + 1 : n$
> $A_{ij} = A_{ij} - A_{ik} \cdot A_{kj}$

There is obvious parallellism in the innermost loop, since each A_{ij} can be updated independently. If A is stored by column, as is the case in Fortran, then since the inner loop combines rows of A, it accesses memory entries (at least) n locations apart. As described in Section 2, this does not respect locality. Algorithm 9 is also called kij-LU decomposition, because of the nesting order of its loops. All the rest of 3! permutations of i, j and k lead to valid algorithms, some of which access columns of A in the innermost loop. Algorithm 10 is one of these, and is used in the LINPACK routine **sgefa** (Dongarra *et al.*, 1979):

Algorithm 10 Column oriented Gaussian elimination (kji-LU decomposition)

> for $k = 1 : n - 1$
>   { choose l so $|A_{lk}| = \max_{k \leq i \leq n} |A_{ik}|$, swap A_{lk} and A_{kk} }
>   for $i = k + 1 : n$
> $A_{ik} = A_{ik}/A_{kk}$
>   for $j = k + 1 : n$
>   { swap A_{lj} and A_{kj} }
>   for $i = k + 1 : n$
> $A_{ij} = A_{ij} - A_{ik} \cdot A_{kj}$

The inner loop of Algorithm 10 can be performed by a single call to the Level 1 BLAS operation **saxpy**. To achieve higher performance, we modify this code first to use the Level 2 and then the Level 3 BLAS in its innermost loops. Again, 3! versions of these algorithms are possible, but we just describe the ones used in the LAPACK library (Anderson *et al.*, 1992). To make the use of BLAS clear, we use matrix/vector operations instead of loops:

Algorithm 11 Gaussian elimination using Level 2 BLAS

> for $k = 1 : n - 1$
>   { choose l so $|A_{lk}| = \max_{k \leq i \leq n} |A_{ik}|$, swap rows l and k of A }
> $A(k + 1 : n, k) = A(k + 1 : n, k)/A_{kk}$
> $A(k + 1 : n, k + 1 : n) = A(k + 1 : n, k + 1 : n)$
> $- A(k + 1 : n, k) \cdot A(k, k + 1 : n)$

The parallellism in the inner loop is evident: most work is performed is a single rank-1 update of the trailing $n - k \times n - k$ submatrix

$$A(k + 1 : n, k + 1 : n),$$

where each entry of

$$A(k + 1 : n, k + 1 : n)$$

can be updated in parallel. Other permutations of the nested loops lead to different algorithms, which depend on the BLAS for matrix–vector multiplication and solving a triangular system instead of rank-1 updating (Anderson and Dongarra, 1990; Robert, 1990); which is faster depends on the relative speed of these on a given machine.

To convert to the Level 3 BLAS involves column blocking

$$A = [A^{(1)}, \ldots, A^{(m)}]$$

into $n \times n_b$ blocks, where n_b is the *block size* and $m = n/n_b$. The optimal choice of n_b depends on the memory hierarchy of the machine in question: our approach is to compute the LU decomposition of each $n \times n_b$ subblock of A using Algorithm 11 in the fast memory, and then use Level 3 BLAS to update the rest of the matrix:

Algorithm 12 Gaussian elimination using Level 3 BLAS (we assume n_b divides n)

> for $l = 1 : m$
> > $k = (l - 1) \cdot n_b + 1$
> > Use Algorithm 11 to factorize $PA^{(l)} = LU$ in place
> > Apply P to prior columns $A(1 : n, 1 : k - 1)$ and later columns
> > > $A(1 : n, k + n_b : n)$
> >
> > Update block row of U:
> > > Replace $A(k : k + n_b - 1, k + n_b : n)$ by the solution X of
> > > $TX = A(k : k + n_b - 1, k + n_b : n)$, where T is the lower
> > > triangular matrix in $A(k : k + n_b - 1, k : k + n_b - 1)$
> >
> > $A(k + n_b : n, k + n_b : n) = A(k + n_b : n, k + n_b : n) -$
> > > $A(k + n_b : n, k : k + n_b - 1) \cdot A(k : k + n_b - 1, k + n_b : n)$

Most of the work is performed in the last two lines, solving a triangular system with many right-hand sides, and matrix multiplication. Other similar algorithms may be derived by conformally partitioning L, U and A, and equating partitions in $A = LU$. Algorithms 11 and 12 are available as, respectively, subroutines `sgetf2` and `sgetrf` in LAPACK (Anderson *et al.*, 1992).

We illustrate these points with the slightly different example of Cholesky decomposition, which uses a very similar algorithm: Table 3 shows the

Table 3. *Speed of Cholesky on a Cray YMP*

	1 PE	8 PEs
Maximum speed	330	2640
LINPACK (Cholesky with BLAS 1), $n = 500$	72	72
Matrix–vector multiplication	311	2285
Matrix–matrix multiplication	312	2285
Triangular solve (one right-hand side)	272	584
Triangular solve (many right-hand sides)	309	2398
LAPACK (Cholesky with BLAS 3), $n = 500$	290	1414
LAPACK (Cholesky with BLAS 3), $n = 1000$	301	2115

speeds in megaflops of the various BLAS and algorithms on one and eight
processors of a Cray YMP.

4.2. Gaussian elimination on a distributed memory machine

As described earlier, layout strongly influences the algorithm. We show the
algorithm for a block scatter mapping in both dimensions, and then discuss
how other layouts may be handled. The algorithm is essentially the same
as Algorithm 12, with communication inserted as necessary. The block size
n_b equals b_2, which determines the layout in the horizontal direction.

Communication is required in Algorithm 11 to find the pivot entry at each
step and swap rows if necessary; then each processor can perform the scaling
and rank-1 updates independently. The pivot search is a *reduction* operation, as described in Section 2. After the block column is fully factorized,
the pivot information must be *broadcast* so other processors can permute
their own data, as well as permute among different processors.

In Algorithm 12, the $n_b \times n_b$ L matrix stored on the diagonal must be
spread rightward to other processors in the same row, so they can compute
their entries of U. Finally, the processors holding the rest of L below the
diagonal must *spread* their submatrices to the right, and the processors
holding the new entries of U just computed must *spread* their submatrices
downward, before the final rank-n_b update in the last line of Algorithm 12
can take place.

The optimal choice of the block sizes b_1 and b_2 depends on the cost of
the communication *versus* that of the computation. For example, if the
communication required to do pivot search and swapping of rows is expensive
then b_1 should be large. The execution time is a function of dimension n,
block sizes b_1 and b_2, processor counts p_1 and p_2, and the cost of computation
and communication (from Section 2, we know how to model these). Given
this function, it may be minimized as a function of b_1, b_2, p_1 and p_2. Some

theoretical analyses of this sort for special cases may be found in Robert (1990) and the references therein. See also Dongarra and Ostrouchov (1990) and Dongarra and van de Geijn (1991a). As an example of the performance that can be attained in practice, on an Intel Delta with 512 processors the speed of LU ranged from a little over 1 gigaflop for $n = 2000$ to nearly 12 gigaflops for $n = 25000$.

Even if the layout is not block scatter as described so far, essentially the same algorithm may be used. As described in Section 3.3, many possible layouts are related by permutation matrices. So simply performing the algorithm just described with (optimal) block sizes b_1 and b_2 on the matrix A as stored is equivalent to performing the LU decomposition of $P_1 A P_2$ where P_1 and P_2 are permutation matrices. Thus at the cost of keeping track of these permutations (a possibly nontrivial software issue), a single algorithm suffices for a wide variety of layouts.

Finally, we need to solve the triangular systems $Ly = b$ and $Ux = y$ arising from the LU decomposition. On a shared memory machine, this is accomplished by two calls to the Level 2 BLAS. Designing such an algorithm on a distributed memory machine is harder, because the fewer floating point operations performed ($\mathcal{O}(n^2)$ instead of $\mathcal{O}(n^3)$) make it harder to mask the communication (see Eisenstat, Heath, Henkel and Romine, 1988; Heath and Romine, 1988; Li and Coleman, 1988; Romine and Ortega, 1988).

4.3. Clever but impractical parallel algorithms for solving $Ax = b$

The theoretical literature provides us with a number of apparently fast but ultimately unattractive algorithms for solving $Ax = b$. These may be unattractive because they need many more parallel processors than is reasonable, ignore locality, are numerically unstable, or any combination of these reasons. We begin with an algorithm for solving $n \times n$ triangular linear systems in $\mathcal{O}(\log^2 n)$ parallel steps. Suppose T is lower triangular with unit diagonal (the diagonal can be factored out in one parallel step). For each i from 1 to $n - 1$, let T_i equal the identity matrix except for column i where it matches T. Then it is simple to verify $T = T_1 T_2 \cdots T_{n-1}$ and so $T^{-1} = T_{n-1}^{-1} \cdots T_2^{-1} T_1^{-1}$. One can also easily see that T_i^{-1} equals the identity except for the subdiagonal of column i, where it is the negative of T_i. Thus it takes no work to compute the T_i^{-1}, and the work involved is to compute the product $T_{n-1}^{-1} \cdots T_1^{-1}$ in $\log_2 n$ parallel steps using a tree. Each parallel step involves multiplying $n \times n$ matrices (which are initially quite sparse, but fill up), and so takes about $\log_2 n$ parallel substeps, for a total of $\log_2^2 n$. The error analysis of this algorithm (Sameh and Brent, 1977) yields an error bound proportional to $\kappa(T)^3 \varepsilon$ where $\kappa(T) = \|T\| \cdot \|T^{-1}\|$ is the condition number and ε is machine precision; this is in contrast to the error bound $\kappa(T)\varepsilon$ for the usual algorithm. The error bound for the parallel algorithm may be pessimistic – the worst example we have found has an error growing like $\kappa(T)^{1.5}\varepsilon$ – but shows that there is a tradeoff between parallellism and stability. Also, to achieve the maximum speedup $\mathcal{O}(n^3)$ processors are required, which is unrealistic for large n.

We can use this algorithm to build an $\mathcal{O}(\log^2 n)$ algorithm for the general problem $Ax = b$ (Csanky, 1977), but this algorithm is so unstable as to be entirely useless in floating point (in IEEE double precision floating point, it achieves no precision in inverting $3I$, where I is an identity matrix of size 60 or larger). There are four steps:

1   Compute the powers of A (A^2, A^3, ... , A^{n-1}) by repeated squaring ($\log_2 n$ matrix multiplications of $\log_2 n$ steps each).
2   Compute the traces $s_i = \mathrm{tr}(A^i)$ of the powers in $\log_2 n$ steps.
3   Solve the Newton identities for the coefficients a_i of the characteristic polynomial; this is a triangular system of linear equations whose matrix entries and right-hand side are known integers and the s_i (we can do this in $\log_2^2 n$ steps as described above).
4   Compute the inverse using Cayley–Hamilton Theorem (in about $\log_2 n$ steps).

For a survey of other theoretical algorithms, see Bertsekas and Tsitsiklis (1989) and Karp and Ramachandran (1990).

4.4. Solving banded systems

These problems do not lend themselves as well to the techniques described above, especially for small bandwidth. The reason is that proportionately less and less parallel work is available in updating the trailing submatrix, and in the limiting case of tridiagonal matrices, the parallel algorithm derived as above and the standard serial algorithm are nearly identical. If the bandwidth is wide enough, however, the techniques of the previous sections still apply (Du Croz et al., 1990; Fox et al., 1988).

The problem of solving banded linear systems with a narrow band has been studied by many authors, see for instance the references in Gallivan et al. (1990) and Ortega (1988). We will only sketch some of the main ideas and we will do so for rather simple problems. The reader should keep in mind that these ideas can easily be generalized for more complicated situations, and many have appeared in the literature.

Most of the parallel approaches perform more arithmetic operations than standard (sequential) Gaussian elimination (typically 2.5 times as many), twisted factorization being the only exception. In twisted factorization the Gaussian elimination process is carried out in parallel from both sides. This method was first proposed in Babuška (1972) for tridiagonal systems $Tx = b$ as a means to compute a specified component of x more accurately. For a tridiagonal matrix twisted factorization leads to the following decomposition of T:

$$
\begin{pmatrix}
\star & \star & & & & & & & \\
\star & \star & \star & & & & & & \\
& \star & \star & \star & & & & & \\
& & \star & \ddots & \ddots & & & & \\
& & & \ddots & \ddots & \ddots & & & \\
& & & & \ddots & \star & \star & & \\
& & & & & \star & \star & \star & \\
& & & & & & \star & \star &
\end{pmatrix}
\tag{4.1}
$$

$$
=
\begin{pmatrix}
\star & & & & & & & \\
\star & \star & & & & & & \\
& \star & \ddots & & & & & \\
& & \ddots & \ddots & & & & \\
& & \star & \star & \star & & & \\
& & & \ddots & \ddots & & & \\
& & & & \star & \star & & \\
& & & & & \star &
\end{pmatrix}
\begin{pmatrix}
\star & \star & & & & & & \\
& \star & \star & & & & & \\
& & \star & \ddots & & & & \\
& & & \ddots & \star & & & \\
& & & & \star & & & \\
& & & & \star & \ddots & & \\
& & & & & \ddots & \star & \\
& & & & & & \star & \star
\end{pmatrix},
$$

or $T = PQ$, where we have assumed that no zero diagonal element is created in P or Q. Such decompositions exist if A is symmetric positive definite, or if A is an M-matrix, or when A is diagonally dominant. The twisted factorization and subsequent forward and back substitutions with P and Q take as many arithmetic operations as the standard factorization, and can be carried out with twofold parallellism by working from both ends of the matrix simultaneously. For an analysis of this process for tridiagonal systems, see van der Vorst (1987a). Twisted factorization can be combined with any of the following techniques, often doubling the parallellism.

The other techniques we will discuss can all be applied to general banded systems, for which most were originally proposed, but for ease of exposition we will illustrate them just with a lower unit bidiagonal system $Lx = b$. A straightforward parallellization approach is to eliminate the unknown x_{i-1} from equation i using equation $i-1$, for all i in parallel. This leads to a new system in which each x_i is coupled only with x_{i-2}. Thus, the original system splits in two independent lower bidiagonal systems of half the size, one for the odd-numbered unknowns, and one for the even-numbered unknowns. This process can be repeated recursively for both new systems, leading to an algorithm known as *recursive doubling* (Stone, 1973). In Algorithm 2

(Section 2.2) it was presented as a special case of parallel prefix. It has been analysed and generalized for banded systems in Dubois and Rodrigue (1977). Its significance for modern parallel computers is limited, which we illustrate with the following examples.

Suppose we perform a single step of recursive doubling. This step can be done in parallel, but it involves slightly more arithmetic than the serial elimination process for solving $Lx = b$. The two resulting lower bidiagonal systems can be solved in parallel. This implies that on a two-processor system the time for a single step of recursive doubling will be slightly more than the time for solving the original system with only one processor. If we have n processors (where n is the dimension of L), then the elimination step can be done in very few time steps, and the two resulting systems can be solved in parallel, so that we have a speedup of about 2. However, this is not very practical, since during most of the time $n - 2$ processors are idle, or formulated differently, the efficiency of the processors is rather low.

If we use n processors to apply this algorithm recursively instead of splitting into just two systems, we can solve in $\mathcal{O}(\log n)$ steps, a speedup of $\mathcal{O}(n/\log n)$, but the efficiency decreases like $\mathcal{O}(1/\log n)$. This is theoretically attractive but inefficient. Because of the data movement required, it is unlikely to be fast without system support for this communication pattern.

A related approach, which avoids the two subsystems, is to eliminate only the odd-numbered unknowns x_{i-1} from the even-numbered equations i. Again, this can be done in parallel, or in vector mode, and it results in a new system in which only the even-numbered unknowns are coupled. After having solved this reduced system, the odd-numbered unknowns can be computed in parallel from the odd-numbered equations. Of course, the trick can be repeated for the subsystem of half size, and this process is known as *cyclic reduction* (Lambiotte and Voigt, 1974; Heller, 1978). Since the amount of serial work is halved in each step by completely parallel (or vectorizable) operations, this approach has been successfully applied on vector supercomputers, especially when the vector speed of the machine is significantly greater than the scalar speed (Ortega, 1988; de Groen, 1991; Schlichting and van der Vorst, 1987). For distributed memory computers the method requires too much data movement for the reduced system to be practical.

However, the method is easily generalized to one with more parallellism. Cyclic reduction can be viewed as an approach in which the given matrix L is written as a lower block bidiagonal matrix with 2×2 blocks along the diagonal. In the elimination process all $(2, 1)$ positions in the diagonal blocks are eliminated in parallel. An obvious idea is to subdivide the matrix into larger blocks, i.e. we write L as a block bidiagonal matrix with $k \times k$ blocks along the diagonal (for simplicity we assume that n is a multiple of k). In practical cases k is chosen so large that the process is not repeated for the resulting subsystems, as for cyclic reduction (where $k = 2$). This approach is referred to as a *divide-and-conquer* approach. For banded triangular systems it was first suggested by Chen, Kuck and Sameh (1978), for tridiagonal systems it was proposed by Wang (1981).

To illustrate, let us apply one parallel elimination step to the lower bidiagonal system $Lx = b$ to eliminate all subdiagonal elements in all diagonal blocks. This yields a system $\tilde{L}x = \tilde{b}$, where for $k = 4$ and $n = 16$ we get

$$
\tilde{L} = \begin{pmatrix}
1 & & & & & & & & & & & & & & & \\
 & 1 & & & & & & & & & & & & & & \\
 & & 1 & & & & & & & & & & & & & \\
 & & & 1 & & & & & & & & & & & & \\
 & & & \star & 1 & & & & & & & & & & & \\
 & & & \star & & 1 & & & & & & & & & & \\
 & & & \star & & & 1 & & & & & & & & & \\
 & & & \star & & & & 1 & & & & & & & & \\
 & & & & & & & \star & 1 & & & & & & & \\
 & & & & & & & \star & & 1 & & & & & & \\
 & & & & & & & \star & & & 1 & & & & & \\
 & & & & & & & \star & & & & 1 & & & & \\
 & & & & & & & & & & & \star & 1 & & & \\
 & & & & & & & & & & & \star & & 1 & & \\
 & & & & & & & & & & & \star & & & 1 & \\
 & & & & & & & & & & & \star & & & & 1
\end{pmatrix}
\tag{4.2}
$$

There are two possibilities for the next step. In the original approach (Wang, 1981), the fill-in in the subdiagonal blocks is eliminated in parallel, or vector mode, for each subdiagonal block (note that each subdiagonal block has only one column with nonzero elements). It has been shown in van der Vorst and Dekker (1989) that this leads to very efficient vectorized code for machines such as Cray, Fujitsu, etc.

For parallel computers, the parallellism in eliminating these subdiagonal blocks is relatively fine-grained compared with the more coarse-grained parallellism in the first step of the algorithm. Furthermore, on distributed memory machines the data for each subdiagonal block have to be spread over all processors. In Michielse and van der Vorst (1988) it has been shown that this limits the performance of the algorithm, the speedup being bounded by the ratio of computational speed and communication speed. This ratio is often very low (Michielse and van der Vorst, 1988).

The other approach is first to eliminate successively the last nonzero elements in the subdiagonal blocks $\tilde{L}_{j,j-1}$. This can be done with a short recurrence of length $n/k - 1$, after which all fill-in can be eliminated in parallel. For the recurrence we need some data communication between processors. However, for k large enough with respect to n/k, one can attain speedups close to $2k/5$ for this algorithm on a k processor system (van der Vorst, 1989c). For a generalization of the divide-and-conquer approach for banded systems, see Meier (1985); the data transport aspects for distributed memory machines have been discussed in Michielse and van der Vorst (1988).

There are other variants of the divide-and-conquer approach that move the fill-in into other columns of the subblocks or are more stabile numerically. For example, in Mehrmann (1991) the matrix is split into a block diagonal matrix and a remainder via rank-1 updates.

5. Least squares problems

Most algorithms for finding the x minimizing $\|Ax - b\|_2$ require computing a QR decomposition of A, where Q is orthogonal and R is upper triangular. We will assume A is $m \times n$, $m \geq n$, so that Q is $m \times m$ and R is $m \times n$. For simplicity we consider only QR without pivoting, and mention work incorporating pivoting at the end.

The conventional approach is to premultiply A by a sequence of simple orthogonal matrices Q_i chosen to introduce zeros below the diagonal of A (Golub and Van Loan, 1989). Eventually A becomes upper triangular, and equal to R, and the product $Q_N \cdots Q_1 = Q$. One kind of Q_i often used is a *Givens rotation*, which changes only two rows of A, and introduces a single zero in one of them; it is the identity in all but two rows and columns, where it is $\begin{bmatrix} c & s \\ -s & c \end{bmatrix}$, with $c^2 + s^2 = 1$. A second kind of Q_i is a *Householder reflection*, which can change any number of rows of A, zeroing out all entries but one in the changed rows of one column of A; a Household reflection may be written $I - 2uu^T$, where u is a unit vector with nonzeros only in the rows to be changed.

5.1. Shared memory algorithms

The basic algorithm to compute a QR decomposition using Householder transformations is given in (Golub and Van Loan, 1989):

Algorithm 13 QR decomposition using Level 2 BLAS

> for $k = 1 : n - 1$
> > Compute a unit vector u_k so that $(I - 2u_k u_k^T)A(k+1:m, k) = 0$
> > Update $A = A - 2 * u_k(u_k^T A)$ $(= Q_k A$ where $Q_k = I - 2u_k u_k^T)$

Computing u_k takes $\mathcal{O}(n - k)$ flops and is essentially a Level 1 BLAS operation. Updating A is seen to consist of a matrix–vector multiplication $(w^T = u_k^T A)$ and a rank-1 update $(A - 2u_k w^T)$, both Level 2 BLAS operations. To convert to Level 3 BLAS requires the observation that one can write $Q_b \cdot Q_{b-1} \cdots Q_1 = I - UTU^T$ where $U = [u_1, ..., u_b]$ is $m \times b$, and T is $b \times b$ and triangular (Schreiber and Van Loan, 1989); for historical reasons this is called a *compact WY transformation*. Thus, by analogy with the LU decomposition with column blocking (Algorithm 12), we may first use Algorithm 13 on a block of n_b columns of A, form U and T of the compact WY transformation, and then update the rest of A by forming $A - UTU^T A$, which consists of three matrix–matrix multiplications. This increases the number of floating point operations by a small amount, and is as stable as the usual algorithm:

Algorithm 14 QR decomposition using Level 3 BLAS (same notation as Algorithm 12)

> for $l = 1 : m$
> $\quad k = (l - 1) \cdot n_b + 1$
> $\quad$ Use Algorithm 13 to factorize $A^{(l)} = Q_l R_l$,
> $\quad$ Form matrices U_l and T_l from Q_l
> $\quad$ Multiply $X = U_l^T \cdot A(k : m, k + n_b : n)$
> $\quad$ Multiply $X = T_l X$
> $\quad$ Update $A(k : m, k + n_b : n) = A(k : m, k + n_b : n) - UX$

Algorithm 14 is available as subroutine **sgeqrf** from LAPACK (Anderson *et al.*, 1992). Pivoting complicates matters slightly. In conventional column pivoting at step k we need to pivot (permute columns) so the next column of A to be processed has the largest norm in rows k through m of all remaining columns. This cannot be directly combined with blocking as we have just described it, and so instead pivoting algorithms which only look among locally stored columns if possible have been developed (Bischof and Tang, 1991a,b).

Other shared memory algorithms based on Givens rotations have also been developed (Chu, 1988a; Gentleman and Kung, 1981; Sameh, 1985), although these do not seem superior on shared memory machines. It is also possible to use Level 2 and 3 BLAS in the modified Gram–Schmidt algorithm (Gallivan, Jalby, Meier and Sameh, 1988).

5.2. Distributed memory algorithms

Just as we could map Algorithm 13 (Gaussian elimination with Level 3 BLAS) to a distributed memory machine with blocked and/or scattered layout by inserting appropriate communication, this can also be done for QR with Level 3 BLAS.

An interesting alternative that works with the same data layouts is based on Givens rotations (Chu, 1988a; Pothen, Jha and Vemulapati, 1987). We consider just the first block column in the block scattered layout, where each of a subset of the processors owns a set of p $r \times r$ subblocks of the block column evenly distributed over the column. Each processor reduces its own $p \cdot r \times r$ submatrix to upper triangular form, spreading the Givens rotations to the right for other processors to apply to their own data. This reduces the processor column to $p \ r \times r$ triangles, each owned by a different processor. Now there needs to be communication among the processors in the column. Organizing them in a tree, at each node in the tree two processors, each of which owns an $r \times r$ triangle, share their data to reduce to a single $r \times r$ triangle. The requisite rotations are again spread rightward. So in $\log_2 p$ of these steps, the first column has been reduced to a single $r \times r$ triangle, and the algorithm moves on to the next block column.

Other Givens based algorithms have been proposed, but seem to require more communication than this one (Pothen *et al.*, 1987).

6. Eigenproblems and the singular value decomposition

6.1. *General comments*

The standard serial algorithms for computing the eigendecomposition of a symmetric matrix A, a general matrix B, or the singular value decomposition (SVD) of a general matrix C have the same two-phase structure: apply orthogonal transformations to reduce the matrix to a condensed form, and then apply an iterative algorithm to the condensed form to compute its eigendecomposition or SVD. For the three problems of this section, the condensed forms are symmetric tridiagonal form, upper Hessenberg form and bidiagonal form, respectively. The motivation is that the iteration requires far fewer flops to apply to the condensed form than to the original dense matrix. We discuss reduction algorithms in Section 6.2.

The challenge for parallel computation is that the iteration algorithms for the condensed forms can be much harder to parallellize than the reductions, since they involve nonlinear, sometimes scalar recurrences and/or little opportunity to use the BLAS. For the nonsymmetric eigenproblem, this has led researchers to explore algorithms that are not parallel versions of serial ones. So far none is as stable as the serial one; this is discussed in Section 6.5.

For the symmetric eigenproblem and SVD, the reductions take $\mathcal{O}(n^3)$ flops, but subsequent iterations to find just the eigenvalues or singular values take only $\mathcal{O}(n^2)$ flops; therefore these iterations have not been bottlenecks on serial machines. But on some parallel machines, the reduction algorithms we discuss are so fast that the $\mathcal{O}(n^2)$ part becomes a bottleneck for surprisingly large values of n. Therefore, parallellizing the $\mathcal{O}(n^2)$ part is of interest; we discuss these problems in Section 6.3.

Other approaches to the symmetric eigenproblem and SVD apply to dense matrices instead of condensed matrices. The best known is Jacobi's method. While attractively parallellizable, the convergence rate is sufficiently slower than methods based on tridiagonal and bidiagonal forms that it is seldom competitive. On the other hand, the Jacobi method is sometimes faster and can be much more accurate than these other methods and so still deserves attention; see Section 6.4. Another method that applies to dense symmetric matrices is a variation of the spectral divide-and-conquer method for nonsymmetric matrices, and discussed in Section 6.5.

In summary, reasonably fast and stable parallel algorithms (if not always implementations) exist for the symmetric eigenvalue problem and SVD. However, no highly parallel and stable algorithms currently exist for the nonsymmetric problem; this remains an open problem.

6.2. Reduction to condensed forms

Since the different reductions to condensed forms are so similar, we discuss only reduction to tridiagonal form; for the others see Dongarra *et al.* (1989). At step k we compute a Householder transformation $Q_k = I - 2u_k u_k^T$ so that column k of $Q_k A$ is zero below the first subdiagonal; these zeros are unchanged by forming the similarity transformation $Q_k A Q_k^T$.

Algorithm 15 Reduction to tridiagonal form using Level 2 BLAS (same notation as Algorithm 12)

> for $k = 1 : n - 2$
> > Compute a unit vector u_k so that $(I - 2u_k u_k^T)A(k + 2 : n, k) = 0$
> > Update $A = (I - 2u_k u_k^T)A(I - 2u_k u_k^T)$ by computing
> > $w_k = 2Au_k$
> > $\gamma_k = w_k^T u_k$
> > $v_k = w_k - \gamma_k u_k$
> > $A = A - v_k u_k^T - u_k v_k^T$

The major work is updating $A = A - vu_k^T - u_k v^T$, which is a symmetric rank-2 update, a Level 2 BLAS operation. To incorporate Level 3 BLAS, we emulate Algorithm 14 by reducing a single column-block of A to tridiagonal form, aggregating the Householder transformations into a few matrices, and then updating via matrix multiply:

Algorithm 16 Reduction to tridiagonal form using Level 3 BLAS (same notation as Algorithm 12)

> for $l = 1 : m$
> > $k = (l - 1) \cdot n_b + 1$
> > Use Algorithm 15 to tridiagonalize the first n_b columns of
> > $A(k : n, k : n)$ as follows:
> > > Do not update all of A at each step, just $A^{(l)}$
> > > Compute $w_k = 2Au_k$ as $2(A - \sum_{q=1}^{k-1}(v_q u_q^T + u_q v_q^T))u_k$
> > > Retain $U^{(l)} = [u_1, ..., u_k]$ and $V^{(l)} = [v_1, ..., v_k]$
> > Update $A(k : n, k : n) = A(k : n, k : n) - U^{(l)}V^{(l)T} - V^{(l)}U^{(l)T}$

Algorithms 15 and 16 are available from LAPACK (Anderson *et al.*, 1992) as subroutines ssytd2 and ssytrf, respectively. Hessenberg reduction is sgehrd, and bidiagonal reduction is sgebrd. The mapping to a distributed memory machine follows as with previous algorithms like QR and Gaussian elimination (Dongarra and van de Geijn, 1991).

For parallel reduction of a band symmetric matrix to tridiagonal form, see Bischof and Sun (1992) and Lang (1992).

The initial reduction of a generalized eigenproblem $A - \lambda B$ involves finding orthogonal matrices Q and Z such that QAZ is upper Hessenberg and QBZ is triangular. So far no profitable way has been found to introduce higher level BLAS into this reduction, in contrast to the other reductions previously mentioned. We return to this problem in Section 6.5.

6.3. The symmetric tridiagonal eigenproblem

The basic algorithms to consider are QR iteration (accelerated) bisection and inverse iteration, and divide-and-conquer. Since the bidiagonal SVD is equivalent to finding the nonnegative eigenvalues of a tridiagonal matrix with zero diagonal (Demmel and Kahan, 1990; Golub and Van Loan, 1989), our comments apply to that problem as well.

QR Iteration The classical algorithm is QR iteration, which produces a sequence of orthogonally similar tridiagonal matrices $T = T_0, T_1, T_2, \ldots$ converging to diagonal form. The mapping from T_i to T_{i+1} is usually summarized as (1) computing a *shift* σ_i, an approximate eigenvalue; (2) factoring $T_i - \sigma_i I = QR$; and (3) forming $T_{i+1} = RQ + \sigma_i I$. Once full advantage is taken of the tridiagonal form, this becomes a nonlinear recurrence that processes the entries of T_i from one end to the other, and amounts to updating T repeatedly by forming PTP^T, with P a Givens rotation. If the eigenvectors are desired, the Ps are accumulated by forming PV, where V is initially the identity matrix. As it stands this recurrence is not parallellizable, but by squaring the matrix entries it can be changed into a recurrence of the form (2.1) in Section 2.2 (see Kuck and Sameh, 1977). The numerical stability of this method is not known, but available analyses are pessimistic (Kuck and Sameh, 1977). Furthermore, QR iterations must be done sequentially, with usually just one eigenvalue converging at a time. If one only wants eigenvalues, this method does not appear to be competitive with the alternatives below. When computing eigenvectors, however, it is easy to parallellize: each processor redundantly runs the entire algorithm updating PTP^T, but only computes n/p of the columns of PV, where p is the number of processors and n is the dimension of T. At the end each processor has n/p components of each eigenvector. Since computing the eigenvectors takes $\mathcal{O}(n^3)$ flops but updating T just $\mathcal{O}(n^2)$, we succeed in parallellizing the majority of the computational work.

Bisection and inverse iteration One of the two most promising methods is (accelerated) bisection for the eigenvalues, followed by inverse iteration for the eigenvectors (Ipsen and Jessup, 1990; Lo, Phillipe and Sameh, 1987). If T has diagonal entries $a_1, ..., a_n$ and offdiagonals $b_1, ..., b_{n-1}$, then we can count the number of eigenvalues of T less than σ (Golub and Van Loan, 1989).

Algorithm 17 Counting eigenvalues using Sturm sequences (1)

$\quad count = 0, d = 1, b_0 = 0$
$\quad \text{for } i = 1 : n$
$\qquad d = a_i - \sigma - b_{i-1}^2/d$
$\qquad \text{if } d < 0, count = count + 1$

This nonlinear recurrence may be transformed into a two-term linear recurrence in $p_i = d_1 d_2 \cdots d_i$:

Algorithm 18 Counting eigenvalues using Sturm sequences (2)

$\quad count = 0, p_0 = 1, p_{-1} = 0, b_0 = 0$
$\quad \text{for } i = 1 : n$
$\qquad p_i = (a_i - \sigma)p_{i-1} - b_{i-1}^2 p_{i-2}$
$\qquad \text{if } p_i p_{i-1} < 0, count = count + 1$

In practice, these algorithms need to protected against over/underflow; Algorithm 17 is much easier to protect (Kahan, 1968). Using either of these algorithms, we can count the number of eigenvalues in an interval. The traditional approach is to bisect each interval, say $[\sigma_1, \sigma_2]$, by running Algorithm 17 or 18 at $\eta = (\sigma_1 + \sigma_2)/2$. By continually subdividing intervals containing eigenvalues, we can compute eigenvalue bounds as tight as we like (and roundoff permits). Convergence of the intervals can be accelerated by using a zero-finder such as `zeroin` (Brent, 1973; Lo et al., 1987), Newton's method, Rayleigh quotient iteration (Beattie and Fox, 1989), Laguerre's method or other methods (Li, Zhang and Sun, 1991). To choose η as an approximate zero of d_n or p_n, i.e. an approximate eigenvalue of T.

There is parallellism both within Algorithm 18 and by running Algorithm 17 or 18 simultaneously for many values of σ. The first kind of parallellism uses parallel prefix as described in (2.1) in Section 2.2, and so care needs to be taken to avoid over/underflow. The numerical stability of the serial implementations of Algorithms 17 (Kahan, 1968) and 18 (Wilkinson, 1965) is very good, but that of the parallel prefix algorithm is unknown, although numerical experiments are promising (Swarztrauber, 1992). This requires good support for parallel prefix operations, and is not as easy to parallellize as simply having each processor refine different sets of intervals containing different eigenvalues (Demmel, 1992a).

Within a single processor one can also run Algorithm 17 or 18 for many different σ by pipelining or vectorizing (Simon, 1989). These many σ could come from disjoint intervals or from dividing a single interval into more than two small ones (*multi-section*); the latter approach appears to be efficient only when a few eigenvalues are desired, so that there are not many disjoint intervals over which to parallellize (Simon, 1989). Achieving good speedup requires load balancing, and this is not always possible to do by statically assigning work to processors. For example, having the ith processor out of p find eigenvalues $(i-1)n/p$ through in/p results in redundant work at the beginning, as each processor refines the initial large interval containing all the eigenvalues. Even if each processor is given a disjoint interval containing an equal number of eigenvalues to find, the speedup may be poor if the eigenvalues in one processor are uniformly distributed in their interval and all the others are tightly clustered in theirs; this is because there will only be one interval to refine in each clustered interval, and many in the uniform one. This means we need to rebalance the load dynamically, with busy processors giving intervals to idle processors. The best way to do this depends on the communication properties of the machine. Since the load imbalance is severe and speedup poor only for problems that run quickly in an absolute sense anyway, pursuing uniformly good speedup may not always be important. The eigenvalues will also need to be sorted at the end if we use dynamic load balancing.

Given the eigenvalues, we can compute the eigenvectors by using inverse iteration in parallel on each processor. At the end each processor will hold the eigenvectors for the eigenvalues it stores; this is in contrast to the parallel QR iteration, which ends up with the transpose of the eigenvector matrix stored. If we simply do inverse iteration without communication, the speedup will be nearly perfect. However, we cannot guarantee orthogonality of eigenvectors of clustered eigenvalues (Ipsen and Jessup, 1990), which currently seems to require reorthogonalization of eigenvectors within clusters (other methods are under investigation (Parlett, 1992a)). We can certainly reorthogonalize against eigenvectors of nearby eigenvalues stored on the same processor without communication, or even against those of neighbouring processors with little communication; this leads to a trade-off between orthogonality, on the one hand, and communication and load balance, on the other.

Other ways to count the eigenvalues in intervals have been proposed as well (Krishnakumar and Morf, 1986; Swarztrauber, 1992), although these are more complicated than either Algorithm 17 or 18. There have also been generalizations to the band definite generalized symmetric eigenvalue problem (Ma, Patrick and Szyld, 1989).

Cuppen's divide-and-conquer algorithm The third algorithm is a div-ide-and-conquer algorithm by Cuppen (1981), and later analysed and modified by many others (Barlow, 1991; Dongarra and Sorensen, 1987; Gu and Eisenstat, 1992; Ipsen and Jessup, 1990; Jessup and Sorensen, 1989; Sorensen and Tang, 1991). If T is $2n \times 2n$, we decompose it into a sum

$$T = \begin{bmatrix} T_1 & 0 \\ 0 & T_2 \end{bmatrix} + \rho x x^T$$

of a block diagonal matrix with tridiagonal blocks T_1 and T_2, and a rank-1 matrix $\rho x x^T$ which is nonzero only in the four entries at the intersection of rows and columns n and $n+1$. Suppose we now compute the eigendecompositions $T_1 = Q_1 \Lambda_1 Q_1^T$ and $T_2 = Q_2 \Lambda_2 Q_2^T$, which can be done in parallel and recursively. This yields the partial eigendecomposition

$$\begin{bmatrix} Q_1 & 0 \\ 0 & Q_2 \end{bmatrix} \cdot \left(\begin{bmatrix} \Lambda_1 & 0 \\ 0 & \Lambda_2 \end{bmatrix} + \rho z z^T \right) \cdot \begin{bmatrix} Q_1^T & 0 \\ 0 & Q_2^T \end{bmatrix}$$

where $z = \text{diag}\,(Q_1^T, Q_2^T)x$. So to compute the eigendecomposition of T, we need to compute the eigendecomposition of the matrix $\text{diag}\,(\Lambda_1, \Lambda_2) + \rho z z^T \equiv D + \rho z z^T$, a diagonal matrix plus a rank-1 matrix. We can easily write down the characteristic polynomial of $D + \rho z z^T$, of which the relevant factor is $f(\lambda)$ in the following so-called *secular equation*

$$f(\lambda) \equiv 1 + \rho \sum_{i=1}^{2n} \frac{z_i^2}{d_i - \lambda} = 0.$$

The roots of $f(\lambda) = 0$ are the desired eigenvalues. Assume the diagonal entries d_i of D are sorted in increasing order. After deflating out easy-to-find eigenvalues (corresponding to tiny z_i or nearly identical d_i) we get a function with guaranteed inclusion intervals $[d_i, d_{i+1}]$ for each zero, and which is also monotonic on each interval. This lets us solve quickly using a Newton-like method (although care must be taken to guarantee convergence (Li, 1992)). The corresponding eigenvector for a root λ_j is then simply given by $(D - \lambda_j I)^{-1}z$. This yields the eigendecomposition $D + \rho z z^T = Q \Lambda Q^T$, from which we compute the full eigendecomposition $T = (\text{diag}\,(Q_1, Q_2)Q)\Lambda(\text{diag}\,(Q_1, Q_2)Q)^T$.

This algorithm, while attractive, proved hard to implement stably. The trouble was that to guarantee the computed eigenvectors were orthogonal, $d_i - \lambda_j$ had to be computed with reasonable relative accuracy, which is not guaranteed even if λ_j is known to high precision; cancellation in $d_i - \lambda_j$ can leave a tiny difference with high relative error. Work by several authors (Barlow, 1991; Sorensen and Tang, 1991) led to the conclusion that λ_i had to be computed to double the input precision in order to determine $d_i - \lambda_i$ accurately. When the input is already in double precision (or whatever is the largest precision supported by the machine), then quadruple is needed, which may be simulated using double, provided double is accurate enough (Dekker, 1971; Priest, 1991). Recently, however, Gu and Eisenstat (1992) have found a new algorithm that makes this unnecessary.

There are two types of parallellism available in this algorithm and both must be exploited to speed up the whole algorithm (Dongarra and Sorensen, 1987; Ipsen and Jessup, 1990). Independent tridiagonal submatrices (such as T_1 and T_2) can obviously be solved in parallel. Initially there are a great many such small submatrices to solve in parallel, but after each secular equation solution, there are half as many submatrices of twice the size. To keep working in parallel, we must find the roots of the different secular equations in parallel; there are equally many roots to find at each level. Also, there is parallellism in the matrix multiplication diag $(Q_1, Q_2) \cdot Q$ needed to update the eigenvectors.

While there is a great deal of parallellism available, there are still barriers to full speedup. First, the speed of the serial algorithm depends strongly on there being a great deal of deflation, or roots of the secular equation that can be computed with little work. If several processors are cooperating to solve a single secular equation, they must either communicate to decide which of their assigned roots were deflated and to rebalance the work load of finding nontrivial roots, or else not communicate and risk a load imbalance. This is the same tradeoff as for the bisection algorithm, except that rebalancing involves more data movement (since eigenvectors must be moved). If it turns out, as with bisection, that load imbalance is severe and speedup poor only when the absolute run time is fast anyway, then dynamic load balancing may not be worth it. The second barrier to full speedup is simply the complexity of the algorithm, and the need to do many different kinds of operations in parallel, including sorting, matrix multiplication, and solving the secular equation. The current level of parallel software support on many machines can make this difficult to implement well.

6.4. Jacobi's method for the symmetric eigenproblem and SVD

Jacobi's method has been used for the nonsymmetric eigenproblem, the symmetric eigenproblem, the SVD, and generalizations of these problems to pairs of matrices (Golub and Van Loan, 1989). It works by applying a series of Jacobi rotations (a special kind of Givens rotation) to the left and/or right of the matrix in order to drive it to a desired canonical form, such as the diagonal form for the symmetric eigenproblem. These Jacobi rotations, which affect only two rows and/or columns of the matrix, are chosen to solve the eigenproblem associated with those two rows and/or columns (this is what makes Jacobi rotations special). By repeatedly solving all 2×2 subproblems of the original, one eventually solves the entire problem. The

Jacobi method works reliably on the symmetric eigenvalue problem and SVD, and less so on the nonsymmetric problem. We will consider only the symmetric problem and SVD in this section, and the nonsymmetric Jacobi later.

Until recently Jacobi methods were of little interest on serial machines because they are usually several times slower than QR or divide-and-conquer schemes, and seemed to have the same accuracy. Recently, however, it has been shown that Jacobi's method can be much more accurate than QR in certain cases (Deichmoller, 1991; Demmel and Veselić, 1992; Slapničar, 1992), which makes it of some value on serial machines.

It has also been of renewed interest on parallel machines because of its inherent parallellism: Jacobi rotations can be applied in parallel to disjoint pairs of rows and/or columns of the matrix, so a matrix with n rows and/or columns can have $\lfloor n/2 \rfloor$ Jacobi rotations applied simultaneously (Brent and Luk, 1985). The question remains of the order in which to apply the simultaneous rotations to achieve quick convergence. A number of good parallel orderings have been developed and shown to have the same convergence properties as the usual serial implementations (Luk and Park, 1989; Shroff and Schreiber, 1989); we illustrate one here in the following diagram (P1–P4 denotes Processor 1–Processor 4). Assume we have distributed $n = 8$ columns on $p = 4$ processors, two per processor. We may leave one column fixed, and 'rotate' the others so that after $n - 1$ steps all possible pairs of columns have simultaneously occupied a single processor, so they could have a Jacobi rotation applied to them:

P1:	1,8	1,7	1,6	1,5	1,4	1,3	1,2
P2:	2,7	8,6	7,5	6,4	5,3	4,2	3,8
P3:	3,6	2,5	8,4	7,3	6,2	5,8	4,7
P4:	4,5	3,4	2,3	8,2	7,8	6,7	5,6
	Step 1	Step 2	Step 3	Step 4	Step 5	Step 6	Step 7

This is clearly easiest to apply when we are applying Jacobi rotations only to columns of the matrix, rather than to both rows and columns. Such a one-sided Jacobi is natural when computing the SVD (Hari and Veselić, 1987), but requires some preprocessing for the symmetric eigenproblem (Demmel and Veselić, 1992; Slapničar, 1992); for example, in the symmetric positive definite case one can perform Cholesky on A to obtain $A = LL^T$, apply one-sided Jacobi on L or L^T to get its (partial) SVD, and then square the singular values to get the eigenvalues of A. It turns out it accelerates convergence to do the Cholesky decomposition with pivoting, and then apply Jacobi to the columns of L rather than the columns of L^T (Demmel and Veselić, 1992). It is possible to use the symmetric-indefinite decomposition of an indefinite symmetric matrix in the same way (Slapničar, 1992).

Jacobi done in this style is a fine-grain algorithm, operating on pairs of columns, and so cannot exploit higher level BLAS. One can instead use block Jacobi algorithms (Bischof, 1989; Shroff and Schreiber, 1989), which work on blocks, and apply the resulting orthogonal matrices to the rest of the matrix using more efficient matrix–matrix multiplication.

6.5. The nonsymmetric eigenproblem

Five kinds of parallel methods for the nonsymmetric eigenproblem have been investigated:

1 Hessenberg **QR** iteration (Bai and Demmel, 1989; Davis, Funderlic and Geist, 1987; Dubrulle, 1991; Geist and Davis, 1990; Stewart, 1987; van de Geijn, 1987; 1989; Watkins, 1992; Watkins and Elsner, 1991);
2 Reduction to nonsymmetric tridiagonal form (Dongarra, Geist and Romine, 1990; Geist, 1990; 1991; Geist, Lu and Wachspress, 1989);
3 Jacobi's method (Eberlein, 1962; 1987; Paardekooper, 1989; Sameh, 1971; Shroff, 1991; Stewart, 1985; Veselić, 1979);
4 Hessenberg divide-and-conquer (Chu, 1988b; Chu, Li and Sauer, 1988; Dongarra and Sidani, 1991; Li and Zeng, 1992; Li, Zeng and Cong, 1992; Zeng, 1991);
5 Spectral divide-and-conquer (Bai and Demmel, 1992; Lin and Zmijewski, 1991; Malyshev, 1991).

In contrast to the symmetric problem or SVD, no guaranteed stable and highly parallel algorithm for the nonsymmetric problem exists. As described in Section 6.2, reduction to Hessenberg form can be done efficiently, but so far it has been much harder to deal with a Hessenberg matrix (Dubrulle, 1991; Jessup, 1991). §

Hessenberg QR **iteration** Parallelizing Hessenberg QR is attractive because it would yield an algorithm that is as stable as the quite acceptable serial one. Unfortunately, doing so involves some of the same difficulties as tridiagonal QR: one is faced with either fine-grain synchronization or larger block operations that execute more quickly but also do much more work without accelerating convergence much. The serial method computes one or two shifts from the bottom right corner of the matrix, and then processes the matrix from the upper left by a series of row and column operations (this processing is called *bulge chasing*). One way to introduce parallellism is to spread the matrix across the processors, but communication costs may exceed the modest computational costs of the row and column operations (Davis *et al.*, 1987; Geist and Davis, 1990; Stewart, 1987; van de Geijn and

§ As noted in Section 6.2, we cannot even efficiently reduce to condensed form for the generalized eigenproblem $A - \lambda B$.

Hudson, 1989; van de Geijn, 1987). Another way to introduce parallellism is to compute $k > 2$ shifts from the bottom corner of the matrix (the eigenvalues of the bottom right $k \times k$ matrix, say), which permits us to work on k rows and columns of the matrix at a time using Level 2 BLAS (Bai and Demmel, 1989). Asymptotic convergence remains quadratic (Watkins and Elsner, 1991). The drawbacks to this scheme are twofold. First, any attempt to use Level 3 BLAS introduces rather small (hence inefficient) matrix–matrix operations, and raises the operation count considerably. Second, the convergence properties degrade significantly, resulting in more overall work as well (Dubrulle, 1991). As a result, speedups have been extremely modest. This routine is available in LAPACK as **shseqr** (Anderson *et al.*, 1992).

Yet another way to introduce parallellism into Hessenberg QR is to pipeline several bulge chasing steps (van de Geijn, 1987; Watkins, 1992; Watkins and Elsner, 1991). If we have several shifts available, then as soon as one bulge chase is launched from the upper left corner, another one may be launched, and so on. Since each bulge chase operates on only two or three adjacent rows and columns, we can potentially have $n/2$ or $n/3$ bulge chasing steps going on simultaneously on disjoint rows (and columns). The problem is that in the serial algorithm, we have to wait until an entire bulge chase has been completed before computing the next shift; in the parallel case we cannot wait. Therefore, we must use 'out-of-date' shifts to have enough available to start multiple bulge chases. This destroys the usual local quadratic convergence, but it remains superlinear (van de Geijn, 1987). It has been suggested that choosing the eigenvalues of the bottom right $k \times k$ submatrix may have superior convergence to just choosing a sequence from the bottom 1×1 or 2×2 submatrices (Watkins, 1992). Parallelism is still fine-grain, however.

Reduction to nonsymmetric tridiagonal form This approach begins by reducing B to nonsymmetric tridiagonal form with a (necessarily) nonorthogonal similarity, and then finding the eigenvalues of the resulting nonsymmetric tridiagonal matrix using the tridiagonal LR algorithm (Dongarra *et al.*, 1990; Geist, 1990; 1991; Geist *et al.*, 1989). This method is attractive because finding eigenvalues of a tridiagonal matrix (even nonsymmetric) is much faster than for a Hessenberg matrix (Wilkinson, 1965). The drawback is that reduction to tridiagonal form may require very ill conditioned similarity transformations, and may even break down (Parlett, 1992a). Breakdown can be avoided by restarting the process with different initializing vectors, or by accepting a 'bulge' in the tridiagonal form. This happens with relatively low probability, but keeps the algorithm from being fully reliable. The current algorithms pivot at each step to maintain and monitor stability, and so can be converted to use Level 2 and Level 3 BLAS in a manner analogous to Gaussian elimination with pivoting. This algorithm illustrates how one can trade off numerical stability for speed. Other nonsymmetric eigenvalue algorithms we discuss later make this tradeoff as well.

Jacobi's method As with the symmetric eigenproblem, nonsymmetric Jacobi methods solve a sequence of 2×2 eigenvalue subproblems by applying 2×2 similarity transformations to the matrix. There are two basic kinds of transformations used. Methods that use only orthogonal transformations maintain numerical stability and converge to Schur canonical form, but converge only linearly at best (Eberlein, 1987; Stewart, 1985). If nonorthogonal transformations are used, one can try to drive the matrix to diagonal form, but if it is close to having a nontrivial Jordan block, the required similarity transformation will be very ill conditioned and so stability is lost. Alternatively, one can try to drive the matrix to be *normal* ($AA^T = A^T A$), at which point an orthogonal Jacobi method can be used to drive it to diagonal form; this still does not get around the problem of (nearly) nontrivial Jordan blocks (Eberlein, 1962; Paardekooper, 1989; Sameh, 1971; Shroff, 1991; Veselić, 1979). On the other hand, if the matrix has distinct eigenvalues, asymptotic quadratic convergence is achieved (Shroff, 1991). Using n^2 processors arranged in a mesh, these algorithms can be implemented in time $\mathcal{O}(n \log n)$ per sweep. Again, we trade off control over numerical stability for speed (of convergence).

Hessenberg divide-and-conquer The divide-and-conquer algorithms we consider here involve setting a middle subdiagonal entry of the original upper Hessenberg matrix H to zero, resulting in a block upper Hessenberg matrix S. The eigenproblems for the two Hessenberg matrices on the diagonal of S can be solved in parallel and recursively. To complete the algorithm, one must merge the eigenvalues and eigenvectors of the two halves of S to get the eigendecomposition of H. Two ways have been proposed to do this: homotopy continuation and Newton's method. Parallelism lies in having many Hessenberg submatrices whose eigendecompositions are needed, in being able to solve for n eigenvalues simultaneously, and in the linear algebra operations needed to find an individual eigenvalue. The first two kinds of parallellism are analogous to those in Cuppen's method (Section 6.3). The main drawback of these methods is loss of guaranteed stability and/or convergence. Newton's method may fail to converge, and both Newton and homotopy may appear to converge to several copies of the same root without any easy way to tell if a root has been missed, or if the root really is multiple. The subproblems produced by divide-and-conquer may be much more ill conditioned than the original problem. These drawbacks are discussed in Jessup (1991).

Homotopy methods replace the original Hessenberg matrix H by the one-parameter linear family $H(t) = tS + (1-t)H$, $0 \le t \le 1$. As t increases from 0 to 1, the eigenvalues (and eigenvectors) trace out curves connecting the eigenvalues of S to the desired ones of H. The numerical method follows these curves by standard curve-following schemes, predicting the position of a nearby point on the curve using the derivative of the eigenvalue with respect to t, and then correcting its predicted value using Newton's method.

Two schemes have been investigated. The first (Li et al., 1992) follows eigenvalue/eigenvector pairs. The homotopy function is $h(z, \lambda, t) = [(H(t)z - \lambda z)^T, \|z\|_2^2 - 1]^T$, i.e. the homotopy path is defined by choosing $z(t)$ and $\lambda(t)$ so that $h(z(t), \lambda(t), t) = 0$ along the path. The simplicity of the homotopy means that over 90% of the paths followed are simple straight lines that require little computation, resulting in a speed up of a factor of up to 2 over the serial QR algorithm. The drawbacks are lack of stability and convergence not being guaranteed. For example, when homotopy paths get very close together, one is forced to take smaller steps (and so converge more slowly) during the curve following. Communication is necessary to decide if paths get close. And as mentioned previously, if two paths converge to the same solution, it is hard to tell if the solution really is a multiple root or if some other root is missing. A different homotopy scheme uses only the determinant to follow eigenvalues (Li and Zeng, 1992; Zeng, 1991); here the homotopy function is simply $\det(H(t) - \lambda I)$. Evaluating the determinant of a Hessenberg matrix costs only a triangular solve and an inner product, and therefore is efficient. It shares similar advantages and disadvantages as the previous homotopy algorithm.

Alternatively, one can use Newton's method to compute the eigendecomposition of H from S (Dongarra and Sidani, 1991). The function to which one applies Newton's method is $f(z, \lambda) = [(Hz - \lambda z)^T, e^T z - 1]^T$, where e is a fixed unit vector. The starting values for Newton's method are obtained from the solutions to S.

Spectral divide-and-conquer A completely different way to divide-and-conquer a matrix is using a projection on part of the spectrum. It applies to a dense matrix B. Suppose Q_1 is an $n \times m$ orthogonal matrix spanning a right invariant subspace of B, and Q_2 is an $n \times (n - m)$ matrix constructed so that $Q = [Q_1, Q_2]$ is square and orthogonal. Then Q deflates B as follows:

$$Q^T B Q = \begin{bmatrix} B_{11} & B_{12} \\ 0 & B_{22} \end{bmatrix}.$$

Note that this is equivalent to having Q_2 span a left invariant subspace of B. The eigenvalues of B_{11} are those corresponding to the invariant subspace spanned by Q_1. Provided we can construct Q_1 effectively, we can use this to divide-and-conquer the matrix.

Of course Hessenberg QR iteration fits into this framework, with Q_2 being $n \times 1$ or $n \times 2$, and computed by (implicit) inverse iteration applied to $B - \sigma I$, where σ is a shift. Just splitting so that B_{22} is 1×1 or 2×2 does not permit much parallellism, however; it would be better to split the matrix nearer the middle. Also, it would be nice to be able to split off just that part of the spectrum of interest to the user, rather than computing all eigenvalues as these methods must all do.

There are several approaches to computing Q. They may be motivated by analogy to Hessenberg QR, where Q is the orthogonal part of the QR factorization $QR = B - \sigma I$. If σ is an exact eigenvalue, so that $B - \sigma I$ is singular, then the last column of Q is (generically) a left eigenvector for 0. One can then verify that the last row of $Q^T(B-\sigma I)Q$ is zero, so that we have deflated the eigenvalue at σ. Now consider a more general function $f(B)$; in principle any (piecewise) analytic function will do. Then the eigenvalues of $f(B)$ are just f evaluated at the eigenvalues of B, and $f(B)$ and B have (modulo Jordan blocks) the same eigenvectors. Suppose that the rank of $f(B)$ is $m < n$, so that $f(B)$ has (at least) $n - m$ zero eigenvalues. Factorize $QR = f(B)$. Then the last $n - m$ columns of Q (generally) span the left null space of $f(B)$, i.e. a left invariant subspace of $f(B)$ for the zero eigenvalue. But this is also a left invariant subspace of B so we get

$$Q^T f(B)Q = \begin{bmatrix} \hat{B}_{11} & \hat{B}_{12} \\ 0 & 0 \end{bmatrix} \quad \text{and} \quad Q^T BQ = \begin{bmatrix} B_{11} & B_{12} \\ 0 & B_{22} \end{bmatrix}.$$

The problem thus becomes finding functions $f(B)$ that are easy to evaluate and have large null spaces, or which map selected eigenvalues of B to zero. One such function f is the *sign function* (Bai and Demmel, 1992; Howland, 1983; Kenney and Laub, 1991; Lin and Zmijewski, 1991; Robert, 1980; Stickel, 1991) which maps points with positive real part to $+1$ and those with negative real part to -1; adding 1 to this function then maps eigenvalues in the right half plane to 2 and in the left plane to 0, as desired.

The only operations we can easily perform on (dense) matrices are multiplication and inversion, so in practice f must be a rational function. A globally, asymptotically quadratically convergent iteration to compute the sign function of B is $B_{i+1} = (B_i + B_i^{-1})/2$ (Howland, 1983; Robert, 1980; Stickel, 1991); this is simply Newton's method applied to $B^2 = I$, and can also be seen to be equivalent to repeated squaring (the power method) of the Cayley transform of B. It converges more slowly as eigenvalues approach the imaginary axis, and is in fact nonconvergent if there are imaginary eigenvalues, as may be expected since the sign function is discontinuous there. Other higher order convergent schemes exist, but they can be more expensive to implement as well (Kenney and Laub, 1991; Pandey, Kenney and Laub, 1990). Another scheme that divides the spectrum between the eigenvalues inside and outside the unit circle is given in Malyshev (1991).

If the eigenvalues are known to be real (as when the matrix is symmetric), we need only construct a function f that maps different parts of the real axis to 0 and 1 instead of the entire left and right half planes. This simplifies both the computation of $f(B)$ and the extraction of its null space. See Auslander and Tsao (1992), Bischof and Sun (1992) and Lederman, Tsao and Turnbull (1992) for details.

Of course, we wish to split not just along the imaginary axis or unit circle but other boundaries as well. By shifting the matrix and multiplying by a complex number $e^{i\theta}$ one can split along an arbitrary line in the complex plane, but at the cost of introducing complex arithmetic. By working on a shifted and squared real matrix, one can divide along lines at an angle of $\pi/4$ and retain real arithmetic (Bai and Demmel, 1992; Howland, 1983; Stickel, 1991).

This method is promising because it allows us to work on just that part of the spectrum of interest to the user. It is stable because it applies only orthogonal transformations to B. On the other hand, if it is difficult to find a good place to split the spectrum, convergence can be slow, and the final approximate invariant subspace inaccurate. At this point, iterative refinement could be used to improve the factorization (Demmel, 1987). These methods apply to the generalized nonsymmetric eigenproblem as well (Bai and Demmel, 1992; Malyshev, 1991).

7. Direct methods for sparse linear systems

7.1. Cholesky factorization

In this section we discuss parallel algorithms for solving sparse systems of linear equations by direct methods. Paradoxically, sparse matrix factorization offers additional opportunities for exploiting parallellism beyond those available with dense matrices, yet it is usually more difficult to attain good efficiency in the sparse case. We examine both sides of this paradox: the additional parallellism induced by sparsity, and the difficulty in achieving high efficiency in spite of it. We will see that regularity and locality play a similar role in determining performance in the sparse case as they do for dense matrices.

We couch most of our discussion in terms of the Cholesky factorization, $A = LL^T$, where A is symmetric positive definite (SPD) and L is lower triangular with positive diagonal entries. We focus on Cholesky factorization primarily because this allows us to discuss parallellism in relative isolation, without the additional complications of pivoting for numerical stability. Most of the lessons learned are also applicable to other matrix factorizations, such as LU and QR. We do not try to give an exhaustive survey of research in this area, which is currently very active, instead referring the reader to existing surveys, such as Heath, Ng and Peyton (1991). Our main point in the current discussion is to explain how the sparse case differs from the dense case, and examine the performance implications of those differences.

We begin by considering the main features of sparse Cholesky factorization that affect its performance on serial machines. Algorithm 19 gives a standard, column-oriented formulation in which the Cholesky factor L overwrites the initial matrix A, and only the lower triangle is accessed:

Algorithm 19 Cholesky factorization

> for $j = 1,\ n$
> > for $k = 1,\ j - 1$
> > > for $i = j,\ n$ $\{\mathrm{cmod}(j,k)\}$
> > > > $a_{ij} = a_{ij} - a_{ik} \cdot a_{jk}$
> > $a_{jj} = \sqrt{a_{jj}}$
> > for $k = j + 1,\ n$ $\{\mathrm{cdiv}(j)\}$
> > > $a_{kj} = a_{kj}/a_{jj}$

The outer loop in Algorithm 19 is over successive columns of A. The the current column (indexed by j) is modified by a multiple of each prior column (indexed by k); we refer to such an operation as $\mathrm{cmod}(j,k)$. The computation performed by the inner loop (indexed by i) is a **saxpy**. After all its modifications have been completed, column j is then scaled by the reciprocal of the square root of its diagonal element; we refer to this operation as $\mathrm{cdiv}(j)$. As usual, this is but one of the 3! ways of ordering the triple-nested loop that embodies the factorization.

The inner loop in Algorithm 19 has no effect, and thus may as well be skipped, if $a_{jk} = 0$. For a dense matrix A, such an event is too unlikely to offer significant advantage. The fundamental difference with a sparse matrix is that a_{jk} is in fact very often zero, and computational efficiency demands that we recognize this situation and take advantage of it. Another way of expressing this condition is that column j of the Cholesky factor L does not depend on prior column k if $\ell_{jk} = 0$, which not only provides a computational shortcut, but also suggests an additional source of parallellism that we will explore in detail later.

7.2. Sparse matrices

Thus far we have not said what we mean by a 'sparse' matrix. A good operational definition is that a matrix is sparse if it contains enough zero entries to be worth taking advantage of them to reduce both the storage and work required in solving a linear system. Ideally, we would like to store and operate on only the nonzero entries of the matrix, but such a policy is not necessarily a clear win in either storage or work. The difficulty is that sparse data structures include more overhead (to store indices as well as numerical values of nonzero matrix entries) than the simple arrays used for dense matrices, and arithmetic operations on the data stored in them usually cannot be performed as rapidly either (due to indirect addressing of operands). There is therefore a tradeoff in memory requirements between sparse and dense representations and a tradeoff in performance between the algorithms that use them. For this reason, a practical requirement for a family of matrices to be 'usefully' sparse is that they have only $\mathcal{O}(n)$ nonzero entries, i.e. a (small) constant number of nonzeros per row or column, independent of the matrix dimension. For example, most matrices arising from finite difference or finite element discretizations of PDEs satisfy this condition. In addition to the number of nonzeros, their particular locations, or pattern, in the matrix also has a major effect on how well sparsity can be exploited. Sparsity arising from physical problems usually exhibits some systematic pattern that can be exploited effectively, whereas the same number of nonzeros located randomly might offer relatively little advantage.

In Algorithm 19, the modification of a given column of the matrix by a prior column not only changes the existing nonzero entries in the target column, but may also introduce new nonzero entries in the target column. Thus, the Cholesky factor L may have additional nonzeros, called *fill*, in locations that were zero in the original matrix A. In determining the storage requirements and computational work, these new nonzeros that the matrix gains during the factorization are equally as important as the nonzeros with which the matrix starts out.

The amount of such fill is dramatically affected by the order in which the columns of the matrix are processed. For example, if the first column of the matrix A is completely dense, then all of the remaining columns, no matter how sparse they start out, will completely fill in with nonzeros during the factorization. On the other hand, if a single such dense column is permuted (symmetrically) to become the last column in the matrix, then it will cause no fill at all. Thus, a critical part of the solution process for sparse systems is to determine an ordering for the rows and columns of the input matrix that limits fill to preserve sparsity. Unfortunately, finding an ordering that minimizes fill is a very hard combinatorial problem (NP-complete), but heuristics are available that do a good job of reducing, if not exactly minimizing, fill. These techniques include minimum degree, nested dissection, and various schemes for reducing the bandwidth or profile of a matrix (see, e.g., Duff, Erisman and Reid (1986) and George and Liu (1981) for details on these and many other concepts used in sparse matrix computations).

One of the key advantages of SPD matrices is that such a sparsity preserving ordering can be selected in advance of the numeric factorization, independent of the particular values of the nonzero entries: only the pattern of the nonzeros matters, not their numerical values. This would not be the case, in general, if we also had to take into account pivoting for numerical stability, which obviously would require knowledge of the nonzero values, and would introduce a potential conflict between preserving sparsity and preserving stability. For the SPD case, once the ordering is selected, the locations of all fill elements in L can be anticipated prior to the numeric factorization, and thus an efficient static data structure can be set up in advance to accommodate them (this process is usually called *symbolic factorization*). This feature also stands in contrast to general sparse linear systems, which usually require dynamic data structures to accommodate fill entries as they occur, since their locations depend on numerical information that becomes known only as the numeric factorization process unfolds. Thus, modern algorithms and software for solving sparse SPD systems include a symbolic preprocessing phase in which a sparsity-preserving ordering is computed and a static data structure is set up for storing the entries of L before any floating point computation takes place.

We introduce some concepts and notation that will be useful in our subsequent discussion of parallel sparse Cholesky factorization. An important tool in understanding the combinatorial aspects of sparse Cholesky factorization is the notion of the *graph* of a symmetric $n \times n$ matrix A, which is an undirected graph having n vertices, with an edge between two vertices i and j if the corresponding entry a_{ij} of the matrix is nonzero. We denote the graph of A by $G(A)$. The structural effect of the factorization process can then be described by observing that the elimination of a variable adds fill edges to the corresponding graph so that the neighbours of the eliminated vertex become a clique (i.e. a fully connected subgraph). We also define the *filled graph*, denoted by $F(A)$, as having an edge between vertices i and j, with $i > j$, if $\ell_{ij} \neq 0$ in the Cholesky factor L (i.e. $F(A)$ is simply $G(A)$ with all fill edges added).

We use the notation M_{i*} to denote the ith row, and M_{*j} to denote the jth column, of a matrix M. For a given sparse matrix M, we define

$$\text{Struct}(M_{i*}) = \{k < i \mid m_{ik} \neq 0\}$$

and

$$\text{Struct}(M_{*j}) = \{k > j \mid m_{kj} \neq 0\}.$$

In other words, $\text{Struct}(M_{i*})$ is the sparsity structure of row i of the strict lower triangle of M, while $\text{Struct}(M_{*j})$ is the sparsity structure of column j of the strict lower triangle of M. For the Cholesky factor L, we define the *parent* function as follows:

$$\text{parent}(j) = \begin{cases} \min\{i \in \text{Struct}(L_{*j})\}, & \text{if } \text{Struct}(L_{*j}) \neq \emptyset, \\ j & \text{otherwise.} \end{cases}$$

Thus, $\text{parent}(j)$ is the row index of the first offdiagonal nonzero in column j of L, if any, and has the value j otherwise. Using the parent function, we define the *elimination tree* as a graph having n vertices, with an edge between vertices i and j, for $i > j$, if $i = \text{parent}(j)$. If the matrix is

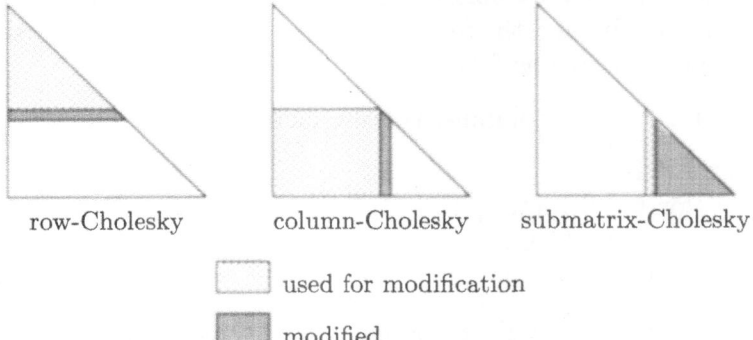

row-Cholesky   column-Cholesky   submatrix-Cholesky

☐   used for modification

■   modified

Fig. 9. Three forms of Cholesky factorization.

irreducible, then the elimination tree is indeed a single tree with its root at vertex n (otherwise it is more accurately termed an *elimination forest*). The elimination tree, which we denote by $T(A)$, is a spanning tree for the filled graph $F(A)$. The many uses of the elimination tree in analysing and organizing sparse Cholesky factorization are surveyed in Liu (1990). We will illustrate these concepts pictorially in several examples below.

7.3. Sparse factorization

There are three basic types of algorithms for Cholesky factorization, depending on which of the three indices is placed in the outer loop:

1   *Row-Cholesky:* Taking i in the outer loop, successive rows of L are computed one by one, with the inner loops solving a triangular system for each new row in terms of the previously computed rows.

2   *Column-Cholesky:* Taking j in the outer loop, successive columns of L are computed one by one, with the inner loops computing a matrix–vector product that gives the effect of previously computed columns on the column currently being computed.

3   *Submatrix-Cholesky:* Taking k in the outer loop, successive columns of L are computed one by one, with the inner loops applying the current column as a rank-1 update to the remaining unreduced submatrix.

These three families of algorithms have markedly different memory reference patterns in terms of which parts of the matrix are accessed and modified at each stage of the factorization, as illustrated in Figure 9, and each has its advantages and disadvantages in a given context.

For sparse Cholesky factorization, row-Cholesky is seldom used for a number of reasons, including the difficulty in providing a row-oriented data structure that can be accessed efficiently during the factorization, and the difficulty in vectorizing or parallellizing the triangular solutions required. We will therefore focus our attention on the column-oriented methods, column-Cholesky and submatrix-Cholesky. Expressed in terms of the column operations cmod and cdiv and the Struct notation defined earlier, sparse column-Cholesky can be stated as follows:

Algorithm 20 Sparse column-Cholesky factorization

> for $j = 1$, n
>> for $k \in \text{Struct}(L_{j*})$
>>> $\text{cmod}(j, k)$
>> $\text{cdiv}(j)$

In column-Cholesky, a given column j of A remains unchanged until the outer loop index reaches that value of j. At that point column j is updated by a nonzero multiple of each column $k < j$ of L for which $\ell_{jk} \neq 0$. After all column modifications have been applied to column j, the diagonal entry ℓ_{jj} is computed and used to scale the completely updated column to obtain the remaining nonzero entries of L_{*j}. Column-Cholesky is sometimes said to be a 'left-looking' algorithm, since at each stage it accesses needed columns to the left of the current column in the matrix. It can also be viewed as a 'demand-driven' algorithm, since the inner products that affect a given column are not accumulated until actually needed to modify and complete that column. For this reason, Ortega (1988) terms column-Cholesky a 'delayed-update' algorithm. It is also sometimes referred to as a 'fan-in' algorithm, since the basic operation is to combine the effects of multiple previous columns on a single target column. The column-Cholesky algorithm is the most commonly used method in commercially available sparse matrix packages.

Similarly, sparse submatrix-Cholesky can be expressed as follows.

Algorithm 21 Sparse submatrix-Cholesky factorization

> for $k = 1$, n
>> $\text{cdiv}(k)$
>> for $j \in \text{Struct}(L_{*k})$
>>> $\text{cmod}(j, k)$

In submatrix-Cholesky, as soon as column k has been computed, its effects on all subsequent columns are computed immediately. Thus, submatrix-Cholesky is sometimes said to be a 'right-looking' algorithm, since at each stage columns to the right of the current column are modified. It can also be viewed as a 'data-driven' algorithm, since each new column is used as soon as it is completed to make all modifications to all the subsequent columns it affects. For this reason, Ortega (1988) terms submatrix-Cholesky an 'immediate-update' algorithm. It is also sometimes referred to as a 'fan-out' algorithm, since the basic operation is for a single column to affect multiple subsequent columns. We will see that these characterizations of the column-Cholesky and submatrix-Cholesky algorithms have important implications for parallel implementations.

We note that many variations and hybrid implementations that lie somewhere between pure column-Cholesky and pure submatrix-Cholesky are possible. Perhaps the most important of these are the multi-frontal methods (see, e.g., Duff *et al.* (1986)), in which updating operations are accumulated in and propagated through a series of *front matrices* until finally being incorporated into the ultimate target columns. Multi-frontal methods have a number of attractive advantages, most of which accrue from the localization of memory references in the front matrices, thereby facilitating the effective use of memory hierarchies, including cache, virtual memory with paging, or explicit out-of-core solutions (the latter was the original motivation for these methods (Irons, 1970)). In addition, since the front matrices are essentially dense, the operations on them can be done using optimized kernels, such as the BLAS, to take advantage of vectorization or any other available architectural features. For example, such techniques have been used to attain very high performance for sparse factorization on conventional vector supercomputers (Ashcraft, Grimes, Lewis, Peyton and Simon, 1987) and on RISC workstations (Rothberg and Gupta, 1989).

7.4. Parallelism in sparse factorization

We now examine in greater detail the opportunities for parallellism in sparse Cholesky factorization and various algorithms for exploiting it. One of the most important issues in designing any parallel algorithm is selecting an appropriate level of *granularity*, by which we mean the size of the computational subtasks that are assigned to individual processors. The optimal choice of task size depends on the tradeoff between communication costs and the load balance across processors. We follow Liu (1986) in identifying three potential levels of granularity in a parallel implementation of Cholesky factorization:

1   *fine-grain*, in which each task consists of only one or two floating point operations, such as a multiply–add pair,
2   *medium-grain*, in which each task is a single column operation, such as cmod or cdiv,
3   *large-grain*, in which each task is the computation of an entire group of columns in a subtree of the elimination tree.

Fine-grain parallellism, at the level of individual floating point operations, is available in either the dense or sparse case. It can be exploited effectively by a vector processing unit or a systolic array, but would incur far too much communication overhead to be exploited profitably on most current generation parallel computers. In particular, the communication latency of these machines is too great for such frequent communication of small messages to be feasible.

Medium-grain parallellism, at the level of operations on entire columns, is also available in either the dense or the sparse case. This level of granularity accounts for essentially all of the parallel speedup in dense factorization on current generation parallel machines, and it is an extremely important source of parallellism for sparse factorization as well. This parallellism is due primarily to the fact that many cmod operations can be computed simultaneously by different processors. For many problems, such a level of granularity provides a good balance between communication and computation, but scaling up to very large problems and/or very large numbers of processors may necessitate that the tasks be further broken up into chunks based on a two-dimensional partitioning of the columns. One must keep in mind, however, that in the sparse case an entire column operation may require only a few floating point operations involving the sparsely populated nonzero elements in the column. For a matrix of order n having a planar graph, for example, the largest embedded dense submatrix to be factored is roughly of order $\sqrt{n}$, and thus a sparse problem must be extremely large before a two-dimensional partitioning becomes essential.

Large-grain parallellism, at the level of subtrees of the elimination tree, is available only in the sparse case. If T_i and T_j are disjoint subtrees of the elimination tree, with neither root node a descendant of the other, then all of the columns corresponding to nodes in T_i can be computed completely independently of the columns corresponding to nodes in T_j, and vice versa, and hence these computations can be done simultaneously by separate processors with no communication between them. For example, each leaf node of the elimination tree corresponds to a column of L that depends on no prior columns, and hence all of the leaf node columns can be completed immediately merely by performing the corresponding cdiv operation on each of them. Furthermore, all such cdiv operations can be performed simultaneously by separate processors (assuming enough processors are available). By contrast, in the dense case all cdiv operations must be performed sequentially (at least at this level of granularity), since there is never more than one leaf node at any given time.

We see from this discussion that the elimination tree serves to characterize the parallellism that is unique to sparse factorization. In particular, the height of the elimination tree gives a rough measure of the parallel computation time, and the width of the elimination tree gives a rough measure of the degree or multiplicity of large-grain parallellism. These measures are only very rough, however, since the medium level parallellism also plays a major role in determining overall performance. Still, we can see that short, bushy elimination trees are more advantageous than tall, slender ones in terms of

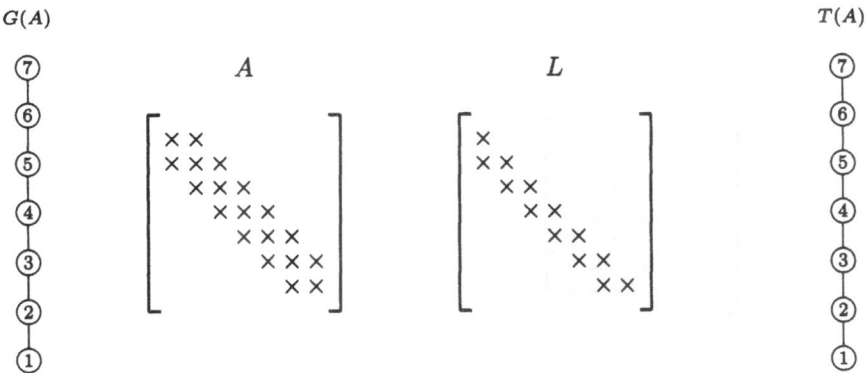

Fig. 10. One-dimensional grid and corresponding tridiagonal matrix (left), with Cholesky factor and elimination tree (right).

the large-grain parallellism available. And just as the fill in the Cholesky factor is very sensitive to the ordering of the matrix, so is the structure of the elimination tree. This suggests that we should choose an ordering to enhance parallellism, and indeed this is possible (see, e.g., Jess and Kees (1982), Lewis, Peyton and Pothen (1989), Liu (1989)), but such an objective may conflict to some degree with preservation of sparsity. Roughly speaking, sparsity and parallellism are largely compatible, since the large-grain parallellism is due to sparsity in the first place. However, these two criteria are by no means coincident, as we will see by example below.

We now illustrate these concepts using a series of simple examples. Figure 10 shows a small one-dimensional mesh with a 'natural' ordering of the nodes, the nonzero patterns of the corresponding tridiagonal matrix A and its Cholesky factor L, and the resulting elimination tree $T(A)$. On the positive side, the Cholesky factor suffers no fill at all and the total work required for the factorization is minimal. However, we see that the elimination tree is simply a chain, and therefore there is no large-grain parallellism available. Each column of L depends on the immediately preceding one, and thus they must be computed sequentially. This behaviour is typical of orderings that minimize the bandwidth of a sparse matrix: they tend to inhibit rather than enhance large-grain parallellism in the factorization. (As previously discussed in Section 4.4, there is in fact little parallellism of any kind to be exploited in solving a tridiagonal system in this natural order. The cmod operations involve only a couple of flops each, so that even the 'medium-grain' tasks are actually rather small in this case.)

Figure 11 shows the same one-dimensional mesh with the nodes reordered by a minimum degree algorithm. Minimum degree is the most effective general purpose heuristic known for limiting fill in sparse factorization (George and Liu, 1989). In its simplest form, this algorithm begins by selecting a

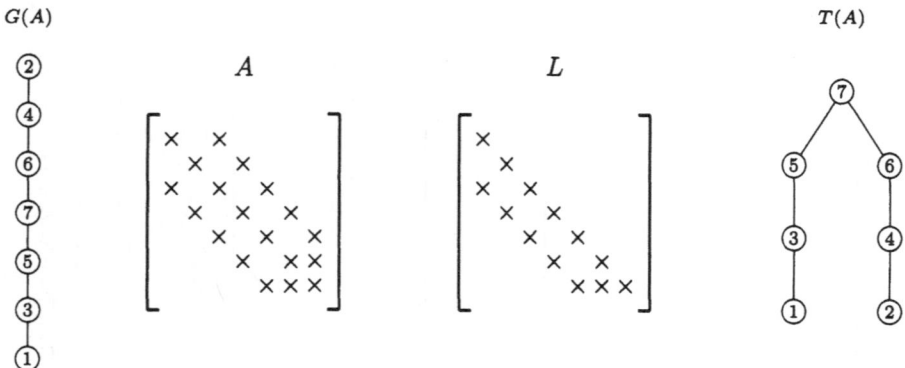

Fig. 11. Graph and matrix reordered by minimum degree (left), with
corresponding Cholesky factor and elimination tree (right).

node of minimum degree (i.e. one having fewest incident edges) in $G(A)$
and numbering it first. The selected node is then deleted and new edges are
added, if necessary, to make its former neighbours into a clique. The process
is then repeated on the updated graph, and so on, until all nodes have been
numbered. We see in Figure 11 that L suffers no fill in the new ordering, and
the elimination tree now shows some large-grain parallellism. In particular,
columns 1 and 2 can be computed simultaneously, then columns 3 and 4,
and so on. This twofold parallellism reduces the tree height (roughly the
parallel completion time) by approximately a factor of two.

At any stage of the minimum degree algorithm, there may be more than
one node with the same minimum degree, and the quality of the ordering produced may be affected by the tie breaking strategy which is used.
In the example of Figure 11, we have deliberately broken ties in the most
favourable way (with respect to parallellism); the least favourable tie breaking would have reproduced the original ordering of Figure 10, resulting in
no parallellism. Breaking ties randomly (which in general is about all one
can do) could produce anything in between these two extremes, yielding
an elimination tree that reveals some large-grain parallellism, but which is
taller and less well balanced than our example in Figure 11. Again, this is
typical of minimum degree orderings. In view of this property, Liu (1989)
has developed an interesting strategy for the further reordering of an initial
minimum degree ordering that preserves fill while reducing the height of the
elimination tree.

Figure 12 shows the same mesh again, this time ordered by nested dissection, a divide-and-conquer strategy (George, 1973). Let S be a set of nodes,
called a *separator*, whose removal, along with all edges incident upon nodes
in S, disconnects $G(A)$ into two remaining subgraphs. The nodes in each
of the two remaining subgraphs are numbered contiguously and the nodes

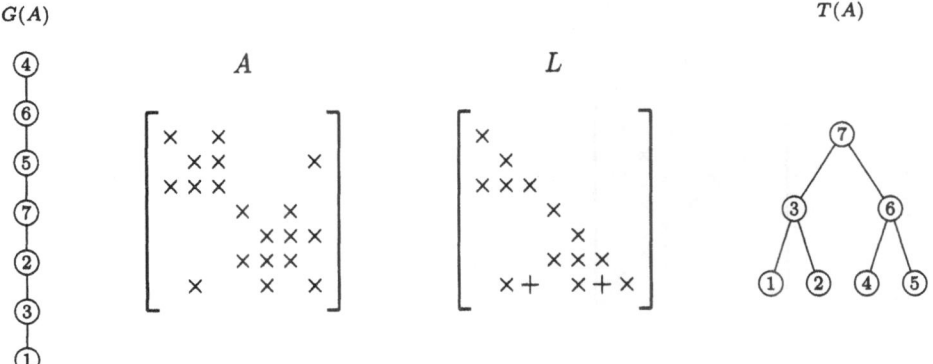

Fig. 12. Graph and matrix reordered by nested dissection (left), with corresponding Cholesky factor and elimination tree (right).

in the separator S are numbered last. This procedure is then applied recursively to split each of the remaining subgraphs, and so on, until all nodes have been numbered. If sufficiently small separators can be found, then nested dissection tends to do a good job of limiting fill, and if the pieces into which the graph is split are of about the same size, then the elimination tree tends to be well balanced. We see in Figure 12 that for our example, with this ordering, the Cholesky factor L suffers fill in two matrix entries (indicated by $+$), but the elimination tree now shows a fourfold large-grain parallellism, and its height has been reduced further. This behaviour is again typical of nested dissection orderings: they tend to be somewhat less successful at limiting fill than minimum degree, but their divide-and-conquer nature tends to identify parallellism more systematically and produce better balanced elimination trees.

Finally, Figure 13 shows the same problem reordered by odd–even reduction. This is not a general purpose strategy for sparse matrices, but it is often used to enhance parallellism in tridiagonal and related systems, so we illustrate it for the sake of comparison with more general purpose methods. In odd–even reduction (see, e.g., Duff *et al.* (1986)), odd node numbers come before even node numbers, and then this same renumbering is applied recursively within each resulting subset, and so on until all nodes are numbered. Although the resulting nonzero pattern of A looks superficially different, we can see from the elimination tree that this method is essentially equivalent to nested dissection for this type of problem.

7.5. Parallel algorithms for sparse factorization

Having developed some understanding of the sources of parallellism in sparse Cholesky factorization, we now consider some algorithms for exploiting it.

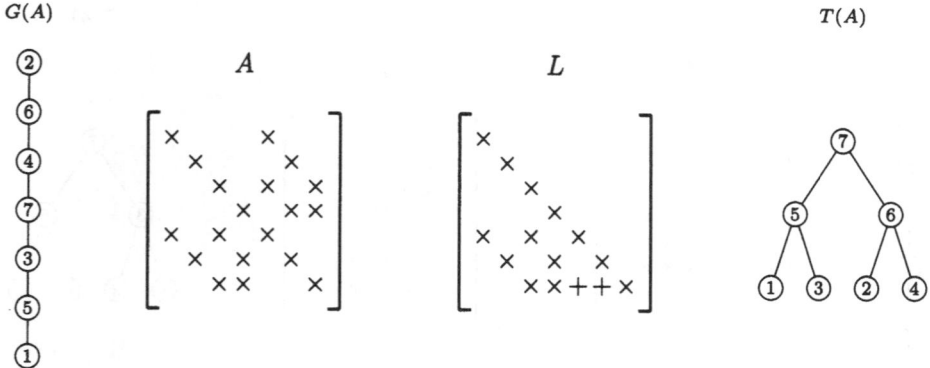

Fig. 13. Graph and matrix reordered by odd–even reduction (left), with corresponding Cholesky factor and elimination tree (right).

In designing any parallel algorithm, one of the most important decisions is how tasks are to be assigned to processors. In a shared memory parallel architecture, the tasks can easily be assigned to processors dynamically by maintaining a common pool of tasks from which available processors claim work to do. This approach has the additional advantage of providing automatic load balancing to whatever degree is permitted by the chosen task granularity. An implementation of this approach for parallel sparse factorization is given in George, Heath, Liu and Ng (1986).

In a distributed memory environment, communication costs often prohibit dynamic task assignment or load balancing, and thus we seek a static mapping of tasks to processors. In the case of column-oriented factorization algorithms, this amounts to assigning the columns of the matrix to processors according to some mapping procedure determined in advance. Such an assignment could be made using the block or wrap mappings, or combinations thereof, often used for dense matrices. However, such simple mappings risk wasting much of the large-grain parallellism identified by means of the elimination tree, and may also incur unnecessary communication. For example, the leaf nodes of the elimination tree can be processed in parallel if they are assigned to different processors, but the latter is not necessarily ensured by a simple block or wrap mapping.

A better approach for sparse factorization is to preserve locality by assigning subtrees of the elimination tree to contiguous subsets of neighbouring processors. A good example of this technique is the 'subtree-to-subcube' mapping often used with hypercube multicomputers (George, Heath, Liu and Ng, 1989). Of course, the same idea applies to other network topologies, such as submeshes of a larger mesh. We will assume that some such mapping is used, and we will comment further on its implications later. Whatever the mapping, we will denote the processor containing column j

by map[j], or, more generally, if J is a set of column numbers, map[J] will denote the set of processors containing the given columns.

One of the earliest and simplest parallel algorithms for sparse Cholesky factorization is the following version of submatrix-Cholesky (George, Heath, Liu and Ng, 1988). Algorithm 22 runs on each processor, with each responsible for its own subset, *mycols*, of columns.

Algorithm 22 Distributed fan-out sparse Cholesky factorization
> for $j \in$ *mycols*
>> if j is a leaf node in $T(A)$
>>> cdiv(j)
>>> send L_{*j} to processors in map(Struct(L_{*j}))
>>> *mycols* = *mycols* $- \{j\}$
>> while *mycols* $\neq \emptyset$
>>> receive any column of L, say L_{*k}
>>> for $j \in$ *mycols* $\cap$ Struct(L_{*k})
>>>> cmod(j, k)
>>>> if column j requires no more cmods
>>>>> cdiv(j)
>>>>> send L_{*j} to processors in map(Struct(L_{*j}))
>>>>> *mycols* = *mycols* $- \{j\}$

In Algorithm 22, any processor that owns a column of L corresponding to a leaf node of the elimination tree can complete it immediately merely by performing the necessary cdiv operation, since such a column depends on no prior columns. The resulting factor columns are then broadcast (fanned-out) to all other processors that will need them to update columns that they own. The remainder of the algorithm is then driven by the arrival of factor columns, as each processor goes into a loop in which it receives and applies successive factor columns, in whatever order they may arrive, to whatever columns remain to be processed. When the modifications of a given column have been completed, then the cdiv operation is done, the resulting factor column is broadcast as before, and the process continues until all columns of L have been computed.

Algorithm 22 potentially exploits both the large-grain parallellism characterized by concurrent cdivs and the medium-grain parallellism characterized by concurrent cmods, but this data-driven approach also has a number of drawbacks that severely limit its efficiency. In particular, performing the column updates one at a time by the receiving processors results in unnecessarily high communication frequency and volume, and in a relatively inefficient computational inner loop. The communication requirements can be reduced by careful mapping and by aggregating updating information over subtrees (see, e.g., George, Liu and Ng (1989), Mu and Rice (1992), Zmijewski (1989)), but even with this improvement, the fan-out algorithm is usually not competitive with other algorithms presented later. The short-comings of the fan-out algorithm motivated the formulation of the following fan-in algorithm for sparse factorization, which is a parallel implementation of column-Cholesky (Ashcraft, Eisenstat and Liu, 1990):

Algorithm 23 Distributed fan-in sparse Cholesky factorization

> for $j = 1$, n
>> if $j \in mycols$ or $mycols \cap \text{Struct}(L_{j*}) \neq \emptyset$
>>> $u = 0$
>>> for $k \in mycols \cap \text{Struct}(L_{j*})$
>>>> $u = u + \ell_{jk} * L_{*k}$ {aggregate column update s}
>>> if $j \in mycols$
>>>> incorporate u into the factor column j
>>>> while any aggregated update column for
>>>>> column j remains, receive in u another
>>>>> aggregated update column for column j, and
>>>>> incorporate it into the factor column j
>>>> cdiv(j)
>> else
>>> send u to processor map[j]

Algorithm 23 takes a demand-driven approach: the updates for a given column j are not computed until needed to complete that column, and they are computed by the sending processors rather than the receiving processor. As a result, all of a given processor's contributions to the updating of the column in question can be combined into a single aggregate update column, which is then transmitted in a single message to the processor containing the target column. This approach not only decreases communication frequency and volume, but it also facilitates a more efficient computational inner loop. In particular, no communication is required to complete the columns corres-ponding to any subtree that is assigned entirely to a single processor. Thus, with an appropriate locality-preserving and load-balanced subtree mapping, Algorithm 23 has a perfectly parallel, communication-free initial phase that is followed by a second phase in which communication takes place over in-creasingly larger subsets of processors as the computation proceeds up the elimination tree, encountering larger subtrees. This perfectly parallel phase, which is due entirely to sparsity, tends to constitute a larger proportion of the overall computation as the size of the problem grows for a fixed number of processors, and thus the algorithm enjoys relatively high efficiencies for sufficiently large problems.

In the fan-out and fan-in factorization algorithms, the necessary infor-mation flow between columns is mediated by factor columns or aggregate update columns, respectively. Another alternative is a *multi-frontal* method, in which update information is mediated through a series of front matrices. In a sense, this represents an intermediate strategy, since the effect of each factor column is incorporated immediately into a front matrix, but its eventual incorporation into the ultimate target column is delayed until actually needed. The principal computational advantage of multi-frontal methods is that the frontal matrices are treated as dense matrices, and hence updating operations on them are much more efficient than the corresponding operations on sparse data structures that require indirect addressing. Unfortunately, although the updating computations employ simple dense arrays, the overall management of the front matrices is relatively complicated. As a consequence, multi-frontal methods are difficult to specify succinctly, so we will not attempt to do so here, but note that multi-frontal methods have been implemented for both shared-memory (e.g., Benner, Montry and Weigand (1987), Duff (1986)) and distributed-memory (e.g., Gilbert and Schreiber (1992), Lucas, Blank and Tieman (1987)) parallel computers, and are among the most effective methods known for sparse factorization in all types of computational environments. For a unified description and comparison of parallel fan-in, fan-out and multi-frontal methods, see Ashcraft, Eisenstat, Liu and Sherman (1990).

In this brief section on parallel direct methods for sparse systems, we have concentrated on numeric Cholesky factorization for SPD matrices. We have omitted many other aspects of the computation, even for the SPD case: computing the ordering in parallel, symbolic factorization and triangular solution. More generally, we have omitted any discussion of LU factorization for general sparse square matrices or QR factorization for sparse rectangular matrices. Instead we have concentrated on identifying the major features that distinguish parallel sparse factorization from the dense case and examining the performance implications of those differences.

8. Iterative methods for linear systems

In this section we discuss parallel aspects of iterative methods for solving large linear systems. For a good mathematical introduction to a class of successful and popular methods, the so-called Krylov subspace methods, see Freund, Golub and Nachtigal (1992). There are many such methods and new ones are frequently proposed. Fortunately, they share enough properties that to understand how to implement them in parallel it suffices to examine carefully just a few.

For the purposes of parallel implementation there are two classes of methods: those with short recurrences, i.e. methods that maintain only a very limited number of search direction vectors, and those with long recurrences. The first class includes CG (Conjugate Gradients), CR (Conjugate Residuals), Bi-CG, CGS (CG squared), QMR (Quasi Minimum Residual),

GMRES(m) for small m (Generalized Minimum Residual), truncated OR-THOMIN (Orthogonal Minimum Residual), Chebychev iteration, and so on. We could further distinguish between methods with fixed iteration parameters and methods with dynamical parameters, but we will not do so; the effects of this aspect will be clear from our discussion. As the archetype for this class we will consider CG; the parallel implementation issues for this method apply to most other short recurrence methods. The second class of methods includes GMRES, GMRES(m) with larger m, ORTHOMIN, OR-THODIR (Orthogonal Directions), ORTHORES (Orthogonal Residuals), and EN (Eirola–Nevanlinna's Rank-1 update method). We consider GM-RES in detail, which is a popular method in this class.

This section is organized as follows. In Section 8.1 we will discuss the parallel aspects of important computational kernels in iterative schemes. From the discussions it should be clear how to combine coarse-grained and fine-grained approaches, for example when implementing a method on a parallel machine with vector processors. The implementation for such machines, in particular those with shared memory, is given much attention in Dongarra *et al.* (1991). In Section 8.2, coarse-grained parallel and data-locality issues of CG will be discussed, while in Section 8.3 the same will be done for GMRES.

8.1. Parallelism in the kernels of iterative methods

The basic time-consuming computational kernels of iterative schemes are usually:

1   inner products,
2   vector updates,
3   matrix–vector products, like Ap_i (for some methods also $A^T p_i$),
4   preconditioning (e.g., solve for w in $Kw = r$).

The inner products can be easily parallellized; each processor computes the inner product of two segments of each vector (local inner products or LIPs). On distributed memory machines the LIPs have to be sent to other processors in order to be reduced to the required global inner product. This step requires communication. For shared memory machines the inner products can be computed in parallel without difficulty. If the distributed memory system supports overlap of communication with computation, then we seek opportunities in the algorithm to do so. In the standard formulation of most iterative schemes this is usually a major problem. We will come back to this in the next two sections. Vector updates are trivially parallellizable: each processor updates its 'own' segment. The matrix–vector products are often easily parallellized on shared memory machines by splitting the matrix into strips corresponding to the vector segments. Each processor takes care of the matrix–vector product of one strip.

For distributed memory machines there may be a problem if each processor has only a segment of the vector in its memory. Depending on the bandwidth of the matrix we may need communication for other elements of the vector, which may lead to some communication problems. However, many sparse matrix problems are related to a network in which only the nearby nodes are connected. In such a case it seems natural to subdivide the network, or grid, into suitable blocks and to distribute these blocks over the processors. When computing the Ap_i each processor needs at most the values of p_i at some nodes in the neighbouring blocks. If the number of connections to these neighbouring blocks is small compared with the number of internal nodes, then the communication time can be overlapped with the computational work. For more detailed discussions on the implementation aspects on distributed memory systems, see de Sturler (1991) and Pommerell (1992).

Preconditioning is often the most problematic part in a parallel environment. Incomplete decompositions of A form a popular class of preconditionings in the context of solving discretized PDEs. In this case the preconditioner $K = LU$, where L and U have a sparsity pattern equal or close to the sparsity pattern of the corresponding parts of A (L is lower triangular, U is upper triangular). For details see Golub and Van Loan (1989), Meijerink and van der Vorst (1977) and Meijerink and van der Vorst (1981). Solving $Kw = r$ leads to solving successively $Lz = r$ and $Uw = z$. These triangular solves lead to recurrence relations that are not easily parallellized. We will now discuss a number of approaches to obtain parallellism in the preconditioning part.

1   *Reordering the computations.* Depending on the structure of the matrix a *frontal approach* may lead to successful parallellism. By inspecting the dependency graph one can select those elements that can be computed in parallel. For instance, if a second-order PDE is discretized by the usual five-point star over a rectangular grid, then the triangular solves can be parallellized if the computation is carried out along diagonals of the grid, instead of the usual lexicographical order. For vector computers this leads to a vectorizable preconditioner (see Ashcraft and Grimes (1988), Dongarra *et al.* (1991), van der Vorst (1989a) and (1989b)). For coarse-grained parallellism this approach is insufficient. By a similar approach more parallellism can be obtained in three-dimensional situations: the so-called *hyperplane approach* (Schlichting and van der Vorst, 1989; van der Vorst, 1989a; 1989b). The disadvantage is that the data need to be redistributed over the processors, since the grid points, which correspond to a hyperplane in the grid, are located quite irregularly in the array. For shared memory machines this also leads to reduced performance because of indirect addressing.

In general one concludes that the data dependency approach is not adequate for obtaining a suitable degree of parallellism.

2   *Reordering the unknowns.* One may also use a *colouring scheme* for reordering the unknowns, so that unknowns with the same colour are not explicitly coupled. This means that the triangular solves can be parallellized for each colour. Of course, communication is required for couplings between groups of different colours. Simple colouring schemes, like red-black ordering for the five-point discretized Poisson operator, seem to have a negative effect on the convergence behaviour. Duff and Meurant (1989) have carried out numerical experiments for many different orderings, which show that the numbers of iterations may increase significantly for other than lexicographical ordering. Some modest degree of parallellism can be obtained, however, with so-called incomplete twisted factorizations (Dongarra *et al.*, 1991; van der Vorst, 1987b; van der Vorst, 1989a). Multi-colour schemes with a large number of colours (e.g., 20 to 100) may lead to little or no degradation in convergence behaviour (Doi, 1991), but also to less parallellism. Moreover, the ratio of computation to communication may be more unfavourable.

3   *Forced parallellism.* Parallelism can also be forced by simply neglecting couplings to unknowns residing in other processors. This is like block Jacobi preconditioning, in which the blocks may be decomposed in incomplete form (Seager, 1986). Again, this may not always reduce the overall solution time, since the effects of increased parallellism are more than undone by an increased number of iteration steps. In order to reduce this effect, it is suggested in Radicati di Brozolo and Robert (1989) to construct incomplete decompositions on slightly overlapping domains. This requires communication similar to that of matrix–vector products. In Radicati di Brozolo and Robert (1989) results are reported on a six-processor shared memory system (IBM3090), and speedups close to six have been observed.

The problems with parallellism in the preconditioner have led to searches for other preconditioners. Often simple diagonal scaling is an adequate preconditioner and this is trivially parallellizable. For results on a Connection Machine, see Berryman, Saltz, Gropp and Mirchandaney (1989). Often this approach leads to a significant increase in iteration steps. Still another approach is to use polynomial preconditioning: $w = p_j(A)r$, i.e. $K^{-1} = p_j(A)$, for some suitable jth degree polynomial. This preconditioner can be implemented by forming only matrix–vector products, which, depending on the structure of A, are easier to parallellize (Saad, 1985b). For p_j one often selects a Chebychev polynomial, which requires some information on the spectrum of A.

Finally we point out the possibility of using the truncated Neumann series for the approximate inverse of A, or parts of L and U. Madsen *et al.* (1976) discuss approximate inversion of A, which from the implementation point of view is equivalent to polynomial preconditioning. In van der Vorst (1982) the use of truncated Neumann series for removing some of the recurrences in the triangular solves is discussed. This approach leads to only fine-grained parallellism (vectorization).

8.2. Parallelism and data locality in preconditioned CG

To use CG to solve $Ax = b$, A must be symmetric and positive definite. In other short recurrence methods, other properties of A may be required or desirable, but we will not exploit these properties explicitly here.

Most often, CG is used in combination with some kind of preconditioning (Freund *et al.*, 1991; Golub and Van Loan, 1989; Hestenes and Stiefel, 1954). This means that the matrix A is implicitly multiplied by an approximation K^{-1} of A^{-1}. Usually, K is constructed to be an approximation of A, and so that $Ky = z$ is easier to solve than $Ax = b$. Unfortunately, a popular class of preconditioners, those based on incomplete factorizations of A, are hard to parallellize. We have discussed some efforts to obtain more parallellism in the preconditioner in Section 8.1. Here we will assume the preconditioner is chosen such that the time to solve $Ky = z$ in parallel is comparable with the time to compute Ap. For CG we also require that the preconditioner K be symmetric positive definite. We exploit this to implement the preconditioner more efficiently.

The preconditioned CG algorithm is as follows.

Algorithm 24 Preconditioned conjugate gradients – variant 1

> $x_0 =$ initial guess; $r_0 = b - Ax_0$;
> $p_{-1} = 0; \beta_{-1} = 0$;
> Solve for w_0 in $Kw_0 = r_0$;
> $\rho_0 = (r_0, w_0)$
> for $i = 0, 1, 2, \ldots.$
> > $p_i = w_i + \beta_{i-1} p_{i-1}$;
> > $q_i = Ap_i$;
> > $\alpha_i = \rho_i / (p_i, q_i)$
> > $x_{i+1} = x_i + \alpha_i p_i$
> > $r_{i+1} = r_i - \alpha_i q_i$;
> > if x_{i+1} accurate enough then quit;
> > Solve for w_{i+1} in $Kw_{i+1} = r_{i+1}$;
> > $\rho_{i+1} = (r_{i+1}, w_{i+1})$;
> > $\beta_i = \rho_{i+1} / \rho_i$;
> end;

If A or K is not very sparse, most work is done in multiplying $q_i = Ap_i$ or solving $Kw_{i+1} = r_{i+1}$, and this is where parallellism is most beneficial. It is also completely dependent on the structures of A and K.

Now we consider parallellizing the rest of the algorithm. Note that updating x_{i+1} and r_{i+1} can only begin after completing the inner product for α_i. Since on a distributed memory machine communication is needed for the inner product, we cannot overlap this communication with useful computation. The same observation applies to updating p_i, which can only begin after completing the inner product for β_{i-1}. Apart from computing Ap_i and solving $Kw_{i+1} = r_{i+1}$, we need to load 7 vectors for 10 vector floating point operations. This means that for this part of the computation only 10/7 floating point operation can be carried out per memory reference on average.

Several authors (Chronopoulos and Gear, 1989; Meurant, 1984a,b; van der Vorst, 1986) have attempted to improve this ratio, and to reduce the number of synchronization points (the points at which computation must wait for communication). In Algorithm 24 there are two such synchronization points, namely the computation of both inner products. Meurant (1984a) (see also Saad (1985b)) has proposed a variant in which there is only one synchronization point, however at the cost of possibly reduced numerical stability, and one additional inner product. In this scheme the ratio between computations and memory references is about 2. We show here yet another variant, proposed by Chronopoulos and Gear (1989).

Algorithm 25 Preconditioned conjugate gradients – variant 2

$x_0 =$ initial guess; $r_0 = b - Ax_0$;
$q_{-1} = p_{-1} = 0; \beta_{-1} = 0$;
Solve for w_0 in $Kw_0 = r_0$;
$s_0 = Aw_0$;
$\rho_0 = (r_0, w_0); \mu_0 = (s_0, w_0)$;
$\alpha_0 = \rho_0/\mu_0$;
for $i = 0, 1, 2, \ldots$.
    $p_i = w_i + \beta_{i-1}p_{i-1}$;
    $q_i = s_i + \beta_{i-1}q_{i-1}$;
    $x_{i+1} = x_i + \alpha_i p_i$;
    $r_{i+1} = r_i - \alpha_i q_i$;
    if x_{i+1} accurate enough then quit;
    Solve for w_{i+1} in $Kw_{i+1} = r_{i+1}$;
    $s_{i+1} = Aw_{i+1}$;
    $\rho_{i+1} = (r_{i+1}, w_{i+1}); \mu_{i+1} = (s_{i+1}, w_{i+1})$;
    $\beta_i = \rho_{i+1}/\rho_i$;
    $\alpha_{i+1} = \rho_{i+1}/(\mu_{i+1} - \rho_{i+1}\beta_i/\alpha_i)$;
end;

In this scheme all vectors need be loaded only once per pass of the loop, which leads to improved data locality. However, the price is $2n$ extra flops per iteration step. Chronopoulos and Gear (1989) claim the method is stable, based on their numerical experiments. Instead of two synchronization points, as in the standard version of CG, we have now only one such synchronization point, as the next loop can be started only when the inner products at the end of the previous loop have been completed. Another slight advantage is that these inner products can be computed in parallel.

Chronopoulos and Gear (1989) propose to improve further the data locality and parallellism in CG by combining s successive steps. Their algorithm is based upon the following property of CG. The residual vectors $r_0, ..., r_i$ form an orthogonal basis (assuming exact arithmetic) for the Krylov subspace spanned by $r_0, Ar_0, ..., A^{i-1}r_0$. Given r_0 through r_j, the vectors $r_0, r_1, ..., r_j, Ar_j, ..., A^{i-j-1}r_j$ also span this subspace. Chronopoulos and Gear propose to combine s successive steps by generating $r_j, Ar_j, ..., A^{s-1}r_j$ first, and then to orthogonalize and update the current solution with this blockwise extended subspace. Their approach leads to slightly more flops than s successive steps of standard CG, and also one additional matrix-vector product every s steps. The implementation issues for vector register computers and distributed memory machines are discussed in great detail in Chronopoulos (1991).

The main drawback in this approach is potential numerical instability: depending on the spectrum of A, the set $r_j, ..., A^{s-1}r_j$ may converge to a vector in the direction of a dominant eigenvector, or in other words, may become dependent for large values of s. The authors claim success in using this approach without serious stability problems for small values of s. Nevertheless, it seems that s-step CG still has a bad reputation (Saad, 1988) because of these problems. However, a similar approach, suggested by Chronopoulos and Kim (1990) for other processes such as GMRES, seems to be more promising. Several authors have pursued this direction, and we will come back to this in Section 8.3.

We consider another variant of CG, in which we may overlap all communication time with useful computations. This is just a reorganized version of the original CG scheme, and is therefore precisely as stable. The key trick is to delay the updating of the solution vector. Another advantage over the previous scheme is that no additional operations are required. We will assume that the preconditioner K can be written as $K = LL^T$. Furthermore, L has a block structure, corresponding to the grid blocks, so that any communication can again be overlapped with computation.

Algorithm 26 Preconditioned conjugate gradients – variant 3

$x_{-1} = x_0 =$ initial guess; $r_0 = b - Ax_0$;
$p_{-1} = 0; \beta_{-1} = 0; \alpha_{-1} = 0$;

$$s = L^{-1}r_0;$$
$$\rho_0 = (s, s)$$
for $i = 0, 1, 2, \ldots$
$$\begin{aligned}
w_i &= L^{-T}s; & (0) \\
p_i &= w_i + \beta_{i-1}p_{i-1}; & (1) \\
q_i &= Ap_i; & (2) \\
\gamma &= (p_i, q_i); & (3) \\
x_i &= x_{i-1} + \alpha_{i-1}p_{i-1}; & (4) \\
\alpha_i &= \rho_i/\gamma; & (5) \\
r_{i+1} &= r_i - \alpha_i q_i; & (6) \\
s &= L^{-1}r_{i+1}; & (7) \\
\rho_{i+1} &= (s, s); & (8) \\
&\text{if } r_{i+1} \text{ small enough then} & (9) \\
&\quad x_{i+1} = x_i + \alpha_i p_i \\
&\quad \text{quit;} \\
\beta_i &= \rho_{i+1}/\rho_i;
\end{aligned}$$
end;

Under the assumptions that we have made, CG can be efficiently parallellized as follows.

1   All compute intensive operations can be done in parallel. Only operations (2), (3), (7), (8), (9), and (0) require communication. We have assumed that the communication in (2), (7), and (0) can be overlapped with computation.

2   The communication required for the reduction of the inner product in (3) can be overlapped with the update for x_i in (4), (which could in fact have been done in the previous iteration step).

3   The reduction of the inner product in (8) can be overlapped with the computation in (0). Also step (9) usually requires information such as the norm of the residual, which can be overlapped with (0).

4   Steps (1), (2), and (3) can be combined: the computation of a segment of p_i can be followed immediately by the computation of a segment of q_i in (2), and this can be followed by the computation of a part of the inner product in (3). This saves on load operations for segments of p_i and q_i.

5   Depending on the structure of L, the computation of segments of r_{i+1} in (6) can be followed by operations in (7), which can be followed by the computation of parts of the inner product in (8), and the computation of the norm of r_{i+1}, required for (9).

6   The computation of β_i can be done as soon as the computation in (8) has been completed. At that moment, the computation for (1) can be started if the requested parts of w_i have been completed in (0).

7   If no preconditioner is used, then $w_i = r_i$, and steps (7) and (0) are

skipped. Step (8) is replaced by $\rho_{i+1} = (r_{i+1}, r_{i+1})$. Now we need some computation to overlap the communication for this inner product. To this end, one might split the computation in (4) in two parts. The first part would be computed in parallel with (3), and the second part with ρ_{i+1}.

More recent work on removing synchroniation points in CG while retaining numerical stability appears in D'Azevedo and Romine (1992) and Eijkhout (1992).

8.3. Parallelism and data locality in GMRES

GMRES, proposed by Saad and Schultz (1985), is a CG-like method for solving general nonsingular linear systems $Ax = b$. GMRES minimizes the residual over the Krylov subspace $\text{span}[r_0, Ar_0, A^2 r_0, ..., A^i r_0]$, with $r_0 = b - Ax_0$. This requires, as with CG, the construction of an orthogonal basis of this space. Since we do not require A to be symmetric, we need long recurrences: each new vector must be explicitly orthogonalized against all previously generated basis vectors. In its most common form GMRES orthogonalizes using Modified Gram–Schmidt (Golub and Van Loan, 1989). In order to limit memory requirements (since all basis vectors must be stored), GMRES is restarted after each cycle of m iteration steps; this is called GMRES(m). A slightly simplified version of GMRES(m) with preconditioning K is as follows (for details, see Saad and Schultz (1985)):

Algorithm 27 GMRES(m)

x_0 is an initial guess; $r = b - Ax_0$;
for $j = 1, 2, ...$
    Solve for w in $Kw = r$;
    $v_1 = w/\|w\|_2$;
    for $i = 1, 2, ..., m$
        Solve for w in $Kw = Av_i$;
        for $k = 1, ..., i$ orthogonalization of w
           $h_{k,i} = (w, v_k)$;   against vs, by modified
           $w = w - h_{k,i} v_k$;   Gram–Schmidt process
        end k;
        $h_{i+1,i} = \|w\|_2$;
        $v_{i+1} = w/h_{i+1,i}$;
    end i;
    Compute x_m using the $h_{k,i}$ and v_i;
    $r = b - Ax_m$;
    if residual r is small enough then quit
    else ($x_0 := x_m$;);
end j;

Another scheme for GMRES, based upon Householder orthogonalization instead of modified Gram–Schmidt, has been proposed in Walker (1988). For some applications the additional computation required by Householder orthogonalization is compensated by improved numerical properties: the better orthogonality saves iteration steps. In van der Vorst and Vuik (1991) a variant of GMRES is proposed in which the preconditioner itself may be an iterative process, which may help to increase parallel efficiency.

Similar to CG and other iterative schemes, the major computations are matrix–vector computations (with A and K), inner products and vector updates. All of these operations are easily parallellizable, although on distributed memory machines the inner products in the orthogonalization act as synchronization points. In this part of the algorithm, one new vector, $K^{-1}Av_j$, is orthogonalized against the previously built orthogonal set v_1, v_2, ... , v_j. In Algorithm 27, this is done using Level 1 BLAS, which may be quite inefficient. To incorporate Level 2 BLAS we can do either Householder orthogonalization or classical Gram–Schmidt twice (which mitigates classical Gram–Schmidt's potential instability (Saad, 1988)). Both approaches significantly increase the computational work and do not remove the synchronization and data-locality problems completely. Note that we cannot, as in CG, overlap the inner product computation with updating the approximate solution, since in GMRES this updating can be done only after completing a cycle of m orthogonalization steps.

The obvious way to extract more parallellism and data locality is to generate a basis v_1, Av_1, ..., $A^m v_1$ for the Krylov subspace first, and to orthogonalize this set afterwards; this is called m-step GMRES(m) (Chronopoulos and Kim, 1990). This approach does not increase the computational work and, in contrast to CG, the numerical instability due to generating a possibly near-dependent set is not necessarily a drawback. One reason is that error cannot build up as in CG, because the method is restarted every m steps. In any case, the resulting set, after orthogonalization, is the basis of some subspace, and the residual is then minimized over that subspace. If, however, one wants to mimic standard GMRES(m) as closely as possible, one could generate a better (more independent) starting set of basis vectors

$$v_1, \; y_2 = p_1(A)v_1, \ldots, y_{m+1} = p_m(A)v_1,$$

where the p_j are suitable degree j polynomials. Newton polynomials are suggested in Bai, Hu and Reichel (1991), and Chebychev polynomials in de Sturler (1991).

After generating a suitable starting set, we still have to orthogonalize it. In de Sturler (1991) modified Gram–Schmidt is used while avoiding communication times that cannot be overlapped. We outline this approach, since it may be of value for other orthogonalization methods. Given a basis for the Krylov subspace, we orthogonalize by

for $k = 2, ..., m + 1$:
　　/* orthogonalize $y_k, ..., y_{m+1}$ w.r.t. v_{k-1} */
　　for $j = k, ..., m + 1$
　　　　$y_j = y_j - (y_j, v_{k-1})v_{k-1}$
　　$v_k = y_k / \|y_k\|_2$

In order to overlap the communication costs of the inner products, we split the j-loop into two parts. Then for each k we proceed as follows.

1. compute in parallel the local parts of the inner products for the first group
2. assemble the local inner products to global inner products
3. compute in parallel the local parts of the inner products for the second group
4. update y_k; compute the local inner products required for $\|y_k\|_2$
5. assemble the local inner products of the second group to global inner products
6. update the vectors $y_{k+1}, ..., y_{m+1}$
7. compute $v_k = y_k / \|y_k\|_2$

From this scheme it is obvious that if the length of the vector segments per processor are not too small, in principle all communication time can be overlapped by useful computations.

For a 150 processor MEIKO system, configured as a 15×10 torus, de Sturler (1991) reports speedups of about 120 for typical discretized PDE systems with $60,000$ unknowns (i.e. 400 unknowns per processor). For larger systems, the speedup increases to 150 (or more if more processors are involved) as expected. Calvetti *et al.* (1991) report results for an implementation of m-step GMRES, using BLAS2 Householder orthogonalization, for a four-processor IBM 6000 distributed memory system. For larger linear systems, they observed speedups close to 2.5.

9. Iterative methods for eigenproblems

The oldest iterative scheme for determining a few dominant eigenvalues and corresponding eigenvectors of a matrix A is the *power method* (Parlett, 1980):

Algorithm 28 Power method

　　select x_0 with $\|x_0\|_2 = 1$
　　$k = 0$
　　repeat
　　　　$k = k + 1$
　　　　$y_k = Ax_{k-1}$
　　　　$\lambda = \|y_k\|_2$
　　　　$x_k = y_k / \lambda$

until x_k converges

If the eigenvalue of A of maximum modulus is well separated from the others, then x_k converges to the corresponding eigenvector and λ converges to the modulus of the eigenvalue. The power method has been superseded by more efficient techniques. However, the method is still used in the form of *inverse iteration* for the rapid improvement of eigenvalue and eigenvector approximations obtained by other schemes. In inverse iteration, the line '$y_k = Ax_{k-1}$' is replaced by 'Solve for y_k in $Ay_k = x_{k-1}$'. Most of the computational effort will be required by this operation, whose (iterative) solution we discussed in Section 8.

All operations in the power method are easily parallellizable, except possibly for the convergence test. There is only one synchronization point: x_k can be computed only after the reduction operation for λ has completed. This synchronization could be avoided by changing the operation $y_k = Ax_{k-1}$ to $y_k = Ay_{k-1}$ (assuming $y_0 = x_0$). This means λ would change every iteration by a factor which converges to the maximum modulus of the eigenvalues of A, and so risks overflow or underflow after enough steps.

The power method constructs basis vectors x_k for the Krylov subspace determined by x_0 and A. It is faster and more accurate to keep all these vectors and then determine stationary points of the Rayleigh quotient over the Krylov subspace. In order to minimize work and improve numerical stability, we compute an orthonormal basis for the Krylov subspace. This can be done by either short recurrences or long recurrences. The short (three-term) recurrence is known as the *Lanczos method*. When A is symmetric this leads to an algorithm with can efficiently compute many, if not all, eigenvalues and eigenvectors (Parlett, 1980).

In fact, the CG method (and Bi-CG) can be viewed as a solution process on top of Lanczos. The long recursion process is known as *Arnoldi's method* (Arnoldi, 1951), which we have seen already as the underlying orthogonalization procedure for GMRES. Not surprisingly, a short discussion on parallellizing the Lanczos and Arnoldi methods would have much in common with our earlier discussions of CG and GMRES.

9.1. The Lanczos method

The Lanczos algorithm is described by the following scheme (Parlett, 1980):

Algorithm 29 Lanczos method

select $r_0 \neq 0$; $q_0 = 0$
$k = 0$
repeat
    $k = k + 1$
    $\beta_{k-1} = \|r_{k-1}\|_2$

$$q_k = r_{k-1}/\beta_{k-1}$$
$$u_k = Aq_k$$
$$s_k = u_k - \beta_{k-1}q_{k-1}$$
$$\alpha_k = (q_k, s_k)$$
$$r_k = s_k - \alpha_k q_k$$

until the eigenvalues of T_k converge (see below)

The q_ks can be saved in secondary storage (they are required for back-transformation of the so-called Ritz vectors below).

The α_m and β_m, for $m = 1, 2, ..., k$, form a tridiagonal matrix T_k:

$$T_k = \begin{pmatrix} \alpha_1 & \beta_1 & & & \\ \beta_1 & \alpha_2 & \beta_2 & & \\ & \beta_2 & \cdot & \cdot & \\ & & \cdot & \cdot & \beta_{k-1} \\ & & & \beta_{k-1} & \alpha_k \end{pmatrix}.$$

The eigenvalues and eigenvectors of T_k are called the *Ritz values* and *Ritz vectors*, respectively, of A with respect to the Krylov subspace of dimension k. The Ritz values converge to eigenvalues of A as k increases, and after backtransformation with the q_ms, the corresponding Ritz vectors approximate eigenvectors of A. For improving these approximations one might consider inverse iteration.

The parallel properties of the Lanczos scheme are similar to those of CG. On distributed memory machines there are two synchronization points caused by the reduction operations for β_{k-1} and α_k. The first synchronization point can, at the cost of n additional flops, be removed by delaying the normalization of the n-vector r_{k-1}. This would lead to the following sequence of statements:

$$u_k = Ar_{k-1}; \quad q_k = r_{k-1}/\beta_{k-1}; \quad s_k = u_k/\beta_{k-1} - \beta_{k-1}q_{k-1}.$$

In this approach, the reduction operation for β_{k-1} can be overlapped with computing Ar_{k-1}. Much in the same spirit as the approach suggested for CG by Chronopoulos and Gear (see algorithm 25), the synchronization point caused by the reduction operation for α_k can be removed by computing As_k right after, and in overlap with, this operation. In that case we reconstruct Ar_{k-1} from recurrence relations for r_{k-1}. This increases the operation count by another n flops, and, even more serious, leads to a numerically less stable algorithm. The Ritz values and Ritz vectors can be computed in parallel by techniques discussed in Section 6. For small k there will not be much to do in parallel, but we also need not compute the eigensystem of T_k for each k, nor check convergence for each k. An elegant scheme for tracking convergence of the Ritz values is discussed in Parlett and Reid (1981). If the Ritz value $\theta_j^{(k)}$, i.e. the jth eigenvalue of T_k, is acceptable as an approximation to some eigenvalue of A, then an approximation $e_j^{(k)}$ to the corresponding eigenvector of A is given by

$$e_j^{(k)} = Q_k y_j^{(k)}, \qquad (9.1)$$

where $y_j^{(k)}$ is the jth eigenvector of T_k, and $Q_k = [q_1, q_2, ..., q_k]$. This is easy to parallellize.

As with CG, one may attempt to improve parallellism in Lanczos by combining s steps in the orthogonalization step. However, the eigensystem of T_k is very sensitive to loss of orthogonality in the q_m vectors. For the standard Lanczos method this loss of orthogonality goes hand in hand with the convergence of Ritz values and leads to multiple eigenvalues of T_k (see Paige (1976) and Parlett (1980)), and so can be accounted for, for instance, by selective reorthogonalization, for which the converged Ritz vectors are required (Parlett, 1980). It is as yet unknown how rounding errors will affect the s step approach, and whether it may lead to inaccurate eigenvalue approximations.

9.2. The Arnoldi method

The Arnoldi algorithm is just the orthogonalization scheme we used before in GMRES:

Algorithm 30 Arnoldi's method

> w is an initial vector with $\|w\|_2 \neq 0$;
> $v_1 = w/\|w\|_2$;
> $k = 0$;
> repeat
> > $k = k + 1$;
> > $w = Av_k$;
> > for $i = 1, ..., k$ orthogonalization of w
> > > $h_{i,k} = (w, v_i)$;   against vs, by modified
> > > $w = w - h_{i,k}v_i$;   Gram–Schmidt process
> > end i;
> > $h_{k+1,k} = \|w\|_2$;
> > $v_{k+1} = w/h_{k+1,k}$;
> until the eigenvalues of $\bar{H}_k$ converge (see below)

The elements $h_{i,j}$ computed after k steps build an upper Hessenberg matrix H_k of dimension $(k+1) \times k$. The eigenvalues and eigenvectors of the upper $k \times k$ part $\bar{H}_k$ of H_k are the Ritz values and Ritz vectors of A with respect to the k dimensional Krylov subspace generated by A and w. The Ritz values $\theta_j^{(k)}$ of $\bar{H}_k$ converge to eigenvalues of A, and the corresponding Ritz vectors $y_j^{(k)}$ can then be backtransformed to approximate eigenvectors of e_j of A via

$$e_j = V_k y_j^{(k)},$$

where $V_k = [v_1, v_2, ..., v_k]$. The parallel solution of the eigenproblem for a Hessenberg matrix is far from easy, for a discussion see Section 6. Normally it is assumed that the order n of A is much larger than the number of Arnoldi steps k. In this case it may be acceptable to solve the eigenproblem for H_k on a single processor. In order to limit k, it has been proposed to carry out the Arnoldi process with a fixed small value for k (a few times larger than the desired number of dominant eigenvalues m), and to repeat the process, very much in the same manner as GMRES(k). At each repetition of the process, it is started with a w that is taken as a combination of the Ritz vectors corresponding to the m dominant Ritz values of the previous cycle. Hopefully, the subspace of m dominant Ritz vectors converges to an invariant subspace of dimension m. This process is known as *subspace iteration*, a generalization of the power method, for details see Saad (1980, 1985a). For a description, as well as a discussion of its performance on the Connection Machine, see Petiton (1992).

A different approach for computing an invariant subspace of order m, based on Arnoldi's process, is discussed in Sorensen (1992). Here one starts with m steps of Arnoldi to create an initial approximation of the invariant subspace of dimension m corresponding to m desired eigenvalues, say the m largest eigenvalues in modulus. Then this subspace is repeatedly expanded by p new vectors, using the Arnoldi process, so that the $m + p$ vectors form a basis for a $m + p$ dimensional Krylov subspace. This information is compressed to the first m vectors of the subset, by a QR algorithm that drives the residual in the projected operator to a small value using p shifts (usually the p unwanted Ritz values of the projected operator). If this expansion and compression process is repeated i times, then the computed m dimensional subspace will be a subset of the $m + i \cdot p$ dimensional Krylov subspace one would get without compression. The hope is that by compressing well, the intersection of the desired invariant subspace with the final m dimensional subspace is close to the intersection with the larger $m + i \cdot p$ dimensional subspace. The benefit of this method is in limiting storage and time spent on the projected Hessenberg eigenproblems to depend on $m + p$ rather than $m + i \cdot p$. An advantage over the previous approach (*subspace iteration*) is in the implicit construction of a suitable starting vector for the $m + p$ dimensional Krylov subspace. For a basis of this subspace only p matrix vector products have to be evaluated for each iteration cycle. In these approaches the eigendecomposition of the projected Hessenberg matrices is still the hardest step to parallellize.

We do not know of successful attempts to combine s successive Krylov subspace vectors v, Av, $A^2 v$, ..., $A^{s-1} v$ (as was proposed in combination

with GMRES). In the case of subspace iteration numerical instability may
not be as severe as for the Lanczos process.

REFERENCES

A. Aho, J. Hopcroft and J. Ullman (1974), *The Design and Analysis of Computer Algorithms*, Addison-Wesley (New York).

E. Anderson and J. Dongarra (1990), 'Evaluating block algorithm variants in LA-PACK', Computer Science Dept. Technical Report CS-90-103, University of Tennessee, Knoxville, TN (LAPACK Working Note 19).

E. Anderson, Z. Bai, C. Bischof, J. Demmel, J. Dongarra, J. Du Croz, A. Greenbaum, S. Hammarling, A. McKenney, S. Ostrouchov and D. Sorensen (1992), *LAPACK Users' Guide, Release 1.0*, SIAM (Philadelphia).

ANSI/IEEE, New York, (1985), *IEEE Standard for Binary Floating Point Arithmetic*, Std 754-1985 edition (New York).

W. E. Arnoldi (1951), 'The principle of minimized iteration in the solution of the matrix eigenproblem', *Quart. Appl. Math.* **9**, 17–29.

C. Ashcraft, S.C. Eisenstat and J.W.-H. Liu (1990), 'A fan-in algorithm for distributed sparse numerical factorization', *SIAM J. Sci. Statist. Comput.* **11**, 593–599.

C. Ashcraft, S.C. Eisenstat, J.W.-H. Liu and A.H. Sherman (1990), 'A comparison of three column-based distributed sparse factorization schemes', Technical Report YALEU/DCS/RR-810, Dept of Computer Science, Yale University, New Haven, CT.

C. Ashcraft and R. Grimes (1988), 'On vectorizing incomplete factorizations and SSOR preconditioners', *SIAM J. Sci. Statist. Comput.* **9**, 122–151.

C. Ashcraft, R. Grimes, J. Lewis, B. Peyton and H. Simon (1987), 'Progress in sparse matrix methods for large linear systems on vector supercomputers', *Int. J. Supercomput. Appl.* **1**(4), 10–30.

L. Auslander and A. Tsao (1992), 'On parallelizable eigensolvers', *Adv. Appl. Math.* **13**, 253–261.

I. Babuška (1972), 'Numerical stability in problems of linear algebra', *SIAM J. Numer. Anal.* **9**, 53–77.

Z. Bai and J. Demmel (1989), 'On a block implementation of Hessenberg multishift QR iteration', *Int. J. High Speed Comput.* **1**(1), 97–112 (also LAPACK Working Note 8).

Z. Bai and J. Demmel (1992), 'Design of a parallel nonsymmetric eigenroutine toolbox', Computer Science Dept preprint, University of California, Berkeley, CA.

Z. Bai, D. Hu and L. Reichel (1991), 'A Newton basis GMRES implementation', Technical Report 91-03, University of Kentucky.

D. H. Bailey, K. Lee and H. D. Simon (1991), 'Using Strassen's algorithm to accelerate the solution of linear systems', *J. Supercomput.* **4**, 97–371.

J. Barlow (1991), 'Error analysis of update methods for the symmetric eigenvalue problem', to appear in *SIAM J. Math. Anal. Appl.* (Tech Report CS-91-21, Computer Science Department, Penn State University.)

C. Beattie and D. Fox (1989), 'Localization criteria and containment for Rayleigh quotient iteration', *SIAM J. Math. Anal. Appl.* **10** (1), 80–93.

K. Bell, B. Hatlestad, O. Hansteen and P. Araldsen (1973), *NORSAM—A programming system for the finite element method, User's Manual, Part I, General description.* Institute for Structural Analysis, NTH, N-7034 Trondheim, Norway.

R. Benner, G. Montry and G. Weigand (1987), 'Concurrent multifrontal methods: shared memory, cache, and frontwidth issues', *Int. J. Supercomput. Appl.* **1** (3), 26–44.

H. Berryman, J. Saltz, W. Gropp and R. Mirchandaney (1989), 'Krylov methods preconditioned with incompletely factored matrices on the CM-2', Technical Report 89-54, NASA Langley Research Center, ICASE, Hampton, VA.

D. Bertsekas and J. Tsitsiklis (1989), *Parallel and Distributed Comptutation: Numerical Methods*, Prentice Hall (New York).

C. Bischof (1989), 'Computing the singular value decomposition on a distributed system of vector processors', *Parallel Comput.* **11** 171–186.

C. Bischof and X. Sun (1992), 'A divide and conquer method for tridiagonalizing symmetric matrices with repeated eigenvalues', MCS Report P286-0192, Argonne National Lab.

C. Bischof and P. Tang (1991a), 'Generalized incremental condition estimation', Computer Science Dept, Technical Report CS-91-132, University of Tennessee, Knoxville, TN (LAPACK Working Note 32).

C. Bischof and P. Tang (1991b), 'Robust incremental condition estimation', Computer Science Dept. Technical Report CS-91-133, University of Tennessee, Knoxville, TN (LAPACK Working Note 33).

R. P. Brent (1973), *Algorithms for Minimization Without Derivatives*, Prentice-Hall (New York).

R. Brent and F. Luk (1985), 'The solution of singular value and symmetric eigenvalue problems on multiprocessor arrays', *SIAM J. Sci. Statist. Comput.* **6**, 69–84.

J. Brunet, A. Edelman and J. Mesirov (1990), 'An optimal hypercube direct n-body solver on the connection machine', in: *Proceedings of Supercomputing '90*, IEEE Computer Society Press (New York), 748–752.

D. Calvetti, J. Petersen and L. Reichel (1991), 'A parallel implementation of the GMRES method', Technical Report ICM-9110-6, Institute for Computational Mathematics, Kent, OH.

L. Cannon (1969), 'A cellular computer to implement the Kalman filter algorithm', PhD thesis, Montana State University, Bozeman, MN.

S. C. Chen, D. J. Kuck and A. H. Sameh (1978), 'Practical parallel band triangular system solvers', *ACM Trans. Math. Software* **4**, 270–277.

A. T. Chronopoulos (1991), 'Towards efficient parallel implementation of the CG method applied to a class of block tridiagonal linear systems', in: *Supercomputing '91*, IEEE Computer Society Press (Los Alamitos, CA), 578–587.

A. T. Chronopoulos and C. W. Gear (1989), 's-step iterative methods for symmetric linear systems', *J. Comput. Appl. Math.* **25**, 153–168.

A. T. Chronopoulos and S. K. Kim (1990), 's-Step Orthomin and GMRES implemented on parallel computers', Technical Report 90/43R, UMSI, Minneapolis.

E. Chu (1988a), 'Orthogonal decomposition of dense and sparse matrices on multiprocessors', PhD thesis, University of Waterloo.

M. Chu (1988b), 'A note on the homotopy method for linear algebraic eigenvalue problems', *Lin. Alg. Appl.* **105**, 225–236.

M. Chu, T.-Y. Li and T. Sauer (1988), 'Homotopy method for general λ-matrix problems', *SIAM J. Math. Anal. Appl.* **9** (4), 528–536.

L. Csanky (1977), 'Fast parallel matrix inversion algorithms', *SIAM J. Comput.* **5** 618–623.

J.J.M. Cuppen (1981), 'A divide and conquer method for the symmetric tridiagonal eigenproblem', *Numer. Math.* **36**, 177–195.

G. Davis, R. Funderlic and G. Geist (1987), 'A hypercube implementation of the implicit double shift QR algorithm', in: *Hypercube Multiprocessors 1987*, SIAM (Philadelphia), 619–626.

E. F. D'Azevedo and C. H. Romine (1992), 'Reducing communication costs in the conjugate gradient algorithm on distributed memory multiprocessors', Technical Report ORNL/TM-12192, Oak Ridge National Laboratory.

P. P. N. de Groen (1991), 'Base p-cyclic reduction for tridiagonal systems of equations', *Appl. Numer. Math.* **8** 117–126.

E. de Sturler (1991), 'A parallel restructured version of GMRES(m)', Technical Report 91-85, Delft University of Technology, Delft.

A. Deichmoller (1991), 'Über die Berechnung verallgemeinerter singulärer Werte mittles Jacobi-ähnlicher Verfahren', PhD thesis, Fernuniversität—Hagen, Hagen, Germany.

E. Dekel, D. Nassimi and S. Sahni (1981), 'Parallel matrix and graph algorithms', *SIAM J. Comput.* **10** (4), 657–675.

T. Dekker (1971), 'A floating point technique for extending the available precision', *Numer. Math.* **18** 224–242.

J. Demmel (1987), 'Three methods for refining estimates of invariant subspaces', *Computing* **38**, 43–57.

J. Demmel (1992a), 'Specifications for robust parallel prefix operations', Technical report, Thinking Machines Corp., Cambridge, MA.

J. Demmel (1992b), 'Trading off parallelism and numerical stability', Computer Science Division Technical Report UCB//CSD-92-702, University of California, Berkeley, CA.

J. Demmel and N. J. Higham (1992), 'Stability of block algorithms with fast Level 3 BLAS', *ACM Trans. Math. Soft.* **18** (3), 274–291.

J. Demmel and W. Kahan (1990), 'Accurate singular values of bidiagonal matrices', *SIAM J. Sci. Statist. Comput.* **11** (5), 873–912.

J. Demmel and K. Veselić (1992), 'Jacobi's method is more accurate than QR', *SIAM J. Mat. Anal. Appl.* **13** (4), 1204–1246.

S. Doi (1991), 'On parallelism and convergence of incomplete LU factorizations', *Appl. Numer. Math.* **7**, 417–436.

J. Dongarra and E. Grosse (1987), 'Distribution of mathematical software via electronic mail', *Commun. ACM* **30**(5), 403–407.

J. Dongarra and S. Ostrouchov (1990), 'LAPACK block factorization algorithms on the Intel iPSC/860', Computer Science Dept Technical Report CS-90-115, University of Tennessee, Knoxville, TN (LAPACK Working Note 24).

J. Dongarra and M. Sidani (1991), 'A parallel algorithm for the non-symmetric eigenvalue problem', Computer Science Dept Technical Report CS-91-137, University of Tennessee, Knoxville, TN.

J. Dongarra and D. Sorensen (1987), 'A fully parallel algorithm for the symmetric eigenproblem', *SIAM J. Sci. Statist. Comput.* **8**(2), 139–154.

J. Dongarra and R. van de Geijn (1991), 'Reduction to condensed form for the eigenvalue problem on distributed memory computers', Computer Science Dept Technical Report CS-91-130, University of Tennessee, Knoxville, TN (LAPACK Working Note 30), to appear in *Parallel Comput.*

J. Dongarra and R. van de Geijn (1991), 'Two dimensional basic linear algebra communication subprograms', Computer Science Dept Technical Report CS-91-138, University of Tennessee, Knoxville, TN (LAPACK Working Note 37).

J. Dongarra, J. Bunch, C. Moler and G. W. Stewart (1979), *LINPACK User's Guide*, SIAM (Philadelphia).

J. Dongarra, J. Du Croz, I. Duff and S. Hammarling (1990), 'A set of Level 3 Basic Linear Algebra Subprograms', *ACM Trans. Math. Soft.* **16** (1), 1–17.

J. Dongarra, J. Du Croz, S. Hammarling and Richard J. Hanson (1988), 'An extended set of fortran basic linear algebra subroutines', *ACM Trans. Math. Soft.* **14** (1), 1–17.

J. Dongarra, I. Duff, D. Sorensen and H. van der Vorst (1991), *Solving Linear Systems on Vector and Shared Memory Computers*, SIAM (Philadelphia).

J. Dongarra, G. A. Geist, and C. Romine (1990b), 'Computing the eigenvalues and eigenvectors of a general matrix by reduction to tridiagonal form', Technical Report ORNL/TM-11669, Oak Ridge National Laboratory (to appear in *ACM TOMS*).

J. Dongarra, S. Hammarling and D. Sorensen (1989), 'Block reduction of matrices to condensed forms for eigenvalue computations', *J. Comput. Appl. Math.* **27**, 215–227 (LAPACK Working Note 2).

J. Du Croz, P. J. D. Mayes and G. Radicati di Brozolo (1990), 'Factorizations of band matrices using Level 3 BLAS', Computer Science Dept Technical Report CS-90-109, University of Tennessee, Knoxville, TN (LAPACK Working Note 21).

P. Dubois and G. Rodrigue (1977), 'An analysis of the recursive doubling algorithm', *High Speed Computer and Algorithm Organization*, (D. J. Kuck and A. H. Sameh, eds), Academic Press (New York).

A. Dubrulle (1991), 'The multishift QR algorithm: is it worth the trouble?', Palo Alto Scientific Center Report G320-3558x, IBM Corp., 1530 Page Mill Road, Palo Alto, CA 94304.

I.S. Duff (1986), 'Parallel implementation of multifrontal schemes', *Parallel Comput.* **3**, 193–204.

I.S. Duff and G.A. Meurant (1989), 'The effect of ordering on preconditioned conjugate gradient', *BIT* **29**, 635–657.

I.S. Duff, A.M. Erisman and J.K. Reid (1986), *Direct Methods for Sparse Matrices*, Oxford University Press (Oxford, UK).

P. Eberlein (1962), 'A Jacobi method for the automatic computation of eigenvalues and eigenvectors of an arbitrary matrix', *J. SIAM* **10**, 74–88.

P. Eberlein (1987), 'On the Schur decomposition of a matrix for parallel computation', *IEEE Trans. Comput.* **36**, 167–174.

V. Eijkhout (1992), 'Qualitative properties of the conjugate gradient and Lanczos methods in a matrix framework', Computer Science Dept. Technical Report CS-92-170, University of Tennessee, Knoxville, TN (LAPACK Working Note 51).

S. Eisenstat, M. Heath, C. Henkel and C. Romine (1988), 'Modified cyclic algorithms for solving triangular systems on distributed memory multiprocessors', *SIAM J. Sci. Statist. Comput.* **9**, 589–600.

G. Fox, S. Hiranandani, K. Kennedy, C. Koelbel, U. Kremer, C.-W. Tseng and M.-Y. Wu (1990), 'Fortran D language specification', Computer Science Department Report CRPC-TR90079, Rice University, Houston, TX.

G. Fox, M. Johnson, G. Lyzenga, S. Otto, J. Salmon and D. Walker (1988), *Solving Problems on Concurrent Processors, vol. I*, Prentice Hall (New York).

R. W. Freund, G. H. Golub and N. M. Nachtigal (1992), 'Iterative solution of linear systems', *Acta Numerica* **1**, 57–100.

K. Gallivan, M. Heath, E. Ng, J. Ortega, B. Peyton, R. Plemmons, C. Romine, A. Sameh and R. Voigt (1990), *Parallel Algorithms for Matrix Computations*, SIAM (Philadelphia).

K. Gallivan, W. Jalby, U. Meier and A. Sameh (1988), 'Impact of hierarchical memory systems on linear algebra algorithm design', *Int. J. Supercomput. Appl.* **2**, 12–48.

K. A. Gallivan, M. T. Heath, E. Ng, *et al.* (1990), *Parallel Algorithms for Matrix Computations*, SIAM (Philadelphia).

K. A. Gallivan, R. J. Plemmons and A. H. Sameh (1990), 'Parallel algorithms for dense linear algebra computations', *SIAM Rev.* **32**, 54–135.

B. S. Garbow, J. M. Boyle, J. J. Dongarra and C. B. Moler (1977), Matrix eigensystem routines – EISPACK guide extension, *Lecture Notes in Computer Science, vol. 51*, Springer-Verlag (Berlin).

G. A. Geist (1990), 'Parallel tridiagonalization of a general matrix using distributed memory multiprocessors', in *Proc. Fourth SIAM Conf. on Parallel Processing for Scientific Computing*, SIAM (Philadelphia), 29–35.

G. A. Geist (1991), 'Reduction of a general matrix to tridiagonal form', *SIAM J. Math. Anal. Appl.* **12**(2), 362–373.

G. A. Geist and G. J. Davis (1990), 'Finding eigenvalues and eigenvectors of unsymmetric matrices using a distributed memory multiprocessor', *Parallel Comput.* **13**(2), 199–209.

G. A. Geist, A. Lu and E. Wachspress (1989), 'Stabilized reduction of an arbitrary matrix to tridiagonal form', Technical Report ORNL/TM-11089, Oak Ridge National Laboratory.

M. Gentleman and H. T. Kung (1981), 'Matrix triangularization by systolic arrays', in *Proc. SPIE 298, Real Time Signal Processing*, San Diego, CA, 19–26.

J.A. George (1973), 'Nested dissection of a regular finite element mesh', *SIAM J. Numer. Anal.* **10**, 345–363.

J.A. George and J.W.-H. Liu (1981), *Computer Solution of Large Sparse Positive Definite Systems*, Prentice-Hall (Englewood Cliffs, NJ).

J.A. George and J.W.-H. Liu (1989), 'The evolution of the minimum degree ordering algorithm', *SIAM Rev.* **31**, 1–19.

J.A. George, M.T. Heath, J.W.-H. Liu and E. Ng (1986), 'Solution of sparse positive definite systems on a shared memory multiprocessor', *Int. J. Parallel Progr.* **15**, 309–325.

J.A. George, M.T. Heath, J.W.-H. Liu and E. Ng (1988), 'Sparse Cholesky factorization on a local-memory multiprocessor', *SIAM J. Sci. Statist. Comput.* **9**, 327–340.

J.A. George, M.T. Heath, J.W.-H. Liu and E. Ng (1989), 'Solution of sparse positive definite systems on a hypercube', *J. Comput. Appl. Math.* **27**, 129–156.

J.A. George, J.W.-H. Liu and E. Ng (1989), 'Communication results for parallel sparse Cholesky factorization on a hypercube', *Parallel Comput.* **10**, 287–298.

J. Gilbert and R. Schreiber (1992), 'Highly parallel sparse cholesky factorization', *SIAM J. Sci. Statist. Comput.* **13**, 1151–1172.

G. Golub and C. Van Loan (1989), *Matrix Computations*, 2nd ed, Johns Hopkins University Press (Baltimore, MD).

A. Gottlieb and G. Almasi (1989), *Highly Parallel Computing*, Benjamin Cummings (Redwood City, CA).

M. Gu and S. Eisenstat (1992), 'A stable and efficient algorithm for the rank-1 modification of the symmetric eigenproblem', Computer Science Dept Report YALEU/DCS/RR-916, Yale University.

V. Hari and K. Veselić (1987), 'On Jacobi methods for singular value decompositions', *SIAM J. Sci. Statist. Comput.* **8**, 741–754.

M.T. Heath and C. Romine (1988), 'Parallel solution of triangular systems on distributed memory multiprocessors', *SIAM J. Sci. Statist. Comput.* **9**, 558–588.

M.T. Heath, E. Ng and B. Peyton (1991), 'Parallel algorithms for sparse linear systems', *SIAM Rev.* **33**, 420–460.

D. Heller (1978), 'Some aspects of the cyclic reduction algorithm for block tridiagonal linear systems', *SIAM J. Numer. Anal.* **13**, 484–496.

M. R. Hestenes and E. Stiefel (1954), 'Methods of conjugate gradients for solving linear systems', *J. Res. Natl Bur. Stand.* **49**, 409–436.

High Peformance Fortran (1991), documentation available via anonymous ftp from titan.cs.rice.edu in directory public/HPFF.

N. J. Higham (1990), 'Exploiting fast matrix multiplication within the Level 3 BLAS', *ACM Trans. Math. Soft.* **16**, 352–368.

C. T. Ho (1990), Optimal communication primitives and graph embeddings on hypercubes, PhD thesis, Yale University.

C. T. Ho, S. L. Johnsson and A. Edelman (1991), 'Matrix multiplication on hypercubes using full bandwidth and constant storage', in *The Sixth Distributed Memory Computing Conf. Proc.*, IEEE Computer Society Press (New York), 447–451.

X. Hong and H. T. Kung (1981), 'I/O complexity: the red blue pebble game', in *Proc. 13th Symposium on the Theory of Computing*, ACM (New York), 326–334.

J. Howland (1983), 'The sign matrix and the separation of matrix eigenvalues', *Lin. Alg. Appl.* **49**, 221–232.

I. Ipsen and E. Jessup (1990), 'Solving the symmetric tridiagonal eigenvalue problem on the hypercube', *SIAM J. Sci. Statist. Comput.* **11**(2), 203–230.

B.M. Irons (1970), 'A frontal solution program for finite element analysis', *Int. J. Numer. Math. Engrg* **2**, 5–32.

J.A.G. Jess and H.G.M. Kees (1982), 'A data structure for parallel L/U decomposition', *IEEE Trans. Comput.* **C-31**, 231–239.

E. Jessup (1991), 'A case against a divide and conquer approach to the nonsymmetric eigenproblem', Technical Report ORNL/TM-11903, Oak Ridge National Laboratory.

E. Jessup and I. Ipsen (1992), 'Improving the accuracy of inverse iteration', *SIAM J. Sci. Statist. Comput.* **13**(2), 550–572.

E. Jessup and D. Sorensen (1989), 'A divide and conquer algorithm for computing the singular value decomposition of a matrix', in *Proc. Third SIAM Conf. on Parallel Processing for Scientific Computing* SIAM (Philadelphia), 61–66. SIAM.

S. L. Johnsson (1987), 'Communication efficient basic linear algebra computations on hypercube architecture', *J. Parallel Distr. Comput.* **4**, 133–172.

S. L. Johnsson (1990), private communication.

S. L. Johnsson and C. T. Ho (1989), 'Matrix multiplication on Boolean cubes using generic communication primitives', in *Parallel Processing and Medium Scale Multiprocessors*, (A. Wouk, ed.), SIAM (Philadelphia), 108–156.

W. Kahan (1968), 'Accurate eigenvalues of a symmetric tridiagonal matrix', Computer Science Dept Technical Report CS41, Stanford University, Stanford, CA, July 1966 (revised June 1968).

R. Karp and V. Ramachandran (1990), 'Parallel algorithms for shared memory machines', in *Handbook of Theoretical Computer Science, vol. A: Algorithms and Complexity*, (J. van Leeuwen, ed.), Elsevier and MIT Press (New York), 869–941.

C. Kenney and A. Laub (1991), 'Rational iteration methods for the matrix sign function', *SIAM J. Math. Anal. Appl.* **21**, 487–494.

A. S. Krishnakumar and Morf (1986), 'Eigenvalues of a symmetric tridiagonal matrix: a divide and conquer approach', *Numer. Math.* **48**, 349–368.

D. Kuck and A. Sameh (1977), 'A parallel QR algorithm for symmetric tridiagonal matrices', *IEEE Trans. Comput.* **C-26**(2).

H. T. Kung (1974), 'New algorithms and lower bounds for the parallel evaluation of certain rational expressions', Technical report, Carnegie Mellon University.

J. J. Lambiotte and R. G. Voigt (1974), 'The solution of tridiagonal linear systems on the CDC-STAR-100 computer', Technical report, ICASE-NASA Langley Research Center, Hampton, VA.

B. Lang (1992), 'Reducing symmetric band matrices to tridiagonal form – a comparison of a new parallel algorithm with two serial algorithms on the iPSC/860', Institut für angewandte mathematik report, Universität Karlsruhe.

C. Lawson, R. Hanson, D. Kincaid and F. Krogh (1979), 'Basic linear algebra subprograms for fortran usage', *ACM Trans. Math. Soft.* **5**, 308–323.

S. Lederman, A. Tsao and T. Turnbull (1992), 'A parallelizable eigensolver for real diagonalizable matrices with real eigenvalues', Report TR-01-042, Supercomputing Research Center, Bowie, MD.

J.G. Lewis, B.W. Peyton and A. Pothen (1989), 'A fast algorithm for reordering sparse matrices for parallel factorization', *SIAM J. Sci. Statist. Comput.* **10**, 1156–1173.

G. Li and T. Coleman (1988), 'A parallel triangular solver on a distributed memory multiprocessor', *SIAM J. Sci. Statist. Comput.* **9**, 485–502.

R. Li (1992), 'Solving the secular equation stably', UC Berkeley Math. Dept Report, in preparation.

T.-Y. Li and Z. Zeng (1992), 'Homotopy-determinant algorithm for solving nonsymmetric eigenvalue problems', *Math. Comput.* to appear.

T.-Y. Li, Z. Zeng and L. Cong (1992), 'Solving eigenvalue problems of nonsymmetric matrices with real homotopies', *SIAM J. Numer. Anal.* **29**(1), 229–248.

T.-Y. Li, H. Zhang and X.-H. Sun (1991), 'Parallel homotopy algorithm for symmetric tridiagonal eigenvalue problem', *SIAM J. Sci. Statist. Comput.* **12**(3), 469–487.

C-C. Lin and E. Zmijewski (1991), 'A parallel algorithm for computing the eigenvalues of an unsymmetric matrix on an SIMD mesh of processors', Department of Computer Science TRCS 91-15, University of California, Santa Barbara, CA.

J.W.-H. Liu (1986), 'Computational models and task scheduling for parallel sparse Cholesky factorization', *Parallel Comput.* **3**, 327–342.

J.W.-H. Liu (1989), 'Reordering sparse matrices for parallel elimination', *Parallel Comput.* **11**, 73–91.

J.W.-H. Liu (1990), 'The role of elimination trees in sparse factorization', *SIAM J. Matrix Anal. Appl.* **11**, 134–172.

S.-S. Lo, B. Phillipe and A. Sameh (1987), 'A multiprocessor algorithm for the symmetric eigenproblem', *SIAM J. Sci. Statist. Comput.* **8**(2), 155–165.

R. Lucas, W. Blank and J. Tieman (1987), 'A parallel solution method for large sparse systems of equations', *IEEE Trans. Comput. Aided Des.* **CAD-6**, 981–991.

F. Luk and H. Park (1989), 'On parallel Jacobi orderings', *SIAM J. Sci. Statist. Comput.* **10**(1), 18–26.

S. C. Ma, M. Patrick and D. Szyld (1989), 'A parallel, hybrid algorithm for the generalized eigenproblem', in *Parallel Processing for Scientific Computing*, (Garry Rodrigue, ed.), SIAM (Philadelphia), ch 16, 82–86.

N. K. Madsen, G. H. Rodrigue and J. I. Karush (1976), 'Matrix multiplication by diagonals on a vector/parallel processor', *Inform. Process. Lett.* **5**, 41–45.

A. N. Malyshev (1991), 'Parallel aspects of some spectral problems in linear algebra', Dept of Numerical Mathematics Report NM-R9113, Centre for Mathematics and Computer Science, Amsterdam.

V. Mehrmann (1991), 'Divide and conquer methods for block tridiagonal systems', Technical Report Bericht Nr. 68, Inst. fuer Geometrie und Prakt. Math., Aachen.

U. Meier (1985), 'A parallel partition method for solving banded systems of linear equations', *Parallel Comput.* **2**, 33–43.

J. A. Meijerink and H. A. van der Vorst (1977), 'An iterative solution method for linear systems of which the coefficient matrix is a symmetric M-matrix', *Math. Comput.* **31**, 148–162.

J. A. Meijerink and H. A. van der Vorst (1981), 'Guidelines for the usage of incomplete decompositions in solving sets of linear equations as they occur in practical problems', *J. Comput. Phys.* **44**, 134–155.

G. Meurant (1984a), 'The block preconditioned conjugate gradient method on vector computers', *BIT* **24**, 623–633.

G. Meurant (1984b), 'Numerical experiments for the preconditioned conjugate gradient method on the CRAY X-MP/2', Technical Report LBL-18023, University of California, Berkeley, CA.

P. H. Michielse and H. A. van der Vorst (1988), 'Data transport in Wang's partition method', *Parallel Comput.* **7**, 87–95.

M. Mu and J.R. Rice (1992), 'A grid based subtree-subcube assignment strategy for solving partial differential equations on hypercubes', *SIAM J. Sci. Statist. Comput.* **13**, 826–839.

J.M. Ortega (1988), *Introduction to Parallel and Vector Solution of Linear Systems*, Plenum Press (New York and London).

M.H.C. Paardekooper (1989), 'A quadratically convergent parallel Jacobi process for diagonally dominant matrices with distinct eigenvalues', *J. Comput. Appl. Math.* **27**, 3–16.

C. C. Paige (1976), 'Error analysis of the Lanczos algorithm for tridiagonalizing a symmetric matrix', *J. Inst. Math. Appl.* **18**, 341–349.

P. Pandey, C. Kenney and A. Laub (1990), 'A parallel algorithm for the matrix sign function', *Int. J. High Speed Comput.* **2**(2), 181–191.

B. N. Parlett (1980), *The Symmetric Eigenvalue Problem*, Prentice-Hall (Englewood Cliffs, NJ).

B. Parlett (1992a), private communication.

B. Parlett (1992b), 'Reduction to tridiagonal form and minimal realizations', *SIAM J. Math. Anal. Appl.* **13**(2), 567–593.

B. Parlett and V. Fernando (1992), 'Accurate singular values and differential QD algorithms', Math. Department PAM-554, University of California, Berkeley, CA.

B. N. Parlett and J. K. Reid (1981), 'Tracking the progress of the Lanczos algorithm for large symmetric eigenproblems', *IMA J. Numer. Anal.* **1**, 135–155.

S. G. Petiton (1992), 'Parallel subspace method for non-Hermitian eigenproblems on the Connection Machine (CM2)', *Appl. Numer. Math.* **10**.

C. Pommerell (1992), Solution of large unsymmetric systems of linear equations, PhD thesis, Swiss Federal Institute of Technology, Zürich.

A. Pothen, S. Jha and U. Vemulapati (1987), 'Orthogonal factorization on a distributed memory multiprocessor', in *Hypercube Multiprocessors 1987*, Knoxville, TN, SIAM (Philadelphia), 587–598.

D. Priest (1991), 'Algorithms for arbitrary precision floating point arithmetic', in *Proc. 10th Symp. Computer Arithmetic, Grenoble, France*, (P. Kornerup and D. Matula, eds), IEEE Computer Society Press (New York), 132–145.

G. Radicati di Brozolo and Y. Robert (1989), 'Parallel conjugate gradient-like algorithms for solving sparse nonsymmetric linear systems on a vector multiprocessor', *Parallel Comput.* **11**, 223–239.

Y. Robert (1990), *The Impact of Vector and Parallel Architectures on the Gaussian Elimination Algorithm*, Wiley (New York).

J. Roberts (1980), 'Linear model reduction and solution of the algebraic Riccati equation', *Int. J. Control* **32**, 677–687.

C. Romine and J. Ortega (1988), 'Parallel solution of triangular systems of equations', *Parallel Comput.* **6**, 109–114.

E. Rothberg and A. Gupta (1989), 'Fast sparse matrix factorization on modern workstations', Technical Report STAN-CS-89-1286, Stanford University, Stanford, California.

Y. Saad (1980), 'Variations on Arnoldi's method for computing eigenelements of large unsymmetric matrices', *Lin. Alg. Appl.* **34**, 269–295.

Y. Saad (1985a), 'Partial eigensolutions of large nonsymmetric matrices', Technical Report, Dept. of Comp. Science, New Haven, CN.

Y. Saad (1985b), 'Practical use of polynomial preconditionings for the conjugate gradient method', *SIAM J. Sci. Statist. Comput.* **6**, 865–881.

Y. Saad (1988), 'Krylov subspace methods on supercomputers', Technical report, RIACS, Moffett Field, CA.

Y. Saad and M. H. Schultz (1985), 'Conjugate Gradient-like algorithms for solving nonsymmetric linear systems', *Math. Comput.* **44**, 417–424.

A. Sameh (1971), '1On Jacobi and Jacobi-like algorithms for a parallel computer', *Math. Comput.* **25**, 579–590.

A. Sameh (1985), 'On some parallel algorithms on a ring of processors', *Comput. Phys. Commun.* **37**, 159–166.

A. Sameh and R. Brent (1977), 'Solving triangular systems on a parallel computer', *SIAM J. Numer. Anal.* **14**, 1101–1113.

J. J. F. M. Schlichting and H. A. van der Vorst (1987), 'Solving bidiagonal systems of linear equations on the CDC CYBER 205', Technical Report NM-R8725, CWI, Amsterdam, the Netherlands.

J. J. F. M. Schlichting and H. A. van der Vorst (1989), 'Solving 3D block bidiagonal linear systems on vector computers', *J. Comput. Appl. Math.* **27**, 323–330.

R. Schreiber and C. Van Loan (1989), 'A storage efficient WY representation for products of Householder transformations', *SIAM J. Sci. Statist. Comput.* **10**, 53–57.

M. K. Seager (1986), 'Parallelizing conjugate gradient for the CRAY X-MP', *Parallel Comput.* **3**, 35–47.

G. Shroff (1991), 'A parallel algorithm for the eigenvalues and eigenvectors of a general complex matrix', *Numer. Math.* **58**, 779–805.

G. Shroff and R. Schreiber (1989), 'On the convergence of the cyclic Jacobi method for parallel block orderings', *SIAM J. Math. Anal. Appl.* **10**, 326–346.

H. Simon (1989), 'Bisection is not optimal on vector processors', *SIAM J. Sci. Statist. Comput.* **10**(1), 205–209.

I. Slapničar (1992), Accurate symmetric eigenreduction by a Jacobi method, PhD thesis, Fernuniversität – Hagen, Hagen, Germany.

B. T. Smith, J. M. Boyle, J. J. Dongarra, B. S. Garbow, Y. Ikebe, V. C. Klema and C. B. Moler (1976), *Matrix Eigensystem Routines – EISPACK Guide: Springer Lecture Notes in Computer Science 6*, Springer-Verlag (Berlin).

D. Sorensen (1992), 'Implicit application of polynomial filters in a k-step Arnoldi method', *SIAM J. Math. Anal. Appl.* **13**(1), 357–385.

D. Sorensen and P. Tang (1991), 'On the orthogonality of eigenvectors computed by divide-and-conquer techniques', *SIAM J. Numer. Anal.* **28**(6), 1752–1775.

G. W. Stewart (1985), 'A Jacobi-like algorithm for computing the Schur decomposition of a non-Hermitian matrix', *SIAM J. Sci. Statist. Comput.* **6**, 853–864.

G. W. Stewart (1987), 'A parallel implementation of the QR algorithm', *Parallel Comput.* **5**, 187–196.

E. Stickel (1991), 'Separating eigenvalues using the matrix sign function', *Lin. Alg. Appl.* **148**, 75–88.

H. S. Stone (1973), 'An efficient parallel algorithm for the solution of a tridiagonal linear system of equations', *J. Assoc. Comput. Mach.* **20**, 27–38.

P. Swarztrauber (1992), 'A parallel algorithm for computing the eigenvalues of a symmetric tridiagonal matrix', *Math. Comput.* to appear.

Thinking Machines Corporation (1987), *Connection Machine Model CM-2 Technical Summary*, IBM.

R. van de Geijn (1987), Implementing the QR algorithm on an array of processors, PhD thesis, University of Maryland, College Park, Computer Science Department Report TR-1897.

R. van de Geijn and D. Hudson (1989), 'Efficient parallel implementation of the nonsymmetric QR algorithm', in *Hypercube Concurrent Computers and Applications*, (J. Gustafson, ed.), ACM (New York).

E. Van de Velde (1992), *Introduction to Concurrent Scientific Computing*, Caltech (Pasadena).

H. A. van der Vorst (1982), 'A vectorizable variant of some ICCG methods', *SIAM J. Sci. Statist. Comput.* **3**, 86–92.

H. A. van der Vorst (1986), 'The performance of Fortran implementations for preconditioned conjugate gradients on vector computers', *Parallel Comput.* **3**, 49–58.

H. A. van der Vorst (1987a), 'Analysis of a parallel solution method for tridiagonal linear systems', *Parallel Comput.* **5**, 303–311.

H. A. van der Vorst (1987b), 'Large tridiagonal and block tridiagonal linear systems on vector and parallel computers', *Parallel Comput.* **5**, 45–54.

H. A. van der Vorst (1989a), 'High performance preconditioning', *SIAM J. Sci. Statist. Comput.* **10**, 1174–1185.

H. A. van der Vorst (1989b), 'ICCG and related methods for 3D problems on vector computers', *Comput. Phys. Commun.* **53**, 223–235.

H. A. van der Vorst (1989c), 'Practical aspects of parallel scientific computing', *Future Gener. Comput. Sys.* **4**, 285–291.

H. A. van der Vorst and K. Dekker (1989), 'Vectorization of linear recurrence relations', *SIAM J. Sci. Statist. Comput.* **10**, 27–35.

H. A. van der Vorst and C. Vuik (1991), 'GMRESR: A family of nested GMRES methods', Technical Report 91-80, Delft University of Technology, Faculty of Technical Mathematics.

K. Veselić (1979), 'A quadratically convergent Jacobi-like method for real matrices with complex conjugate eigenvalues', *Numer. Math.* **33**, 425–435.

H. F. Walker (1988), 'Implementation of the GMRES method using Householder transformations', *SIAM J. Sci. Statist. Comput.* **9**, 152–163.

H. H. Wang (1981), 'A parallel method for tridiagonal equations', *ACM Trans. Math. Soft.* **7**, 170–183.

D. Watkins (1992), 'Shifting strategies for the parallel QR algorithm', Dept of Pure and Applied Mathematics report, Washington State Univ., Pullman, WA.

D. Watkins and L. Elsner (1991), 'Convergence of algorithms of decomposition type for the eigenvalue problem', *Lin. Alg. Appl.* **143**, 19–47.

C.-P. Wen and K. Yelick (1992), 'A survey of message passing systems', Computer science division, University of California, Berkeley, CA.

J. H. Wilkinson (1965), *The Algebraic Eigenvalue Problem*, Oxford University Press (Oxford).

Z. Zeng (1991), Homotopy-determinant algorithm for solving matrix eigenvalue problems and its parallelizations, PhD thesis, Michigan State University.

H. Zima and B. Chapman (1991), *Supercompilers for Parallel and Vectors Computers*, ACM (New York).

E. Zmijewski (1989), 'Limiting communication in parallel sparse Cholesky factorization', Technical Report TRCS89-18, Department of Computer Science, University of California, Santa Barbara, CA.

Acta Numerica (1993), pp. 199–237

Linear stability analysis in the numerical solution of initial value problems

J.L.M. van Dorsselaer

J.F.B.M. Kraaijevanger

M.N. Spijker

Department of Mathematics and Computer Science,
University of Leiden
The Netherlands
E-mail: spijker@rulcri.leidenuniv.nl

This article addresses the general problem of establishing upper bounds for the norms of the nth powers of square matrices. The focus is on upper bounds that grow only moderately (or stay constant) when n, or the order of the matrices, increases. The so-called resolvent condition, occurring in the famous Kreiss matrix theorem, is a classical tool for deriving such bounds.

Recently the classical upper bounds known to be valid under Kreiss's resolvent condition have been improved. Moreover, generalizations of this resolvent condition have been considered so as to widen the range of applications. The main purpose of this article is to review and extend some of these new developments.

The upper bounds for the powers of matrices discussed in this article are intimately connected with the stability analysis of numerical processes for solving initial(-boundary) value problems in ordinary and partial linear differential equations. The article highlights this connection.

The article concludes with numerical illustrations in the solution of a simple initial-boundary value problem for a partial differential equation.

CONTENTS

1. Introduction

1.1. Linear stability analysis

This article deals with step-by-step methods for the numerical solution of linear differential equations. Both initial-boundary value problems in partial differential equations and initial value problems in ordinary differential equations will be included in our considerations.

A crucial question in the step-by-step solution of such problems is whether the method will behave *stably* or not. Here we use the term stability to designate that any numerical errors, introduced at some stage of the calculations, are propagated in a mild fashion – i.e. do not blow up in the subsequent steps of the methods.

Classical tools to assess the stability *a priori*, in the numerical solution of partial differential equations, include *Fourier transformation* and the corresponding famous Von Neumann condition (see the classical work by Richtmyer and Morton (1967)). Further tools of recognized merit for assessing stability, in the solution of ordinary differential equations, comprise the so-called *stability regions* in the complex plane (see e.g. the excellent works by Butcher (1987) and Hairer and Wanner (1991)). During the last 25 years these stability regions have been studied extensively; numerous papers have appeared dealing with the shape and various peculiarities of these regions.

However, these tools are based on the behaviour that the numerical method would have when applied to quite simple test problems. Accordingly, in the case of partial differential equations, Fourier transformation provides a straightforward and reliable stability criterion primarily only for *pure initial value* problems in linear differential equations with *constant* coefficients. Similarly, in the case of ordinary differential equations, stability regions are primarily relevant only to *scalar* equations

$$U'(t) = \lambda U(t) \quad \text{for } t \geq 0, \tag{1.1}$$

with given complex constant λ.

In the pioneering work by F. John (1952) the scope of Fourier transformation had already been widened in that it was used in deriving sufficient conditions for stability in the numerical solution of linear partial differential equations with *variable* coefficients. For subsequent related work, relevant to equations with variable coefficients and to *initial-boundary value* problems, the reader may consult Richtmyer and Morton (1967), Kreiss (1966), Gustafsson, Kreiss and Sundström (1972), Meis and Marcowitz (1981), Thomée (1990), and the references therein.

Clearly, rigorous stability criteria with a wider scope than the simple classical test equations are important – both from a practical and a theoretical point of view. It is equally important to know to what extent stability regions can be relied upon in assessing stability in the numerical solution of differential equations more general than (1.1). The present article reviews and extends some recent developments which are relevant to these two questions. No essential use will be made of Fourier transformation.

1.2. Stability and power boundedness

In this paper we shall deal with numerical processes of the form

$$u_n = Bu_{n-1} + b_n \quad \text{for } n = 1, 2, 3, \ldots, \tag{1.2}$$

with a given square matrix B of order $s \geq 1$ and given s-dimensional vectors b_n. The s-dimensional vectors u_n are computed sequentially from (1.2) starting from a given vector u_0. Processes of the form (1.2) occur in the numerical solution of linear initial value problems that are essentially more general than the simple classical test problems mentioned earlier. The vectors u_n provide numerical approximations to the true solution of the initial (-boundary) value problem under consideration.

As an illustration of (1.2) we consider the initial-boundary value problem

$$u_t(x, t) = a(x)u_{xx}(x, t) + b(x)u_x(x, t) + c(x)u(x, t) + d(x),$$
$$u_x(0, t) = 0, \quad u(1, t) = g(t), \tag{1.3}$$
$$u(x, 0) = f(x),$$

where $0 < x < 1$, $t > 0$ and a, b, c, d, f, g are given functions with

$$a(x) > 0, \quad b(x) \geq 0, \quad c(x) \leq 0.$$

We choose $\Delta t = h > 0$, $\Delta x = 1/s$ and consider the approximation of $u(j/s, nh)$ by quantities u_j^n. The following finite difference scheme has been constructed by standard principles (see Richtmyer and Morton (1967)):

$$h^{-1}(u_j^n - u_j^{n-1}) =$$
$$s^2 a(j/s)\{\theta(u_{j-1}^n - 2u_j^n + u_{j+1}^n) + (1 - \theta)(u_{j-1}^{n-1} - 2u_j^{n-1} + u_{j+1}^{n-1})\}$$
$$+ sb(j/s)\{\theta(u_{j+1}^n - u_j^n) + (1 - \theta)(u_{j+1}^{n-1} - u_j^{n-1})\}$$
$$+ c(j/s)\{\theta u_j^n + (1 - \theta)u_j^{n-1}\} + d(j/s),$$
$$u_{-1}^{n-1} = u_1^{n-1}, \quad u_s^{n-1} = g((n-1)h),$$
$$u_j^0 = f(j/s),$$

where $j = 0, 1, \ldots, s - 1$ and $n = 1, 2, 3, \ldots$. θ denotes a parameter, with $0 \leq \theta \leq 1$, specifying the finite difference method.

Defining vectors u_n by

$$u_n = \begin{pmatrix} u_0^n \\ u_1^n \\ \vdots \\ u_{s-1}^n \end{pmatrix} \simeq \begin{pmatrix} u(0, nh) \\ u(1/s, nh) \\ \vdots \\ u((s-1)/s, nh) \end{pmatrix},$$

one easily verifies that the u_n satisfy a relation of the form (1.2). Here the matrix B satisfies

$$B = (I + (1 - \theta)hA)(I - \theta hA)^{-1}, \qquad (1.4)$$

where I denotes the $s \times s$ identity matrix and $A = (\alpha_{jk})$ an $s \times s$ tridiagonal matrix with its (nonzero) entries given by

$$
\begin{aligned}
\alpha_{j+1,j+1} &= -2s^2 a(j/s) - sb(j/s) + c(j/s) & (0 \le j \le s-1), \\
\alpha_{j+1,j} &= s^2 a(j/s) & (1 \le j \le s-1), \\
\alpha_{j+1,j+2} &= s^2 a(j/s) + sb(j/s) & (1 \le j \le s-2), \\
\alpha_{j+1,j+2} &= 2s^2 a(j/s) + sb(j/s) & (j = 0).
\end{aligned}
\qquad (1.5)
$$

Suppose the numerical calculations based on the general process (1.2) were performed using a perturbed starting vector $\tilde{u}_0$, instead of u_0. We would then obtain approximations that we denote by $\tilde{u}_n$. For instance $\tilde{u}_0$ may stand for a finite-digit representation in a computer of the true u_0, and the $\tilde{u}_n$ then stand for the numerical approximations obtained in the presence of the rounding error $v_0 = \tilde{u}_0 - u_0$.

In the stability analysis of (1.2) the crucial question is whether the difference $v_n = \tilde{u}_n - u_n$ (for $n \ge 1$) can be bounded suitably in terms of the perturbation $v_0 = \tilde{u}_0 - u_0$. Since

$$v_n = \tilde{u}_n - u_n = [B\tilde{u}_{n-1} + b_n] - [Bu_{n-1} + b_n]$$

we have

$$v_n = Bv_{n-1},$$

and consequently

$$v_n = B^n v_0.$$

The last expression makes clear that a central issue in stability analysis is the question of whether given matrices have powers that are uniformly bounded. Accordingly, in the following we focus, for an arbitrary $s \times s$ matrix B, on the stability property

$$\|B^n\| \le M_0 \quad \text{for } n = 0, 1, 2, \ldots, \qquad (1.6)$$

where M_0 is a positive constant. For the time being $\| \cdot \|$ stands for the spectral norm, i.e. for the norm induced by the Euclidean vector norm in $\mathbb{C}^s$ (for an $s \times s$ matrix A we have $\|A\| = \max\{|Ax|/|x| : x \in \mathbb{C}^s$ with $x \ne 0\}$, where $|\cdot|$ denotes the Euclidean norm defined by $|x| = \sqrt{x^*x}$ with x^* standing for the Hermitian adjoint of the column vector $x \in \mathbb{C}^s$).

1.3. Power boundedness and the eigenvalue condition

For any given matrix B one can easily deduce from its Jordan canonical form (see, e.g., Horn and Johnson (1990)) a criterion for the existence of an M_0 with property (1.6). A necessary and sufficient requirement for the existence of such an M_0 is the following *eigenvalue condition*:

All eigenvalues λ of B have modulus $|\lambda| \leq 1$, and the geometric multiplicity of each eigenvalue with modulus 1 is equal to its algebraic multiplicity. (1.7)

However, in the stability analysis of numerical processes one is often interested in property (1.6) for all B belonging to some infinite family $\mathcal{F}$ of matrices. The crucial question then is whether a single finite M_0 exists such that (1.6) holds simultaneously for all B belonging to $\mathcal{F}$. In this situation, (1.7) may only provide a condition that is necessary (and not sufficient) for such an M_0 to exist.

For instance, in the example of Section 1.2 one can only expect great accuracy in the approximations u_j^n to $u(j/s, nh)$ when Δx (and Δt) become very small. Accordingly one is primarily interested in bounds for B^n that are uniformly valid for all B of the form (1.4), (1.5) with arbitrarily small $\Delta x = 1/s$.

An instructive counterexample, illustrating the fact that criterion (1.7) can be misleading for the case of families $\mathcal{F}$, is provided by the $s \times s$ bidiagonal matrices

$$B = \begin{pmatrix} -1/2 & 3/2 & & \mbox{\Large 0} \\ & -1/2 & \ddots & \\ & & \ddots & 3/2 \\ \mbox{\Large 0} & & & -1/2 \end{pmatrix}. \tag{1.8}$$

Matrices of the form (1.8) may be thought of as arising in the numerical solution of the initial-boundary value problem

$$u_t(x, t) = u_x(x, t),$$
$$u(1, t) = 0, \quad u(x, 0) = f(x), \quad \text{where } 0 < x < 1, \ t > 0.$$

Consider the finite difference scheme

$$h^{-1}(u_j^n - u_j^{n-1}) = s(u_{j+1}^{n-1} - u_j^{n-1}),$$
$$u_s^{n-1} = 0, \quad u_j^0 = f(j/s).$$

Here $\Delta t = h > 0$, $\Delta x = 1/s$, and u_j^n approximates $u(j/s, nh)$ for $j = 0, 1, \ldots, s-1$ and $n = 1, 2, 3, \ldots$. Clearly, with the choice $hs = 3/2$ this finite difference scheme can be written in the form (1.2) with B as in (1.8).

For each $s \geq 1$ the matrix B defined by (1.8) satisfies the eigenvalue condition (1.7).

Defining the $s \times s$ shift matrix E by

$$E = \begin{pmatrix} 0 & 1 & & \text{\Large 0} \\ & 0 & \ddots & \\ & & \ddots & 1 \\ \text{\Large 0} & & & 0 \end{pmatrix}, \tag{1.9}$$

we have from (1.8) the expression

$$B = -\tfrac{1}{2}I + \tfrac{3}{2}E,$$

so that

$$B^n = \sum_{k=0}^{n} \binom{n}{k} (-\tfrac{1}{2})^{n-k} (\tfrac{3}{2})^k E^k.$$

Defining x to be the s-dimensional vector whose jth component equals $\xi_j = (-1)^j$, and denoting the jth component of $y = B^n x$ by η_j we easily obtain, from the above expression for B^n,

$$|\eta_j| = \sum_{k=0}^{n} \binom{n}{k} (\tfrac{1}{2})^{n-k} (\tfrac{3}{2})^k = 2^n \quad \text{for } 1 \le j \le s - n.$$

For $s > n$ we thus have

$$\left(\sum_{j=1}^{s} |\eta_j|^2 \right)^{1/2} \ge \sqrt{s - n}\, 2^n,$$

and since

$$\left(\sum_{j=1}^{s} |\xi_j|^2 \right)^{1/2} = \sqrt{s},$$

it follows that $\|B^n\| \ge \sqrt{1 - n/s}\, 2^n$. Denoting the $s \times s$ matrix B by B_s we thus have

$$\|(B_{2n})^n\| \ge 2^{n-1/2} \quad \text{for } n = 1, 2, 3, \dots.$$

Clearly, no M_0 can exist such that (1.6) is valid for all B belonging to $\mathcal{F} = \{B_s : s = 1, 2, 3, \dots\}$.

The fact that the eigenvalue criterion can be a misleading guide to stability was already known in the 1960s, see e.g. Parter (1962). A related, but stronger, necessary requirement for stability is the so-called *Godunov–Ryabenkii stability condition*, a discussion of which can be found, e.g., in Richtmyer and Morton (1967), Morton (1980) and Thomée (1990). The latter condition is not satisfied in example (1.8).

The earlier counterexample is similar to examples in Richtmyer and Morton (1967), Spijker (1985), Kreiss (1990) and Reddy and Trefethen (1992).

Further examples of instability under the eigenvalue condition (1.7) can be found in Griffiths, Christie and Mitchell (1980), Kraaijevanger, Lenferink and Spijker (1987) and Lenferink and Spijker (1991b). The matrices B in these references have s different eigenvalues λ with $|\lambda| < 1$, and occur in the numerical solution of problems of the form (1.3). See Trefethen (1988) and Reddy and Trefethen (1992) for related counterexamples in spectral methods.

We conclude this subsection with the remark that in some special cases the eigenvalue criterion can be reliable. For normal matrices B (i.e. $B^*B = BB^*$ with B^* denoting the Hermitian adjoint of B) the stability estimate (1.6) is valid with $M_0 = 1$ as soon as all eigenvalues of B have a modulus not exceeding 1 (see, e.g., Horn and Johnson (1990)). But, in general, one has to look for conditions different from (1.7).

1.4. Power boundedness and the resolvent condition

The famous *Kreiss matrix theorem* (see, e.g., Kreiss (1962) and Richtmyer and Morton (1967)) relates (1.6) to conditions on B which are more reliable than (1.7). One of these conditions involves the so-called *resolvent* $(\zeta I - B)^{-1}$ of B, and reads as follows:

$$\zeta I - B \text{ is invertible and } \|(\zeta I - B)^{-1}\| \le M_1(|\zeta| - 1)^{-1} \text{ for all complex numbers } \zeta \notin D. \tag{1.10}$$

Here M_1 is a positive constant, I the $s \times s$ identity matrix and

$$D = \{\zeta : \zeta \in \mathbb{C} \text{ and } |\zeta| \le 1\}$$

the closed *unit disk* in the complex plane.

If (1.6) is satisfied, then all eigenvalues of B lie in D, so that for all $\zeta \notin D$ the matrix $\zeta I - B$ is invertible and

$$\|(\zeta I - B)^{-1}\| = \left\|\sum_{k=0}^{\infty} \zeta^{-k-1} B^k\right\| \le \sum_{k=0}^{\infty} |\zeta|^{-k-1} M_0 = M_0(|\zeta| - 1)^{-1}.$$

Hence (1.6) implies (1.10) with $M_1 = M_0$. The Kreiss matrix theorem asserts that, conversely, (1.10) implies (1.6) with M_0 depending only on M_1 and the dimension s, but otherwise independent of the matrix B.

The Kreiss matrix theorem has often been used in the stability analysis of numerical methods for solving initial value problems for partial differential equations. In the classical situation the matrices B are obtained by Fourier transformation of the numerical solution operators, and they stand essentially for the so-called amplification matrices (see, e.g., Richtmyer and Morton (1967)). These matrices are of a *fixed* finite order s. On the other hand, the implication of (1.6) by (1.10) can also be used without Fourier transformation, with B standing for the numerical solution operator in (1.2).

In this situation we may be dealing with a family of matrices B of finite –
but not *uniformly* bounded – orders s. Therefore, of particular interest is
the dependence of the stability constant M_0 in (1.6) on the dimension s (see
also Tadmor (1981)).

Various authors (see, e.g., Morton (1964), Miller and Strang (1966), Tadmor (1981), LeVeque and Trefethen (1984) and Spijker (1991)) have studied
the size of (the optimal) M_0 as a function of M_1 and s, and recently some
open problems in this field were solved. Moreover, the implication of (1.6) by
(1.10) as previously discussed has recently been generalized in several directions. More general norms than the spectral norm have been dealt with and
the resolvent condition (1.10) has been adapted to domains different from
the unit disk D. In the latter case the matrices B in (1.6) and (1.10) are not
the same, but are related to each other by a given (rational) transformation.

1.5. Scope of the rest of the article

In the rest of this article we review and extend some of the recent results just mentioned, and illustrate them in the numerical solution of initial
(-boundary) value problems.

In Section 2 we still deal with resolvent condition (1.10) with respect to
the unit disk D, but we consider general norms on the vector space of all
$s \times s$ matrices. In this situation we focus on the best upper bounds for $\|B^n\|$
that are possible under condition (1.10).

In Section 3.1 we relate estimates like (1.6) more explicitly to the stability analysis of numerical methods for the solution of ordinary and partial
differential equations. In Section 3.2 we show that in this analysis it is useful to consider resolvent conditions with respect to regions $V \subset \mathbb{C}$ that are
different from the unit disk D. The focus will be on regions V that are
contained in the stability regions corresponding to the numerical methods
under consideration. We give a review of stability estimates from the literature based on resolvent conditions with respect to such V. Section 3.3
provides various comments on these estimates. We confine our considerations throughout to so-called one-step methods. For related stability results
pertinent to (linear) multistep methods we refer to Crouzeix (1987), Grigorieff (1991), Lubich (1991), Lubich and Nevanlinna (1991) and Reddy and
Trefethen (1990, 1992).

Section 4 deals with various concepts and problems that are related to
(generalized) resolvent conditions. In Section 4.1 the resolvent condition is
related to the concept of ϵ-pseudospectra recently used by Trefethen and
others (see e.g. Trefethen (1992) and Reddy and Trefethen (1990, 1992)). In
Section 4.2 it is related to the so-called M-numerical range introduced by
Lenferink and Spijker (1990). Part of the material presented here is used in
some proofs given in Section 2. In Section 4.3 we consider the problem of

bounding the exponential function of a matrix under the assumption that the matrix satisfies a resolvent condition (with respect to the complex left half plane).

In Section 5 we focus on the range of applications of the stability results reviewed in Section 3. Moreover, a numerical illustration is presented involving the solution of a partial differential equation.

2. Stability estimates under resolvent conditions with respect to the unit disk

2.1. The classical situation for arbitrary $M_1 \geq 1$

We start by reviewing classical upper bounds for $\|B^n\|$ that were derived from (1.10), with $\| \cdot \|$ standing for the spectral norm.

As already mentioned in the introduction, the Kreiss matrix theorem asserts, for the spectral norm, that resolvent condition (1.10) implies power boundedness (1.6) with a stability constant M_0 depending only on M_1 and the dimension s. According to Tadmor (1981), the original proof by Kreiss (1962) yields an upper bound $\|B^n\| \leq M_0$ with

$$M_0 \simeq (M_1)^{s^s},$$

which is far from sharp. After successive improvements by various authors (Morton, 1964; Miller and Strang, 1966), it was Tadmor (1981) who succeeded in proving a bound that is linear in s,

$$\|B^n\| \leq 32e\pi^{-1}sM_1.$$

LeVeque and Trefethen (1984) lowered this upper bound to $2esM_1$, and conjectured that the latter bound can be improved further to

$$\|B^n\| \leq esM_1 \quad \text{for } n = 0, 1, 2, \ldots. \tag{2.1}$$

Moreover, these authors showed by means of a counterexample that the factor e in (2.1) cannot be replaced by any smaller constant – if the upper bound is to be valid for arbitrary factors M_1 in (1.10) and arbitrarily large integers s.

Smith (1985) proved a result which, combined with the arguments of LeVeque and Trefethen (1984), leads to the improved upper bound $\|B^n\| \leq \pi^{-1}(\pi+2)esM_1$, which is an improvement over the upper bound $2esM_1$ but still weaker than conjecture (2.1). The conjecture was finally proved to be true by Spijker (1991) (see also Wegert and Trefethen (1992)).

In addition to the upper bound (2.1), which is linear in s and independent of n, it is possible to derive an upper bound from (1.10) that is linear in n and independent of s. By the Cauchy integral formula (see, e.g., Conway

(1985)) we have

$$B^n = \frac{1}{2\pi i} \int_\Gamma \zeta^n (\zeta I - B)^{-1} d\zeta, \tag{2.2}$$

where the contour of integration Γ is any positively oriented circle $|\zeta| = 1 + \epsilon$ with $\epsilon > 0$. Choosing $\epsilon = 1/n$ it readily follows from (1.10) and (2.2) that

$$\|B^n\| \leq (1 + 1/n)^n (n+1) M_1 \leq e(n+1) M_1 \quad \text{for } n = 1, 2, 3, \ldots \tag{2.3}$$

(see also Reddy and Trefethen (1990) and Lubich and Nevanlinna (1991)).

In the next subsection we will discuss a generalization of the upper bounds (2.1), (2.3) to norms different from the spectral norm. We will also investigate the sharpness of these bounds in the general case.

2.2. Stability estimates for arbitrary $M_1 \geq 1$ and arbitrary norms

In this subsection we consider a generalization of the upper bounds (2.1), (2.3) to the case where $\| \cdot \|$ is an *arbitrary norm* on $\mathbb{C}^{s,s}$, the vector space of all complex $s \times s$ matrices. If the norm is submultiplicative (i.e. $\|AB\| \leq \|A\| \|B\|$ for all $A, B \in \mathbb{C}^{s,s}$) the norm is called a *matrix norm*. Norms for which $\|I\| = 1$ are called *unital*.

Theorem 2.1   Let $s \geq 1$, $B \in \mathbb{C}^{s,s}$ and $\| \cdot \|$ denote an arbitrary norm on the vector space $\mathbb{C}^{s,s}$.

(a) If (1.6) holds for some M_0, then (1.10) holds with $M_1 = M_0$;
(b) If (1.10) holds for some M_1, then

$$\|B^n\| \leq (1 + 1/n)^n \min(s, n+1) M_1 \quad \text{for } n = 1, 2, 3, \ldots. \tag{2.4}$$

Proof.   1. The proof of (a) is the same as the proof in Section 1.4 for the spectral norm. Since the proof of (2.3) as given in Section 2.1 also remains valid for arbitrary norms, the proof of (b) is complete if we can show that

$$\|B^n\| \leq (1 + 1/n)^n s M_1 \quad \text{for } n = 1, 2, 3, \ldots. \tag{2.5}$$

In order to prove (2.5) we now consider arbitrary but fixed $n \geq 1$ and B satisfying (1.10).

2. A well known corollary to the Hahn–Banach theorem (see, e.g., Chapter 3 in Rudin (1973), or Chapter 5 in Horn and Johnson (1990)) states that, corresponding to any normed vector space X and vector $y \in X$, there exists a linear transformation $F : X \to \mathbb{C}$ with

$$F(y) = \|y\| \quad \text{and} \quad |F(x)| \leq \|x\| \quad \text{for all } x \in X.$$

Applying this result with $X = \mathbb{C}^{s,s}, y = B^n$ we see that a linear $F : \mathbb{C}^{s,s} \to \mathbb{C}$ exists with

$$|F(A)| \leq \|A\| \quad \text{for all } s \times s \text{ matrices } A, \tag{2.6}$$

$$F(B^n) = \|B^n\|. \tag{2.7}$$

Combination of (2.7) and (2.2) yields

$$\|B^n\| = \frac{1}{2\pi i} \int_\Gamma \zeta^n R(\zeta)\, d\zeta,$$

where Γ is the positively oriented circle $|\zeta| = 1 + 1/n$ and R is the rational function defined by $R(\zeta) = F((\zeta I - B)^{-1})$. Integration by parts gives

$$\|B^n\| = \frac{-1}{2\pi i(n+1)} \int_\Gamma \zeta^{n+1} R'(\zeta)\, d\zeta \le \frac{1}{2\pi n}(1 + 1/n)^n \int_\Gamma |R'(\zeta)|\, |d\zeta|. \tag{2.8}$$

3. Let E_{jk} stand for the $s \times s$ matrix with entry in the jth row and kth column equal to 1, and all other entries 0. Denoting the entries of the matrix $(\zeta I - B)^{-1}$ by $r_{jk}(\zeta)$ we thus have

$$(\zeta I - B)^{-1} = \sum_{j,k} r_{jk}(\zeta) E_{jk},$$

and therefore also

$$R(\zeta) = \sum_{j,k} r_{jk}(\zeta) F(E_{jk}).$$

We define a rational function to be of order s if its numerator and denominator are polynomials of a degree not exceeding s. By Cramer's rule, the $r_{jk}(\zeta)$ are rational functions of order s with the same denominator. Hence $R(\zeta)$ is also of order s.

It was proved by Spijker (1991) that, for any rational function $R(\zeta)$ which has no poles on the circle Γ and is of order s, the following inequality holds:

$$\int_\Gamma |R'(\zeta)|\, |d\zeta| \le 2\pi s \max_\Gamma |R(\zeta)|. \tag{2.9}$$

The proof of (2.5) now easily follows by a combination of (2.8), (2.9), (2.6) and (1.10). $\square$

We remark that this proof of (2.5) is essentially based on ideas taken from LeVeque and Trefethen (1984) and Lenferink and Spijker (1991a). For an interesting discussion and generalization of inequality (2.9) we refer to Wegert and Trefethen (1992).

In the following theorem we focus on the sharpness of the bound (2.4) in the case $n = s - 1$.

Theorem 2.2 Let $s \ge 2$ and an arbitrary norm $\|\cdot\|$ on $\mathbb{C}^{s,s}$ be given. Then

$$\sup\{\|B^{s-1}\|/M_1(B): B \in \mathbb{C}^{s,s}, M_1(B) < \infty\} = \left(1 + \frac{1}{s-1}\right)^{s-1} s. \tag{2.10}$$

where $M_1(B)$ denotes the smallest M_1 such that (1.10) holds (we define $M_1(B) = \infty$ if (1.10) is not fulfilled for any M_1).

Proof. Define $B \in \mathbb{C}^{s,s}$ by $B = \gamma E$, where $\gamma > 0$ is large and the $s \times s$ matrix E is defined by (1.9). We have

$$
M_1(B) = \sup_{|\zeta|>1} (|\zeta| - 1)\|(\zeta I - B)^{-1}\| = \sup_{|\zeta|>1} \frac{|\zeta| - 1}{|\zeta|} \left\| \sum_{j=0}^{s-1} \left(\frac{\gamma}{\zeta}E\right)^j \right\|
$$

$$
\leq \sum_{j=0}^{s-1} \mu_j \gamma^j \|E^j\| = \mu_{s-1}\gamma^{s-1}\|E^{s-1}\| \left(1 + \mathcal{O}(\gamma^{-1})\right),
$$

where

$$
\mu_j = \sup_{|\zeta|>1} (|\zeta| - 1)|\zeta|^{-j-1} = \max_{0 \leq x \leq 1} (1 - x)x^j = j^j(j + 1)^{-j-1},
$$

so that

$$
\|B^{s-1}\|/M_1(B) \geq 1/\mu_{s-1} + \mathcal{O}(\gamma^{-1})
$$

$$
= \left(1 + \frac{1}{s-1}\right)^{s-1} s + \mathcal{O}(\gamma^{-1}) \quad \text{(as } \gamma \to \infty\text{)}.
$$

It follows that the left-hand member of (2.10) is not smaller than the right-hand member. In view of (2.4) the proof is complete. $\square$

Corollary 2.3 For each $s \geq 1$, let a norm $\|\cdot\| = \|\cdot\|^{(s)}$ be given on $\mathbb{C}^{s,s}$. Then there exist matrices $B_s \in \mathbb{C}^{s,s}$ for $s = 1, 2, 3, \ldots$, such that $M_1(B_s) < \infty$ and

$$
\|(B_s)^{s-1}\|^{(s)} \sim esM_1(B_s) \quad \text{(as } s \to \infty\text{)}, \tag{2.11}
$$

where $M_1(B_s)$ has the same meaning as in Theorem 2.2.

Proof. Immediate from Theorem 2.2. $\square$

This corollary was proved by LeVeque and Trefethen (1984) for the spectral norm. Our proof of Theorem 2.2 is essentially based on ideas taken from that paper.

In view of (2.4), the estimate (2.1) is valid for general norms $\|\cdot\|$ on $\mathbb{C}^{s,s}$. By virtue of Corollary 2.3, this general version of (2.1) is sharp in the sense of (2.11). However, it should be emphasized that this does not resolve the sharpness question for given *fixed* M_1, since $M_1(B_s)$ in (2.11) may depend on s. In the next two subsections we will focus on the situation where M_1 is a given fixed number.

2.3. About the best stability estimates for $M_1 = 1$

In the special situation where the resolvent condition (1.10) holds with $M_1 = 1$, the upper bound (2.4) can be improved in various ways. First we concentrate on arbitrary matrix norms on $\mathbb{C}^{s,s}$, and at the end of this subsection we focus on matrix norms $\|\cdot\|_p$ induced by the *pth Hölder norm on* $\mathbb{C}^s$ (for $p = 1, 2, \infty$).

Theorem 2.4 Let $s \geq 1$, $B \in \mathbb{C}^{s,s}$ and $\|\cdot\|$ denote an arbitrary matrix norm on $\mathbb{C}^{s,s}$. If (1.10) holds with $M_1 = 1$, then

$$\|B^n\| \leq n!n^{-n}e^n \leq \sqrt{2\pi(n+1)} \quad \text{for } n = 1, 2, 3, \ldots. \tag{2.12}$$

Proof. Property (1.10) with $M_1 = 1$ implies that

$$\|e^{zB}\| \leq e^{|z|} \quad \text{for all } z \in \mathbb{C}.$$

This can be seen from Theorem 4.7, to be presented in Section 4, or more directly by using

$$e^{zB} = \lim_{k \to \infty} \left[I - \frac{z}{k}B\right]^{-k} = \lim_{k \to \infty} \left(\frac{k}{z}\right)^k \left[\frac{k}{z}I - B\right]^{-k}$$

and applying (1.10) with $\zeta = k/z$.

We have

$$B^n = \frac{n!}{2\pi i} \int_\Gamma z^{-n-1} e^{zB} \, dz,$$

where Γ is the positively oriented circle with radius n and centre 0. Therefore $\|B^n\| \leq n!n^{-n}e^n$. From Stirling's formula

$$n! = (n/e)^n \sqrt{2\pi n} \exp[\theta_n(12n)^{-1}] \quad \text{with } 0 < \theta_n < 1$$

(see, e.g., Abramowitz and Stegun (1965)), we finally obtain

$$\|B^n\| \leq \sqrt{2\pi n} \exp[(12n)^{-1}] \leq \sqrt{2\pi(n+1)}.$$

$\square$

This proof of (2.12) is essentially based on ideas taken from Bonsall and Duncan (1980) (see also Bonsall and Duncan (1971)). Another proof can be given along the lines of Lubich and Nevanlinna (1991) (Theorem 2.1) or McCarthy (1992).

The next theorem shows that the upper bound for $\|B^n\|$ in (2.12) is sharp. For the elegant proof, which is beyond the scope of this paper, we refer to McCarthy (1992).

Theorem 2.5 Let $s \geq 2$ be given. Then there exists a vector norm on $\mathbb{C}^s$ such that the $s \times s$ shift matrix E, defined by (1.9), has the following two properties with respect to the corresponding induced matrix norm $\|\cdot\|$:

(a) E satisfies the resolvent condition (1.10) with $M_1 = 1$;

(b) $\|E^n\| = n! e^n n^{-n} \geq \sqrt{2\pi n}$ for $n = 1, 2, \ldots, s - 1$.

According to the following theorem the stability estimate (2.12) can be substantially improved for the case of some important matrix norms.

Theorem 2.6 Let $s \geq 1$, $Q \in \mathbb{C}^{s,s}$ invertible, and $p = 1, 2$ or ∞. Let the norm $\|\cdot\|$ on $\mathbb{C}^{s,s}$ be defined by $\|A\| = \|QAQ^{-1}\|_p$ (for all $A \in \mathbb{C}^{s,s}$). Then (1.10) with $M_1 = 1$ implies (1.6) with $M_0 = 1$ (if $p = 1$ or ∞) or $M_0 = 2$ (if $p = 2$).

Proof. Since the result for general invertible Q easily follows from the result for $Q = I$, it is sufficient to consider the latter case only.

Let $p = \infty$. Suppose $B = (\beta_{jk})$ satisfies (1.10) with $\|\cdot\| = \|\cdot\|_\infty$, $M_1 = 1$. Clearly (4.11) holds with $W = D$, $M = 1$. By Theorem 4.7 relation (4.10) holds as well. In view of the expression for $\tau_1[B]$ (with $\|\cdot\| = \|\cdot\|_\infty$) given at the end of Section 4.2, we conclude that each disk with its centre at β_{jj} and radius $\rho_j = \sum_{k \neq j} |\beta_{jk}|$ lies in the unit disk D. Consequently, $|\beta_{jj}| + \rho_j \leq 1$, and

$$\|B\|_\infty = \max_{1 \leq j \leq s} (|\beta_{jj}| + \rho_j) \leq 1,$$

so that (1.6) holds with $M_0 = 1$.

For $p = 1$ the proof follows from the result for $p = \infty$ and the fact that $\|A\|_1 = \|A^T\|_\infty$ for all $A \in \mathbb{C}^{s,s}$ (where A^T denotes the transpose of A).

For $p = 2$ the value $M_0 = 2$ was stated, e.g., in Reddy and Trefethen (1992) and McCarthy (1992). The proof runs as follows. It can be seen by a straightforward calculation (or directly from the material in Bonsall and Duncan (1980) or Lenferink and Spijker (1990)) that the numerical range $\{x^* B x : x \in \mathbb{C}^s$ with $x^* x = 1\}$ is contained in the unit disk D. The proof continues by applying Berger's inequality (see, e.g., Pearcy (1966), Richtmyer and Morton (1967 p. 89), Bonsall and Duncan (1980) or Horn and Johnson (1990)). This inequality reads

$$r(A^n) \leq [r(A)]^n \quad \text{for } n = 1, 2, 3, \ldots,$$

where A is any $s \times s$ matrix, and $r(A)$ denotes the so-called numerical radius of A defined by

$$r(A) = \max\{|x^* A x| : x \in \mathbb{C}^s \text{ with } x^* x = 1\}.$$

Since $r(B) \leq 1$, there follows

$$r(B^n) \leq 1.$$

By splitting B^n into a sum $B^n = A_1 + iA_2$ with Hermitian A_1, A_2, and by noting that for any Hermitian A the relation $\|A\|_2 = r(A)$ is valid, we finally obtain

$$\|B^n\| \leq \|A_1\| + \|A_2\| = r(A_1) + r(A_2) \leq 2r(B^n) \leq 2.$$

□

2.4. About the best stability estimates for fixed $M_1 > 1$

Theorem 2.1 shows that if resolvent condition (1.10) is satisfied with fixed M_1, then $\|B^n\|$ can grow at most linearly with n or s. Corollary 2.3 reveals that the corresponding upper bound is sharp – if we allow M_1 to be variable.

For the special case $M_1 = 1$, however, this linear growth with n or s is too pessimistic, as can be seen from Theorems 2.4 and 2.6 in the previous subsection.

For fixed values $M_1 > 1$ the question also arises as to whether the upper bound (2.4) can be improved. We do not know of any positive results in this direction. In the following we shall therefore present negative results only – in the form of lower bounds for $\|B^n\|$.

A negative result we have seen already is Theorem 2.5, which is also relevant for any fixed $M_1 > 1$. It shows that $\|B^n\|$ may grow at the rate $\sqrt{n}$ or $\sqrt{s}$.

The following two theorems show what growth rates can be achieved for the three important matrix norms $\|\cdot\|_p$, $p = 1, 2, \infty$.

Theorem 2.7 Let $p = 1$ or $p = \infty$. Then there exist $C > 0$ and $M > 1$ such that

$$\sup_{s,B} \|B^n\|_p \geq C\sqrt{n} \quad \text{for } n = 0, 1, 2, \ldots,$$

where the supremum is over all integers $s \geq 1$ and all matrices $B \in \mathbb{C}^{s,s}$ satisfying the resolvent condition (1.10) with $M_1 = M$ and $\|\cdot\| = \|\cdot\|_p$.

Proof. The proof for the case $p = \infty$ easily follows from a straightforward adaptation of Example 2.2 in Lubich and Nevanlinna (1991) to the finite-dimensional case.

More precisely, let ϕ denote a Möbius transformation that maps the unit disk onto itself and is not just a rotation (such ϕ exist, see, e.g., Henrici (1974)). We define $B_s = \phi(E_s)$, where E_s stands for the $s \times s$ matrix E defined by (1.9). From the material in Lubich and Nevanlinna (1991) it follows that every B_s satisfies the resolvent condition (1.10) with $\|\cdot\| = \|\cdot\|_\infty$ and a constant M_1 independent of s, and that

$$\lim_{s \to \infty} \|B_s^n\|_\infty \geq C\sqrt{n} \quad \text{for } n = 0, 1, 2, \ldots,$$

where C is a positive constant. This proves the theorem for the case $p = \infty$.

For $p = 1$ the result is obtained by noting that $\|A\|_1 = \|A^T\|_\infty$ for all $A \in \mathbb{C}^{s,s}$. □

Theorem 2.8 Let $M > \pi + 1$ be given. Then there exist a constant $C > 0$ and matrices $B_s \in \mathbb{C}^{s,s}$ for $s = 2, 4, 6, \ldots$, such that all B_s satisfy (1.10) with $M_1 = M$, $\|\cdot\| = \|\cdot\|_2$, and

$$\|(B_s)^{s/2}\|_2 \geq C(\log s)^{1/2} / \log \log s.$$

Proof. It was shown by McCarthy and Schwartz (1965) that for each $M > \pi + 1$ there exist a constant $\gamma > 0$ and $s \times s$ matrices $E_{s,j}$ (for all even positive s and $j = 1, 2, \ldots, s$) with the following properties:

$$(E_{s,j})^2 = E_{s,j} \neq 0, \quad E_{s,j} E_{s,k} = 0 \ (j \neq k), \quad \sum_{j=1}^{s} E_{s,j} = I, \tag{2.13}$$

$$\left\| \sum_{j \text{ odd}} E_{s,j} \right\|_2 \geq \gamma (\log s)^{1/2} / \log \log s, \tag{2.14}$$

$$B_s = \sum_{j=1}^{s} e^{2\pi i j/s} E_{s,j} \text{ satisfies (1.10) with} \atop M_1 = M \text{ and } \|\cdot\| = \|\cdot\|_2. \tag{2.15}$$

For even s we have

$$(B_s)^{s/2} = \sum_{j=1}^{s} (-1)^j E_{s,j} = I - 2 \sum_{j \text{ odd}} E_{s,j}.$$

In view of (2.14) this implies

$$\|(B_s)^{s/2}\|_2 \geq 2\gamma (\log s)^{1/2} / \log \log s - 1 \quad \text{for } s = 2, 4, 6, \ldots.$$

Since all $(B_s)^{s/2} \neq 0$ there exists a constant C with the property stated in the theorem. $\square$

For additional interesting examples for the matrix norms $\|\cdot\|_p$ with $p = 2$, ∞ we refer to McCarthy (1992).

We also mention that after completion of the present article new results related to this were found by Kraaijevanger (1992) for the matrix norm $\|\cdot\|_\infty$.

3. Stability estimates under resolvent conditions with respect to general regions V

3.1. Linear stability analysis and stability regions

Consider an initial value problem for a system of s *ordinary differential equations* of the form

$$\begin{aligned} U'(t) &= AU(t) + b(t) \quad (t \geq 0), \\ U(0) &= u_0. \end{aligned} \tag{3.1}$$

Here A is a given constant $s \times s$ matrix, and u_0, $b(t)$ are given vectors in $\mathbb{C}^s$. The vector $U(t) \in \mathbb{C}^s$ is unknown for $t > 0$.

In this section we analyse the stability of numerical processes for approximating $U(t)$. This analysis will also be relevant to classes of numerical processes for solving *partial differential equations*.

To elucidate this relevance, we assume an initial-boundary value problem to be given for a linear partial differential equation with variable coefficients in the differential operator (which depend on the space variable x but not on the time variable t). Applying the method of *semi-discretization*, where discretization is applied to the space variable x only, one arrives at an initial value problem for a large system of the form (3.1). In this case the matrix A, the inhomogeneous term $b(t)$, and the vector u_0 are determined by the original initial-boundary value problem and by the process of semi-discretization. The solution $U(t)$ to (3.1) then provides an approximation to the solution of the original initial-boundary value problem. For an example we refer to problem (1.3); by replacing the derivatives with respect to x in (1.3) by the same *finite difference* quotients as referred toin Section 1.2, one arrives at an initial value problem (3.1) with the tridiagonal $s \times s$ matrix $A = (\alpha_{jk})$ given by (1.5). We note that problems (3.1) arise not only when the semi-discretization relies on the introduction of finite differences, but also when it is based on *spectral* approximations (see Gottlieb and Orszag (1977) and Canuto, Hussaini, Quarteroni and Zang (1988)) or on the *finite element* method (see, e.g., Oden and Reddy (1976) and Strang and Fix (1973)).

Many step-by-step methods for the numerical solution of ordinary differential equations, like Runge–Kutta methods or Rosenbrock methods (see Butcher (1987) and Hairer and Wanner (1991)), reduce – when applied to (3.1) – to processes of the form

$$u_n = \varphi(hA)u_{n-1} + b_n \quad \text{for } n = 1, 2, 3, \ldots. \tag{3.2}$$

Here $\varphi(\zeta) = P(\zeta)/Q(\zeta)$ is a rational function, depending only on the underlying step-by-step method. $P(\zeta), Q(\zeta)$ are polynomials, without common zeros, such that $\varphi(0) = \varphi'(0) = 1$. Further, $h = \Delta t > 0$ denotes the *stepsize*, and we define $\varphi(hA) = P(hA)Q(hA)^{-1}$ when $Q(hA)$ is invertible. The vectors $b_n \in \mathbb{C}^s$ are related to $b(t)$, and $u_n \simeq U(nh)$ are calculated successively from (3.2). It is worth noting that many numerical processes in partial differential equations which are *not* constructed with the process of semi-discretization in mind are still of the form (3.2), and can *a posteriori* be conceived as relying on semi-discretization. For instance, it follows from (1.4) that the process constructed in Section 1.2 can be written in the form (3.2) with

$$\varphi(\zeta) = (1 + (1 - \theta)\zeta)(1 - \theta\zeta)^{-1}$$

and $A = (\alpha_{jk})$ satisfying (1.5).

Since (3.2) is a special case of (1.2), the stability analysis of (3.2) amounts to investigating the growth of matrices B^n with

$$B = \varphi(hA).$$

In this analysis it is useful to introduce the *stability region* S, defined by

$$S = \{\zeta : \zeta \in \mathbb{C} \text{ with } Q(\zeta) \neq 0 \text{ and } |\varphi(\zeta)| \leq 1\}. \tag{3.3}$$

Consider the following requirement on hA with regard to S,

$\sigma[hA] \subset S$, and for each $\zeta \in \partial S$ which is a zero of the
minimal polynomial of hA with multiplicity $m > 1$,   (3.4)
the derivatives $\varphi^{(j)}(\zeta)$ vanish for $j = 1, 2, \ldots, m - 1$.

Here $\sigma[hA]$ denotes the *spectrum* (set of eigenvalues) of hA, and ∂S the *boundary* of S. For the concept of minimal polynomial see, e.g., Horn and Johnson (1990). The spectral mapping theorem (see Conway (1985) or Rudin (1973)) states that, if $Q(\zeta) \neq 0$ for all $\zeta \in \sigma[hA]$, then

$$\sigma[\varphi(hA)] = \{\varphi(\zeta) : \zeta \in \sigma[hA]\}.$$

Hence, the condition $\sigma[hA] \subset S$ in (3.4) is equivalent to the condition $\sigma[B] \subset D$ in (1.7) with $B = \varphi(hA)$. Further, from the Jordan canonical form of hA it can be deduced that the condition regarding $\zeta \in \partial S$ in (3.4) is equivalent to the condition on the geometric multiplicities in (1.7). Consequently, (3.4) is equivalent to (1.7). It follows that (3.4) is a necessary and sufficient condition in order that a finite M_0 exists with stability property (1.6) for $B = \varphi(hA)$.

We note that most functions $\varphi(\zeta)$ of practical interest have nonvanishing derivatives $\varphi'(\zeta)$ on the whole of ∂S. In this case (3.4) simply reduces to $\sigma[hA] \subset S$ and the condition that all $\zeta \in \partial S \cap \sigma[hA]$ are zeros of the minimal polynomial of hA with multiplicity 1.

In general (3.4) has similar advantages (it is relatively simple to verify, and reliable for normal matrices) and disadvantages (quite unreliable for families of matrices that are not normal) as the eigenvalue condition (1.7). In the rest of this section we adapt (3.4) to conditions on hA that reliably predict stability – also for nonnormal matrices and norms $\| \cdot \|$ on $\mathbb{C}^{s,s}$ different from the spectral norm. An advantage of these conditions on hA over a resolvent condition on $B = \varphi(hA)$ (as dealt with in Section 2) lies in the circumstance that, in general, hA has a simpler structure than B, and that knowledge available about S can be exploited.

3.2. Reviewing stability estimates from the literature

In the literature various stability results can be found, which are essentially based on the use of resolvent conditions of the form

$\zeta I - hA$ is invertible and $\|(\zeta I - hA)^{-1}\| \leq M_1 \, \mathrm{d}(\zeta, V)^{-1}$
for all complex numbers $\zeta \notin V$.   (3.5)

Here, V is a closed subset of the stability region S (see (3.3)), M_1 is a constant, $\| \cdot \|$ denotes a norm on $\mathbb{C}^{s,s}$ and $d(\zeta, V) = \min\{|\zeta - \eta| : \eta \in V\}$ is the distance from ζ to V. Under additional assumptions, to be stated below, it is shown in the literature that (3.5) implies a stability estimate

$$\|\varphi(hA)^n\| \leq M_1 g(n,s) \quad \text{for } n = 1, 2, 3, \ldots, \tag{3.6}$$

where the function g *only depends on* φ *and* V (and not on h, A, M_1 or $\| \cdot \|$).

In the following we list some of these stability results. We assume throughout that (3.5) is satisfied with closed $V \subset S$ and a norm $\| \cdot \|$ on $\mathbb{C}^{s,s}$. In each separate case we formulate the relevant additional assumptions and the resulting function g.

For any $W \subset \mathbb{C}$ we denote by ∂W the *boundary* of W, and write $\mathbb{C}^- = \{\zeta : \zeta \in \mathbb{C} \text{ with } \mathrm{Re}\,\zeta \leq 0\}$.

1. In Lenferink and Spijker (1991a) (Theorem 2.2) estimate (3.6) is proved with $g(n,s) \equiv \gamma s$ where γ depends only on φ and V. The additional assumptions are: V is bounded and convex; $\varphi'(\zeta) \neq 0$ on $\partial V \cap \partial S$; and ∂V lies on an algebraic curve.

2. In Lenferink and Spijker (1991b) (Lemma 3.3) estimate (3.6) is proved with $g(n,s) \equiv \gamma n$ where γ depends only on φ and V. The additional assumptions are: V is bounded and convex; and $\| \cdot \|$ is induced by a vector norm on $\mathbb{C}^s$.

3. In Reddy and Trefethen (1992) (Theorem 7.1) estimate (3.6) is proved with $g(n,s) \equiv \gamma \min(n,s)$ where γ depends only on φ. The additional assumptions are: $V = S$, S is bounded; $\varphi'(\zeta) \neq 0$ on ∂S; and $\| \cdot \|$ is a weighted spectral norm (i.e. $\|B\| = \|QBQ^{-1}\|_2$ for all $B \in \mathbb{C}^{s,s}$, where Q is an invertible matrix).

4. In Lubich and Nevanlinna (1991) (Theorem 3.1) estimate (3.6) is proved with $g(n,s) \equiv \gamma \min(n,s)$ where γ depends only on φ. The additional assumptions are: $V = \mathbb{C}^-$ and $\| \cdot \|$ is induced by a vector norm on $\mathbb{C}^s$.

5. From the material in the important paper by Brenner and Thomée (1979) it follows that (3.6) holds with $g(n,s) \equiv \gamma\sqrt{n}$ where γ depends only on φ. The additional assumptions are: $V = \mathbb{C}^-$, $M_1 = 1$ and $\| \cdot \|$ is induced by a vector norm on $\mathbb{C}^s$.

6. For $\delta \geq 0$ the *wedge* $W(\delta)$ is defined by $W(\delta) = \{\zeta : \zeta = 0 \text{ or } |\arg\zeta - \pi| \leq \delta\}$. In Lenferink and Spijker (1991b) (Lemma 3.1) estimate (3.6) is proved with $g(n,s) \equiv \gamma$ where γ depends only on φ and V. The additional assumptions are: V is a bounded convex subset of $W(\alpha)$, where $0 \leq \alpha < \pi/2$, $V \subset \mathrm{int}(S) \cup \{0\}$; and $\| \cdot \|$ is induced by a vector norm on $\mathbb{C}^s$.

7. In Crouzeix, Larsson, Piskarev and Thomée (1991) (Theorem 5) estimate (3.6) is proved with $g(n,s) \equiv \gamma$ where γ depends only on φ and

V. The additional assumptions, slightly adapted in order to fit in our framework, are: $V = W(\alpha)$, $S \supset W(\beta)$, $0 \le \alpha < \beta \le \pi/2$ and $\| \cdot \|$ is induced by a vector norm on $\mathbb{C}^s$. For related material see Palencia (1991, 1992) and Lubich and Nevanlinna (1991).

8   For $\rho > 0$ the *disk* $D(\rho)$ is defined by $D(\rho) = \{\zeta : \zeta \in \mathbb{C} \text{ and } |\zeta + \rho| \le \rho\}$. In Lubich and Nevanlinna (1991) (Theorem 3.4) estimate (3.6) is proved with $g(n, s) \equiv \gamma \sqrt{1 + n r_0}$ where γ depends only on φ. The additional assumptions are: $r_0 > 0$, $V = D(r_0)$, $S \supset \mathbb{C}^-$ and $\| \cdot \|$ is induced by a vector norm on $\mathbb{C}^s$. (The assumption $S \supset \mathbb{C}^-$ can be relaxed, see Lubich and Nevanlinna (1991).)

9   The quantity $r = \sup\{\rho : \rho > 0 \text{ and } D(\rho) \subset S\}$ is called the *stability radius* of the step-by-step method (3.2) (see, e.g., Kraaijevanger *et al.* (1987)). In Lenferink and Spijker (1991b) (Sections 2.3 and 2.4) it was noted that, for $0 < r < \infty$, estimate (3.6) holds with $g(n, s) \equiv \gamma \sqrt{n}$ where γ only depends on φ. The additional assumptions are: $M_1 = 1$, $\| \cdot \| = \| \cdot \|_\infty$ and $V = D(r)$. Next consider $r \in (0, \infty]$ and $0 < r_0 < r$. If (3.5) holds with $V = D(r_0)$, then, again under the assumptions $M_1 = 1$, $\| \cdot \| = \| \cdot \|_\infty$, inequality (3.6) even holds with $g(n, s) \equiv \gamma$, where γ depends only on φ and r_0 (see Kraaijevanger *et al.* (1987) and Lenferink and Spijker (1991b)).

10   In Brenner and Thomée (1979) and Lubich and Nevanlinna (1991) more refined estimates of the form (3.6) were derived for functions φ satisfying special conditions. For example, from Lubich and Nevanlinna (1991) (Theorem 3.2) it follows that, in the situation of point 4, an estimate (3.6) with $g(n, s) \equiv \gamma \min(n^\alpha, s)$, $\alpha < 1$, is possible for functions φ with $|\varphi(\zeta)|$ not identically 1 on the imaginary axis. We refer to Brenner and Thomée (1979) and Lubich and Nevanlinna (1991) for more details.

3.3. *Various comments on stability estimates from the literature*

Remark 3.1   Results 1, 2, 6 and 8 in the last subsection were proved by using integral representations of the form

$$\varphi(hA)^n = \frac{1}{2\pi i} \int_\Gamma \varphi(\zeta)^n (\zeta I - hA)^{-1} \, d\zeta,$$

where Γ is a proper curve in the complex plane surrounding V, and by estimating the integral (see, e.g., the proof of Theorem 2.1). Results 5, 7 and 10 were proved by using related, but different, integral representations for $\varphi(hA)^n$.

Results 3 and 4 were obtained by first proving that resolvent condition (3.5) for hA implies a resolvent condition (1.10) for $B = \varphi(hA)$ (with a different constant M_1) and then applying (a version of) Theorem 2.1 to this matrix B.

Finally, the proof of Result 9 relies on an expansion of $\varphi(hA)^n$ in a power series

$$\varphi(hA)^n = \gamma_0 I + \gamma_1 (hA + \rho I) + \gamma_2 (hA + \rho I)^2 + \cdots \qquad (3.7)$$

with $\rho = r$ or r_0, and on bounding the terms of the series using the fact that the resolvent condition (3.5) (with $M_1 = 1$, $\|\cdot\| = \|\cdot\|_\infty$ and $V = D(\rho)$) implies a so-called *circle condition* $\|hA + \rho I\|_\infty \le \rho$. The latter implication, which is in fact an equivalence, follows immediately from Theorem 2.6 (with $B = \rho^{-1}(hA + \rho I)$), and was stated in Lenferink and Spijker (1991b) (Section 2.4). In Kraaijevanger *et al.* (1987), Nevanlinna (1984) and Spijker (1985) this circle condition was combined with (3.7) to yield the desired stability bounds.

Remark 3.2 We note that Results 2, 3, 4, 6, 7 and 8, although formulated in Kraaijevanger *et al.* (1987), Nevanlinna (1984) and Spijker (1985) for special norms, are valid as well for arbitrary norms $\|\cdot\|$ on $\mathbb{C}^{s,s}$. This can be seen by a straightforward adaptation of the proofs in Kraaijevanger *et al.* (1987), Nevanlinna (1984) and Spijker (1985).

Further, it is easy to see that Result 9 is also valid for norms $\|\cdot\|$ defined by $\|B\| = \|QBQ^{-1}\|_p$ (for all $B \in \mathbb{C}^{s,s}$), where Q is an invertible matrix and $p = 1$ or ∞.

Remark 3.3 In all of the above, the resolvent condition (3.5) occurs as a *sufficient condition* for stability estimates of the form (3.6). Reddy and Trefethen (1992) (Theorem 7.1) succeeded in showing (for the weighted spectral norm, see Result 3) that (3.5) is also a *necessary condition* for stability. In fact, they showed – for any matrix hA belonging to a specific family $\mathcal{F}$ defined in their paper – that, in general, strong stability (i.e. $\|\varphi(hA)^n\| \le M_0$ for all $n \ge 0$) implies the resolvent condition (3.5) with $V = S$ and $M_1 = \gamma M_0$. Here γ depends only on φ and $\mathcal{F}$.

Remark 3.4 Modifications of Results 3, 5 and 9 can be proved if we relax slightly the assumption $V \subset S$ for the set V in the resolvent condition (3.5). This can be useful in applications (see Section 5).

(a) Let S be bounded and $\varphi'(\zeta) \ne 0$ on ∂S. Further, let $\beta > 0$ and $h > 0$ be given. Suppose that the resolvent condition (3.5) is (only) satisfied with respect to the set $V = S + \beta h D$ (but not necessarily with respect to the smaller set $V = S$ itself).

It follows from Reddy and Trefethen (1992) (Theorem 8.2) that there exist positive constants $\gamma_1, \gamma_2, \gamma_3$ (only depending on φ) such that these assumptions imply the stability estimate

$$\|\varphi(hA)^n\| \le M_1 \gamma_1 e^{\gamma_2 \beta n h} \min(n, s) \quad \text{for } n = 1, 2, 3, \ldots \qquad (3.8)$$

whenever $\beta h \leq \gamma_3$. This was proved in Reddy and Trefethen (1992) for the weighted spectral norm (defined in Result 3). The proof in that paper can be adapted in a straightforward way to arbitrary norms on $\mathbb{C}^{s,s}$.

(b) Let $r \leq \infty$ have the same meaning as in Result 9, and let $M_1 = 1$, $\| \cdot \| = \| \cdot \|_\infty$. Further, let $0 < r_0 < \infty$, $r_0 \leq r$ and $\beta > 0$ be given. Then there exists a constant $h_0 > 0$ such that φ is analytic on $W = D(r_0) + \beta h_0 D = \{\zeta : \zeta \in \mathbb{C} \text{ and } |\zeta + r_0| \leq r_0 + \beta h_0\}$. Suppose that $0 < h \leq h_0$ and (3.5) is (only) satisfied with $V = D(r_0) + \beta h D$. These assumptions imply the stability estimate

$$\|\varphi(hA)^n\| \leq \gamma_1 e^{\gamma_2 \beta n h} \sqrt{n} \quad \text{for } n = 1, 2, 3, \ldots, \tag{3.9}$$

where the constants γ_1, γ_2 depend only on φ, r_0 and βh_0 (and not on h, n, s or A). The proof is again based on the expansion (3.7) (with $\rho = r_0$), and can be given in two steps. First we apply Theorem 2.6 (with $B = (r_0 + \beta h)^{-1}(hA + r_0 I)$) to obtain $\|hA + r_0 I\| \leq r_0 + \beta h$ and then use (3.7) and estimates for the $|\gamma_k|$ (see Spijker (1985), Corollary 4.3) to derive (3.9). Further, it is easy to see that this result is also valid for norms $\| \cdot \|$ defined by $\|B\| = \|QBQ^{-1}\|_p$ (for all $B \in \mathbb{C}^{s,s}$), where Q is an invertible matrix and $p = 1$ or ∞.

(c) An estimate of the form (3.9) can also be proved if we replace the condition $V = \mathbb{C}^-$ in Result 5 by $V = \mathbb{C}^- + \beta h D$. We refer to Brenner and Thomée (1979) (Theorem 1) for more details.

Remark 3.5 Some of the arguments recently used in Kreiss (1990) and Kreiss and Wu (1992) are closely related to the above, and can be interpreted as yielding a result of the form (3.6). The assumptions on φ which are made in Kreiss (1990) and Kreiss and Wu (1992) in order to derive stability estimates comprise:

The half disk $\{\zeta : \text{Re } \zeta \leq 0, |\zeta| \leq R_1\}$ is contained in S, $\qquad$ (3.10)

φ is a polynomial which does not transform any
two different points ζ with Re $\zeta = 0, |\zeta| < R_1$ $\qquad$ (3.11)
into one and the same image point z with $|z| = 1$.

Here R_1 is a given positive constant. Assume the $s \times s$ matrix hA satisfies

$$\|hA\| \leq R < R_1, \tag{3.12}$$
$$\|(\zeta I - hA)^{-1}\| \leq K_1 (\text{Re } \zeta)^{-1} \quad \text{for all } \zeta \in \mathbb{C} \text{ with Re } \zeta > 0. \tag{3.13}$$

Although the setting in Kreiss (1990) and Kreiss and Wu (1992) is different in appearance from the one we use here, Theorem 3.2 in Kreiss and Wu (1992) essentially states that (3.10)–(3.13) imply

$$\|(e^\zeta I - \varphi(hA))^{-1}\| \leq K_0 (\text{Re } \zeta)^{-1} \quad \text{for all } \zeta \text{ with Re } \zeta > 0. \tag{3.14}$$

We now show that this conclusion is related to the stability results described earlier. First of all, the assumptions (3.12), (3.13) imply our resolvent condition (3.5) with $V = \{\zeta : \text{Re } \zeta \leq 0, |\zeta| \leq R\}$ and $M_1 = \sqrt{2}K_1$. Further, (3.14) can be proved to be equivalent to a resolvent condition of the form (1.10) with $B = \varphi(hA)$. Therefore, by Theorem 2.1, (3.14) implies a result of the form (3.6).

The stability estimates which are focused on in Kreiss (1990) and Kreiss and Wu (1992) are pertinent to l_2 norms, and essentially different from (1.6) or (3.6). In fact, the estimates (3.6) are relevant to stability with respect to perturbations in the initial value u_0 of process (3.2), whereas the estimates in Kreiss (1990) and Kreiss and Wu (1992) are relevant to *stability with respect to perturbations in the vectors b_n of (3.2)*. In Kreiss (1990) and Kreiss and Wu (1992) this stability concept, referred to as *stability in a generalized sense*, is argued to be equivalent to an inequality of the form (3.14) (see Kreiss and Wu (1992) (Theorem 3.1)). Moreover, an analogous concept (of stability in a generalized sense) for the continuous problem (3.1) is stated to be equivalent to a resolvent condition of the form (3.13).

4. Various related concepts and problems

4.1. ϵ-pseudospectra

The useful concept of ϵ-pseudospectra has been introduced and studied by Landau (1975), Varah (1979), Reddy and Trefethen (1990, 1992), Trefethen (1992) and others. The focus in these papers is on the (weighted) spectral norm. The main purpose of this subsection is to extend the notion of ϵ-pseudospectra to the situation of general matrix norms, and to relate it to the resolvent condition (3.5).

Let $\|\cdot\|$ denote an arbitrary matrix norm on $\mathbb{C}^{s,s}$. Let B be an $s \times s$ matrix and $\epsilon > 0$. Consider for a given complex number λ the situation where

$$\text{there exists an } s \times s \text{ matrix } E \text{ with } \|E\| \leq \epsilon \text{ such that } \lambda \in \sigma[B + E]. \quad (4.1)$$

Analogously to Reddy and Trefethen (1990, 1992), Reichel and Trefethen (1992) and Trefethen (1992) we give the following definition.

Definition 4.1 The set of all complex numbers λ satisfying (4.1) is called the ϵ-*pseudospectrum* of B and is denoted by $\sigma_\epsilon[B]$.

We emphasize that – unlike the spectrum $\sigma[B]$ – the pseudospectrum $\sigma_\epsilon[B]$ depends on the norm $\|\cdot\|$.

The concept of an ϵ-pseudospectrum can be related to the following properties:

$$\text{There exists an } s \times s \text{ matrix } E \text{ with } \|E\| = \epsilon \qquad (4.2)$$
$$\text{such that } \lambda \in \sigma[B + E];$$

There exists an $s \times s$ matrix U with $\|U\| = 1$
such that $\|(B - \lambda I)U\| \leq \epsilon$;   (4.3)

$B - \lambda I$ is singular,
or $B - \lambda I$ is invertible with $\|(B - \lambda I)^{-1}\| \geq \epsilon^{-1}$.   (4.4)

We have

Theorem 4.2   (a) Let $\|\cdot\|$ be a matrix norm on $\mathbb{C}^{s,s}$. Then (4.1) and (4.2) are equivalent (provided $s \geq 2$). Moreover (4.2) implies (4.3), and (4.3) implies (4.4). If $\|I\| = 1$ then (4.3) and (4.4) are equivalent.
  (b) Let $\|\cdot\|$ be induced by a vector norm $|\cdot|$ on $\mathbb{C}^s$, $s \geq 2$. Then properties (4.1) – (4.4) are equivalent to each other. Moreover, they are equivalent to the requirement that

there exists a vector $u \in \mathbb{C}^s$ with $|u| = 1$ such that $|(B - \lambda I)u| \leq \epsilon$.   (4.5)

Proof.   (a) First we prove the equivalence of (4.1) and (4.2). The implication of (4.1) by (4.2) is trivial. To prove the reverse implication we assume there exists a matrix E with $\|E\| < \epsilon$ and a vector $u \neq 0$ such that $(B + E - \lambda I)u = 0$. When $s \geq 2$ we can choose a matrix C with $C \neq 0$ and $Cu = 0$. Define the matrix $E(t) = E + tC$ for $t \geq 0$. There exists a positive t_0 such that $\|E(t_0)\| = \epsilon$ and $\lambda \in \sigma[B + E(t_0)]$, which proves (4.2).
  Assume (4.2). Define $V = [u, 0, 0, \ldots, 0] \in \mathbb{C}^{s,s}$ where $u \in \mathbb{C}^s$ is an eigenvector of $B + E$ corresponding to the eigenvalue λ. Defining $U = \|V\|^{-1}V$ we arrive at $\|U\| = 1$ and $(B + E)U = \lambda U$. Hence $\|(B - \lambda I)U\| = \|EU\| \leq \epsilon$, which proves (4.3).
  Assume (4.3). For invertible $B - \lambda I$ we get with $E = (B - \lambda I)U$ the relation $\|E\| \leq \epsilon$, and therefore $1 = \|(B - \lambda I)^{-1}E\| \leq \|(B - \lambda I)^{-1}\|\epsilon$, which proves (4.4).
  Assume (4.4) and $\|I\| = 1$. If $B - \lambda I$ is singular then (4.3) holds with $U = [u, 0, 0, \ldots, 0]$, where $u \in \mathbb{C}^s$ is in the null space of $B - \lambda I$ and is chosen such that $\|U\| = 1$. If $B - \lambda I$ is invertible then (4.3) holds with $U = \|(B - \lambda I)^{-1}\|^{-1}(B - \lambda I)^{-1}$.
  (b) Assume (4.3). Choosing $v \in \mathbb{C}^s$ with $|v| = 1$, $|Uv| = 1$, we have $|(B - \lambda I)Uv| \leq \epsilon$. With $u = Uv$ we arrive at (4.5).
  Assume (4.5). Taking $X = \mathbb{C}^s$ and $y = u$ in the corollary to the Hahn–Banach theorem formulated in the proof of Theorem 2.1, we see that there exists a linear transformation $F : \mathbb{C}^s \to \mathbb{C}$ with

$$F(u) = 1 \quad \text{and} \quad |F(x)| \leq |x| \quad \text{for all } x \in \mathbb{C}^s.$$

Defining the matrix E by

$$Ex = -F(x)(B - \lambda I)u \quad \text{for all } x \in \mathbb{C}^s,$$

it follows that $Eu = -(B - \lambda I)u$ and $\|E\| = |(B - \lambda I)u| \leq \epsilon$, which proves (4.1).

In view of (a) the proof is complete. □

Remark 4.3 (a) In part (a) of Theorem 4.2 the assumption $\|I\| = 1$ is essential for the equivalence of (4.3) and (4.4). This can be seen as follows. Let $s \geq 1$, $\epsilon = \|I\|^{-1}$, $\lambda = 0$ and $B = I$. Then (4.4) is always satisfied but (4.3) holds if and only if $\|I\| = 1$.

(b) For arbitrary matrix norms, (4.1) can be a stronger condition than (4.3), even if $\|I\| = 1$. This can be seen from the following example. On $\mathbb{C}^{2,2}$ we define a matrix norm by

$$\|A\| = \max\{\|A\|_1, \|A\|_\infty\} \quad \text{for all } A = (a_{ij}) \in \mathbb{C}^{2,2}$$

(see, e.g., Horn and Johnson (1990, p. 308)) and choose

$$\lambda = 0, \quad \epsilon = \tfrac{1}{2} \quad \text{and} \quad B = \begin{pmatrix} 1 & 1 \\ 0 & 1 \end{pmatrix}.$$

One easily verifies that condition (4.3) is satisfied by taking

$$U = \begin{pmatrix} \tfrac{1}{2} & -\tfrac{1}{2} \\ 0 & \tfrac{1}{2} \end{pmatrix}.$$

But, a straightforward calculation reveals that $B + E$ is invertible for all $s \times s$ matrices E with $\|E\| \leq 1/2$, showing that Condition (4.1) is violated.

Remark 4.4 If $\| \cdot \|$ is the spectral norm, then Conditions (4.1)–(4.5) are all equivalent to the requirement that $B - \lambda I$ has a singular value σ with $\sigma \leq \epsilon$ (see Reddy and Trefethen (1990), Trefethen (1992) and Varah (1979)).

Following Reddy and Trefethen (1990, 1992) we now formulate a theorem which shows that the resolvent condition (3.5) can be nicely interpreted in terms of the ϵ-pseudospectra of the matrix hA.

Theorem 4.5 Let the norm $\| \cdot \|$ on $\mathbb{C}^{s,s}$ be induced by a vector norm on $\mathbb{C}^s$, and let V, h, A, M_1 be as in Section 3.2. Then the resolvent condition (3.5) is equivalent to the requirement that for all $\epsilon > 0$ the set $\sigma_\epsilon[hA]$ is contained in $V + M_1 \epsilon D = \{\zeta : \zeta = \xi + \eta \text{ with } \xi \in V, |\eta| \leq M_1 \epsilon\}$.

The theorem can be proved in a straightforward way by using the fact that, according to Theorem 4.2(b), Properties (4.1) and (4.4) with $B = hA$ are equivalent in the situation of the theorem.

Following the ideas of Trefethen (1992), the concept of ϵ-pseudospectra can also be used to *determine numerically* regions V and constants M_1 such that (3.5) holds. In order to explain how this can be done we assume $\| \cdot \|$

to be induced by a vector norm on $\mathbb{C}^s$, write $B = hA$, choose a fixed $\epsilon > 0$, denote the boundary of $\sigma_\epsilon[B]$ by Γ_ϵ and its length by $|\Gamma_\epsilon|$. The set

$$V = \sigma_\epsilon[B] \tag{4.6}$$

can be determined numerically, e.g., by checking for a large set of complex numbers λ whether (4.4) is satisfied. A corresponding constant M_1 can be computed from the formula

$$M_1 = |\Gamma_\epsilon|(2\pi\epsilon)^{-1}. \tag{4.7}$$

In order to establish (4.7) we note that for $\zeta \notin V$ we have

$$(\zeta I - B)^{-1} = \frac{1}{2\pi i} \int_{\Gamma_\epsilon} (\zeta - \lambda)^{-1}(\lambda I - B)^{-1} \, d\lambda$$

and therefore

$$\|(\zeta I - B)^{-1}\| \le \frac{|\Gamma_\epsilon|}{2\pi} \max_{\lambda \in \Gamma_\epsilon} |(\zeta - \lambda)^{-1}|\epsilon^{-1} = M_1 \, d(\zeta, V)^{-1}.$$

It should be noted that both V and M_1 depend on ϵ so that it may pay to evaluate (4.6) and (4.7) for various values of ϵ.

We refer to Trefethen (1992) for closely related considerations and many further interesting applications of ϵ-pseudospectra.

4.2. The M-numerical range

When applying the stability results discussed in Sections 3.2 and 3.3, one may want to *prove* rigorously resolvent conditions of the form (3.5). In the following we show that the concept of the M-numerical range, introduced by Lenferink and Spijker (1990), can be helpful. The M-numerical range, to be defined below, can be viewed as a generalization of the *classical numerical range* (for an $s \times s$ matrix B),

$$\{x^*Bx : x \in \mathbb{C}^s \text{ with } x^*x = 1\}.$$

The resolvent condition (3.5) will be seen to be satisfied when V contains the M_1-numerical range of hA.

Let $\| \cdot \|$ be a matrix norm on $\mathbb{C}^{s,s}$, and M a constant with $M \ge \|I\|$. Assume B is a given $s \times s$ matrix. We focus on disks

$$D[\gamma, \rho] = \{\zeta : \zeta \in \mathbb{C} \text{ with } |\zeta - \gamma| \le \rho\}$$

with arbitrary $\gamma \in \mathbb{C}$, $\rho \ge 0$ such that

$$\|(B - \gamma I)^k\| \le M\rho^k \quad \text{for } k = 1, 2, 3, \ldots. \tag{4.8}$$

Definition 4.6 The *M-numerical range of B with respect to the norm* $\|\cdot\|$ is the set $\tau_M[B]$ defined by

$$\tau_M[B] = \bigcap D[\gamma, \rho], \tag{4.9}$$

where the intersection is over all disks $D[\gamma, \rho]$ with property (4.8).

Let W be a nonempty, closed and convex subset of $\mathbb{C}$. If ξ belongs to the boundary ∂W of W and

$$\mathrm{Re}\{e^{-i\theta}(\zeta - \xi)\} \leq 0 \quad \text{for all } \zeta \in W,$$

where θ is a real constant, then θ is called a *normal direction* to W at ξ.

In order to formulate a basic theorem about the M-numerical range we consider the following four conditions on B:

$$\tau_M[B] \subset W, \tag{4.10}$$

$\zeta I - B$ is invertible and $\|(\zeta I - B)^{-k}\| \leq M \cdot \mathrm{d}(\zeta, W)^{-k}$
for all $\zeta \notin W$ and $k = 1, 2, 3, \ldots,$ $\qquad$ (4.11)

$\|\exp[te^{-i\theta}(B - \xi I)]\| \leq M$ for all $t \geq 0, \xi \in \partial W$
and normal directions θ to W at ξ, $\qquad$ (4.12)

there is a unital matrix norm $\|\cdot\|'$ on $\mathbb{C}^{s,s}$ with
corresponding 1-numerical range $\tau_1'[B] \subset W$ and $\qquad$ (4.13)
$M^{-1}\|A\| \leq \|A\|' \leq M\|A\|$ (for all $s \times s$ matrices A).

The following theorem was proved by Lenferink and Spijker (1990).

Theorem 4.7 Properties (4.10)–(4.13) are equivalent to each other.

Clearly, $\tau_M[B]$ is the smallest nonempty, closed and convex set $W \subset \mathbb{C}$ with property (4.10). Therefore, Theorem 4.7 reveals three new characterizations of the M-numerical range. We see that $\tau_M[B]$ equals the smallest nonempty, closed and convex set $W \subset \mathbb{C}$ with property (4.11), and the same holds with regard to properties (4.12) and (4.13).

It is clear that (4.11) is fulfilled for any set W with

$$\tau_M[B] \subset W \subset \mathbb{C}.$$

In view of Definition 4.6 we thus can make the two following observations.

(I) If V is any closed subset of $\mathbb{C}$ with $\tau_M[hA] \subset V$, then (3.5) is fulfilled with $M_1 = M$.

(II) In order to construct a set V as in (I) we only have to determine a finite number of pairs γ_j, ρ_j such that $B = hA$ satisfies (4.8) for all $\gamma = \gamma_j$, $\rho = \rho_j$. Clearly the set $V = \bigcap_j D[\gamma_j, \rho_j]$ is as required.

In Lenferink and Spijker (1990) (Theorem 3.1) and Lenferink and Spijker

(1991b) stability estimates were derived essentially along the lines of the observations (I), (II).

We finally note that for $M = 1$ the set (4.9) coincides with the so-called *algebra numerical range* (see, e.g., Bonsall and Duncan (1980) and Lenferink and Spijker (1990)). In this case some simple expressions for $\tau_1[B]$ are known. For $\| \cdot \| = \| \cdot \|_p$ with $p = 1, 2, \infty$ these expressions are as follows.

- Let $\| \cdot \| = \| \cdot \|_\infty$. Then $\tau_1[B]$ is equal to the convex hull of the union of the Gerschgorin disks $D[\gamma_j, \rho_j]$ defined by $\gamma_j = \beta_{jj}$ and $\rho_j = \sum_{k \neq j} |\beta_{jk}|$, where β_{jk} denote the entries of B (see, e.g., Lenferink and Spijker (1990) (Section 3.1.1)).

- Let $\| \cdot \| = \| \cdot \|_1$. Then $\tau_1[B]$ is easily seen to be equal to the 1-numerical range of B^T with respect to $\| \cdot \|_\infty$.

- Let $\| \cdot \| = \| \cdot \|_2$. Then $\tau_1[B]$ equals the classical numerical range $\{x^*Bx : x \in \mathbb{C}^s \text{ with } x^*x = 1\}$ – see the papers mentioned earlier.

4.3. Bounds on the exponential function of a matrix

In Section 3 we focused on stability of the time stepping process (3.2). In this subsection we shall investigate stability of the underlying initial value problem (3.1) itself.

Suppose the initial value u_0 in (3.1) is replaced by a slightly perturbed vector $\tilde{u}_0$ and $\tilde{U}(t)$ is the solution to (3.1) with initial value $\tilde{u}_0$. In analogy to Section 1.2, (3.1) is said to be *stable* if a small perturbation $v_0 = \tilde{u}_0 - u_0$ always yields errors $V(t) = \tilde{U}(t) - U(t)$ (for $t > 0$) that are also small. Therefore, the stability analysis of the initial value problem (3.1) amounts to bounding $V(t)$ (for $t > 0$) suitably in terms of v_0. Since $V(t) = e^{tA}v_0$ we consider the stability property

$$\|e^{tA}\| \leq M_0 \quad \text{for all } t \geq 0, \tag{4.14}$$

where M_0 is a positive constant and $\| \cdot \|$ a norm on $\mathbb{C}^{s,s}$.

By using the Jordan canonical form of A it can be easily seen that there exists a finite M_0 with the stability property (4.14) if and only if the following eigenvalue condition is satisfied:

> All eigenvalues λ of A have a real part $\operatorname{Re} \lambda \leq 0$,
> and the geometric multiplicity of each eigenvalue λ (4.15)
> with $\operatorname{Re} \lambda = 0$ is equal to its algebraic multiplicity.

Similar to the situation for the eigenvalue conditions (1.7) and (3.4), eigenvalue condition (4.15) can be reliable (e.g. for normal matrices) or misleading (for families of matrices that are not normal). A reliable criterion for the stability property (4.14), in general situations, can be based on the resolvent of A, and reads

$\zeta I - A$ is invertible and $\|(\zeta I - A)^{-1}\| \leq M_1 (\operatorname{Re} \zeta)^{-1}$
for all ζ with $\operatorname{Re} \zeta > 0$. $\qquad$ (4.16)

In the following we shall discuss the relation between the stability property (4.14) and the resolvent condition (4.16).

By using the formula

$$(\zeta I - A)^{-1} = \int_0^\infty e^{-\zeta t} e^{tA} \, dt \quad \text{for all } \zeta \text{ with } \operatorname{Re} \zeta > 0$$

one can easily see that (4.14) implies (4.16) with $M_1 = M_0$. Conversely, (4.16) implies (4.14) with M_0 depending only on M_1 and the dimension s, but otherwise independent of A. Various authors have studied the size of the optimal M_0 as a function of M_1 and s for the spectral norm or other special norms (see Miller (1968), Laptev (1975), Gottlieb and Orszag (1977) and LeVeque and Trefethen (1984)). The following theorem sharpens and generalizes their results to the case of arbitrary norms on $\mathbb{C}^{s,s}$.

Theorem 4.8 Let $s \geq 1$, $A \in \mathbb{C}^{s,s}$ and $\| \cdot \|$ denote an arbitrary norm on $\mathbb{C}^{s,s}$. If (4.16) holds for some M_1, then

$$\|e^{tA}\| \leq es M_1 \quad \text{for all } t \geq 0. \qquad (4.17)$$

Proof. The proof is analogous to that of Theorem 2.1, and is based on the representation of e^{tA} for $t > 0$ as

$$e^{tA} = \frac{1}{2\pi i} \int_{\operatorname{Re} \zeta = t^{-1}} e^{t\zeta} (\zeta I - A)^{-1} \, d\zeta$$

(see also LeVeque and Trefethen (1984)). $\square$

The sharpness of the bound (4.17) is considered in the following theorem, which generalizes a result by LeVeque and Trefethen (1984) for the spectral norm to the case of arbitrary norms on $\mathbb{C}^{s,s}$.

Theorem 4.9 Let $s \geq 2$ and an arbitrary norm $\| \cdot \|$ on $\mathbb{C}^{s,s}$ be given. Then we have for all $t > 0$

$$\sup\{\|e^{tA}\|/M_1(A) : A \in \mathbb{C}^{s,s}, M_1(A) < \infty\} \geq$$
$$\frac{s^s}{(s-1)!} e^{-(s-1)} > e(2\pi)^{-1/2}(s-1)^{1/2}, \qquad (4.18)$$

where $M_1(A)$ denotes the smallest M_1 such that (4.16) holds (we define $M_1(A) = \infty$ if (4.16) is not fulfilled for any M_1).

Proof. Let $t > 0$ be given. Define $A \in \mathbb{C}^{s,s}$ by $A = -\alpha I + \gamma E$ where $\alpha > 0$ will be specified later, $\gamma > 0$ is large and E is the matrix defined by (1.9). After some calculations similar to those in the proof of Theorem 2.2

we obtain the relations

$$M_1(A) \leq s^{-s}(s-1)^{s-1}\alpha^{1-s}\gamma^{s-1}\|E^{s-1}\|\left(1+\mathcal{O}(\gamma^{-1})\right),$$

$$\|e^{tA}\| = e^{-\alpha t}\frac{t^{s-1}}{(s-1)!}\gamma^{s-1}\|E^{s-1}\|\left(1+\mathcal{O}(\gamma^{-1})\right)$$

and hence

$$\|e^{tA}\|/M_1(A) \geq e^{-\alpha t}(\alpha t)^{s-1}s^s(s-1)^{1-s}/(s-1)! + \mathcal{O}(\gamma^{-1}) \qquad \text{(as } \gamma \to \infty).$$

If we choose $\alpha = (s-1)/t$, the right-hand side of the inequality tends to $s^s e^{-(s-1)}/(s-1)!$ as $\gamma \to \infty$, which is strictly larger than $e(2\pi)^{-1/2}(s-1)^{1/2}$ by Stirling's formula (see e.g. the proof of Theorem 2.4). $\square$

Note that the upper bound $\|e^{tA}\|/M_1(A) \leq es$ of Theorem 4.8 and the lower bound (4.18) of Theorem 4.9 differ by a factor $\sim \sqrt{2\pi s}$. This is a less satisfactory situation than in Section 2.2, where the upper bound $\|B^n\|/M_1(B) \leq es$ was shown to be essentially sharp. Further, Theorem 4.9 does not shed any light on the sharpness question for *fixed* constants M_1, since arbitrarily large $M_1(A)$ are allowed in (4.18).

For the special situation where (4.16) holds with $M_1 = 1$, the upper bound (4.17) can be improved. This is the content of the following theorem, which is a well-known result in semigroup theory (see, e.g., Pazy (1983) or Theorem 4.7 above).

Theorem 4.10 Let $M_1 = 1$ and $\|\cdot\|$ be a matrix norm. Then (4.16) implies (4.14) with $M_0 = 1$.

In the remainder of this section we will answer the question whether – in addition to the upper bound (4.17) – there exists an upper bound depending only on t and M_1. This would be analogous to the situation in Section 2, where $\|B^n\|$ was not only bounded by esM_1, but also by $e(n+1)M_1$ (see Theorem 2.1). Clearly, this question is equivalent to the existence of a function g such that

$$\|e^{tA}\| \leq g(t, M_1) \quad \text{for all } t \geq 0,$$

whenever resolvent condition (4.16) is fulfilled. The nonexistence of such a function g is proved in Theorem 4.11.

Theorem 4.11 The matrices

$$A_s = \begin{pmatrix} -1 & -2 & \cdots & -2 \\ & -1 & \ddots & \vdots \\ & & \ddots & -2 \\ \mathbf{0} & & & -1 \end{pmatrix} \in \mathbb{C}^{s,s}, \quad s \geq 1,$$

satisfy the resolvent condition (4.16) with $\|\cdot\| = \|\cdot\|_\infty$ and $M_1 = 2$. Moreover we have

$$\lim_{s \to \infty} \|e^{tA_s}\|_\infty = \infty \quad \text{for all } t > 0. \tag{4.19}$$

Proof. For $A = A_s$ we have $A = -(I + E)(I - E)^{-1}$, where E is the matrix defined by (1.9). Hence we arrive for $\zeta \in \mathbb{C}$ with $\text{Re}\,\zeta > 0$ at

$$(\zeta I - A)^{-1} = \frac{1}{\zeta + 1} \left\{ I - \frac{2}{\zeta + 1} \sum_{j=1}^{s-1} \left(\frac{\zeta - 1}{\zeta + 1}\right)^{j-1} E^j \right\},$$

from which we obtain

$$(\text{Re}\,\zeta)\|(\zeta I - A)^{-1}\|_\infty \leq \frac{\text{Re}\,\zeta}{|\zeta + 1|} \left\{ 1 + \frac{2}{|\zeta + 1|} \sum_{j=1}^{\infty} \left|\frac{\zeta - 1}{\zeta + 1}\right|^{j-1} \right\}$$

$$= (\text{Re}\,\zeta)|\zeta + 1|^{-1}\{1 + 2(|\zeta + 1| - |\zeta - 1|)^{-1}\} \leq 2,$$

implying (4.16) with constant $M_1 = 2$.

In order to prove (4.19) we fix $t > 0$ and define the complex function f by $f(\zeta) = \exp[-t(1 + \zeta)(1 - \zeta)^{-1}]$ (for all $\zeta \neq 1$). The function f is analytic on $\mathbb{C} \setminus \{1\}$ and can therefore be represented on the open unit disk by a power series

$$f(\zeta) = \sum_{n=0}^{\infty} a_n(t)\zeta^n.$$

Since $f(e^{i\theta}) = \exp[-it/\tan(\frac{1}{2}\theta)]$ (for small positive θ), we see that the limit $\lim_{\theta \to 0} f(e^{i\theta})$ does not exist, implying that

$$\sum_{n=0}^{\infty} |a_n(t)| = \infty. \tag{4.20}$$

The proof of (4.19) is completed by combining (4.20) with

$$e^{tA} = f(E) = \sum_{n=0}^{s-1} a_n(t)E^n, \quad \|e^{tA}\|_\infty = \sum_{n=0}^{s-1} |a_n(t)|.$$

$\square$

We remark that after completion of the present paper new results related to this were found by Kraaijevanger (1992) for the maximum norm $\|\cdot\|_\infty$.

5. Applications and examples

5.1. Range of applications

It is clear from Sections 1.2 and 3.1 that the stability estimates discussed in Sections 2 and 3 are relevant to numerical processes for solving linear differential equations which are essentially more general than the classical test problems mentioned in Section 1.1. The results of Sections 2 and 3 have a potential for clarifying actual stability problems in cases where Fourier transformation techniques are unlikely to be successful. Such cases comprise linear differential equations with nonsmooth variable coefficients, spectral methods, and finite difference or finite element methods with highly irregular geometries.

Still, at first sight, most of the stability results in Sections 2 and 3 may be considered to be quite weak in that the upper bounds for $\|B^n\|$ do not remain bounded as $n \to \infty$ or $s \to \infty$. However, in computational practice troublesome instability usually manifests itself by an exponential growth of the error. Evidently such growth is not possible when the upper bounds of Sections 2 and 3 are in force – these upper bounds grow at the rate of some power of s or n. In fact, various authors have allowed such polynomial growth in their definition of stability – e.g. Strang (1960), Forsythe and Wasow (1960) and Gottlieb and Orszag (1977).

In Section 1.2 we indicated that bounds on $\|B^n\|$ are useful when analysing the propagation of *rounding errors* $v_0 = \tilde{u}_0 - u_0$. But the stability estimates of Sections 2 and 3 are also relevant to the question of how fast the so-called *global discretization errors*

$$d_n = U(nh) - u_n \tag{5.1}$$

approach zero when $h = \Delta t \to 0$. Here $U(t)$, u_n satisfy (3.1) and (3.2), respectively. We define the *local discretization error* e_n by $e_n = h^{-1}r_n$, where r_n denotes the residual in the right-hand member of (3.2) when u_n and u_{n-1} in that formula are replaced by $U(nh)$ and $U((n-1)h)$, respectively. Writing $B = \varphi(hA)$ we then have $d_n = Bd_{n-1} + he_n$ and therefore

$$d_n = h \sum_{j=1}^{n} B^{n-j} e_j. \tag{5.2}$$

From this representation it is evident that the stability estimates from Sections 2 and 3, in combination with bounds on the local discretization errors, can be used to derive bounds on the errors (5.1). We note that the same holds true when in the numerical solution of a given partial differential equation, with solution $u(x,t)$, one replaces the vector $U(nh)$ in (5.1) by a suitable projection in $\mathbb{C}^s$ of the true $u(x,t)$. Of course e_n should then be defined accordingly.

If $nh = t > 0$ is fixed, and the bounds on $\|B^n\|$ grow with some power of n (or s), then a straightforward application of (5.2) yields bounds on the global errors that are of a *lower* order than the local discretization errors. But, Strang (1960) has already shown that, even in the presence of such polynomial growth, it may be possible to establish bounds on the global discretization errors that are of the *same* order as the local errors – provided the problem itself is sufficiently smooth.

For subsequent extensions of Strang's result to linear and nonlinear problems, see, e.g., Strang (1964), Spijker (1972), Brenner and Thomée (1979), Thomée (1990), and the references therein. In the case of nonlinear problems the basic assumptions in these papers include the requirement that a linearization of the actual numerical process is stable (in the sense that polynomial growth is allowed). Therefore the stability analysis of linear processes (as in the present paper) may contribute to the stability analysis of numerical processes for nonlinear differential equations, see also López-Marcos and Sanz-Serna (1988).

We finally comment on the relevance of the bounds on $\| \exp(tA) \|$ obtained in Section 4.3. Similar to the situation for $\| \varphi(hA)^n \|$ discussed above, these bounds are not only relevant for studying the effect of initial perturbations $v_0 = \tilde{u}_0 - u_0$ (such as rounding errors) on the solution $U(t)$ of initial value problem (3.1), but also for analysing the global discretization error

$$d(t) = \tilde{U}(t) - U(t)$$

when (3.1) is obtained by semi-discretization of a partial differential equation. Here $\tilde{U}(t)$ denotes a suitable projection in $\mathbb{C}^s$ of the solution to the partial differential equation. Defining the corresponding local discretization error $e(t)$ to be the residual appearing in the right-hand side of the differential equation in (3.1) when $U(t)$ is replaced by $\tilde{U}(t)$, we readily obtain $d'(t) = Ad(t) + e(t)$, so that

$$d(t) = \int_0^t \exp((t-\tau)A)e(\tau)\, d\tau.$$

From this representation, which is a continuous analogue of (5.2), one can derive bounds on the global errors $d(t)$ by combining bounds on the local errors $e(t)$ and the bounds on $\| \exp(tA) \|$ obtained in Section 4.3.

5.2. *Examples pertinent to the theory of Section 3*

In order to illustrate some of the preceding notions and theorems we consider the simple initial-boundary value problem

$$\begin{aligned} u_t(x,t) &= (a(x)u(x,t))_x + g(x,t), \\ u(x,0) &= f(x), \quad u(1,t) = 0, \quad \text{where } 0 < x < 1,\ t > 0. \end{aligned} \tag{5.3}$$

Here a, g, f denote given functions with $a(x) \geq 0$. The values $u(x,t)$ are considered unknown for $0 \leq x < 1$, $t > 0$.

We select an integer $s \geq 1$ and define $\Delta x = 1/s$. Approximating $(au)_x$ by the forward difference quotient (see, e.g., Richtmyer and Morton (1967))

$$(a(x)u(x,t))_x \simeq (\Delta x)^{-1}\{a(x + \Delta x)u(x + \Delta x, t) - a(x)u(x,t)\},$$

problem (5.3) is transformed into a semi-discrete problem of the form (3.1) with $A = (\alpha_{jk})$, where

$$\begin{cases} \alpha_{jj} & = & -sa((j-1)/s) & (j = 1, 2, \ldots, s), \\ \alpha_{j,j+1} & = & sa(j/s) & (j = 1, 2, \ldots, s-1), \\ \alpha_{jk} & = & 0 & \text{otherwise.} \end{cases}$$

Further,

$$\begin{aligned} b(t) & = & (g(0,t), g(1/s,t), \ldots, g((s-1)/s,t))^{\mathrm{T}}, \\ u_0 & = & (f(0), f(1/s), \ldots, f((s-1)/s))^{\mathrm{T}}, \end{aligned}$$

and the jth component $U_j(t)$ of the solution $U(t)$ to (3.1) approximates the solution $u(x,t)$ to (5.3) at $(x,t) = ((j-1)/s, t)$ (for $j = 1, 2, \ldots, s$).

In the following we focus on conditions that guarantee the stability of the fully discrete numerical process (3.2). We will derive upper bounds for $\|\varphi(hA)^n\|_p$ in the cases $p = 1$ and $p = \infty$. For simplicity we assume that the ratio $\mu = h/\Delta x$ is fixed. Further we introduce the constants

$$\alpha = \max_{0 \leq x \leq 1} a(x), \quad \beta = \max_{0 \leq x \leq 1} a'(x).$$

Case 1: $p = 1$. For the norm $\|\cdot\|_1$ one easily verifies that the matrix hA satisfies

$$\|hA + \alpha\mu I\|_1 \leq \alpha\mu.$$

Applying part (a) of Theorem 2.1 to the matrix $B = I + (\alpha\mu)^{-1}hA$ we see that hA satisfies the resolvent condition (3.5) with

$$M_1 = 1 \quad \text{and} \quad V = \{\zeta : |\zeta + \alpha\mu| \leq \alpha\mu\}.$$

Suppose that $\alpha\mu \leq r$, where r is the stability radius, which was defined in Section 3.2 (Result 9) to be the radius of the largest disk in the complex left half-plane which is tangent to the imaginary axis at the origin and lies in the stability region S (defined by (3.3)). Then it follows from Remark 3.2 and the material in Section 3.2 that

$$\|\varphi(hA)^n\|_1 \leq \gamma \min(s, \sqrt{n}) \quad \text{for } n = 1, 2, 3, \ldots,$$

where γ depends only on φ.

Under the more stringent condition $\alpha\mu < r$ it follows from Result 9 (with $r_0 = \alpha\mu$) and Remark 3.2 that we even have

$$\|\varphi(hA)^n\|_1 \leq \gamma \quad \text{for } n = 1, 2, 3, \ldots,$$

where γ depends only on φ and $\alpha\mu$. *Case 2:* $p = \infty$. For the norm $\|\cdot\|_\infty$ one easily verifies that the matrix hA satisfies

$$\|hA + \alpha\mu I\|_\infty \leq \alpha\mu + \beta h.$$

When $\beta \leq 0$ we can proceed as in Case 1. In the following we assume that $\beta > 0$.

An application of part (a) of Theorem 2.1 to the matrix $B = (\alpha\mu + \beta h)^{-1}(hA + \alpha\mu I)$ shows that hA satisfies the resolvent condition (3.5) with

$$M_1 = 1 \quad \text{and} \quad V = \{\zeta : |\zeta + \alpha\mu| \leq \alpha\mu + \beta h\}.$$

Suppose that $\alpha\mu \leq r$ and $r < \infty$. Let the stability region S be bounded and $\varphi'(\zeta) \neq 0$ on ∂S. Then it follows from Remark 3.4 (parts (a) and (b)) that we have

$$\|\varphi(hA)^n\|_\infty \leq \gamma_1 e^{\gamma_2 \beta nh} \min(s, \sqrt{n}) \quad \text{for} \quad n = 1, 2, 3, \ldots$$

whenever $\beta h \leq \gamma_3$. Here $\gamma_1, \gamma_2, \gamma_3 > 0$ only depend on φ.

In case $r = \infty$ we can apply the general result mentioned in Remark 3.4 (part (c)) so as to obtain a similar stability estimate.

Further illustrations of the theory of Section 3 can be found, e.g., in Lenferink and Spijker (1991b) and Reddy and Trefethen (1992). In Kraaijevanger *et al.* (1987) an example was presented pertinent to problem (1.3).

5.3. Numerical illustrations

In order to give a numerical illustration of the material of Section 5.2 we consider the classical fourth-order Runge–Kutta method (see, e.g., Butcher (1987)). Applying this method to the semi-discrete problem (3.1) as specified in Section 5.2, one arrives at a fully discrete process (3.2) with

$$\varphi(\zeta) = 1 + \zeta + \frac{\zeta^2}{2!} + \frac{\zeta^3}{3!} + \frac{\zeta^4}{4!}. \tag{5.4}$$

The corresponding stability radius r is equal to

$$r = 1.393 \tag{5.5}$$

(rounded to four decimal places). For later use we note that it follows from the definition of r that

$$\text{the interval } [-2r, 0] \text{ is contained in } S. \tag{5.6}$$

We consider the matrix A, defined in Section 5.2, with three different choices for the function $a(x)$, viz.

$$a_1(x) = 1, \quad a_2(x) = 1 - x^{10}, \quad a_3(x) = 1 - x.$$

Using the notations of the preceding subsection, we have for all of these functions that

$$\alpha = 1, \quad \beta \leq 0.$$

For given s and function $a(x)$ we shall measure the stability of the corresponding numerical process by the quantity

$$c(\mu, a) = \sup_{n \geq 0} \|\varphi(hA)^n\|_\infty.$$

Table 1. *Values of* $c(\mu, a_i)$ *for* $s = 80$

h	μ	$c(\mu, a_1)$	$c(\mu, a_2)$	$c(\mu, a_3)$
0.0125	1	1	1	1
0.0150	1.2	1.12	1.12	1.08
0.0175	1.4	2.26	2.26	1.30
0.0200	1.6	3.47×10^9	2.58×10^7	4.20
0.0225	1.8	4.93×10^{19}	7.56×10^{15}	1.73×10^2
0.0250	2.0	9.06×10^{50}	2.68×10^{25}	4.17×10^4

For $s = 80$ we have listed some values of $c(\mu, a_i)$ in Table 1. In the table we see good stability up to $\mu = 1.4$. This is perfectly in agreement with the conditions of Section 5.2 since, in view of (5.5), the requirement

$$\alpha\mu \leq r$$

amounts to $\mu \leq 1.393$. For $\mu > 1.4$ we see large values in the table, indicating strong instability. It is worth noting that for all $\mu \leq 2.0$ requirement (3.4) is still fulfilled, since for these μ we have

$$\sigma[hA] \subset S \setminus \partial S.$$

This inclusion follows from (5.5) and (5.6) and the fact that, for our functions a_i,

$$\sigma[hA] \subset [-\mu, 0).$$

The numerical results thus confirm the reliability of the stability criteria discussed in Section 5.2, and the failing of the eigenvalue condition (3.4).

For further numerical illustrations related to the material of Sections 2 and 3 we refer to Trefethen (1988), Lenferink and Spijker (1991b) and Reddy and Trefethen (1992). For a numerical illustration pertinent to problem (1.3) see Kraaijevanger *et al.* (1987).

Acknowledgements

The authors wish to thank L.N. Trefethen and S.C. Reddy for many stimulating discussions on the topic of this paper. They are also indebted to L.N. Trefethen and A. Iserles for editorial comments on a preliminary version of this paper, and to J. Groeneweg for performing the computations displayed in Section 5.3. This research has been supported by the Netherlands Organization for Scientific Research (N.W.O.). This research has been made possible by the award of a fellowship of the Royal Netherlands Academy of Arts and Sciences (K.N.A.W.) to Kraaijevanger.

REFERENCES

M. Abramowitz and I.A. Stegun (1965), *Handbook of Mathematical Functions*, Dover (New York).

F.F. Bonsall and J. Duncan (1971), *Numerical Ranges of Operators on Normed Spaces and of Elements of Normed Algebras*, Cambridge University Press (Cambridge).

F.F. Bonsall and J. Duncan (1980), 'Numerical ranges', in *Studies in Functional Analysis* (R.G. Bartle, ed.), The Mathematical Association of America, 1–49.

Ph. Brenner and V. Thomée (1979), 'On rational approximations of semigroups', *SIAM J. Numer. Anal.* **16**, 683–694.

J.C. Butcher (1987), *The Numerical Analysis of Ordinary Differential Equations*, John Wiley (Chichester).

C. Canuto, M.Y. Hussaini, A. Quarteroni and T.A. Zang (1988), *Spectral Methods in Fluid Dynamics*, Springer (New York).

J.B. Conway (1985), *A Course in Functional Analysis*, Springer (New York).

M. Crouzeix (1987), 'On multistep approximation of semigroups in Banach spaces', *J. Comput. Appl. Math.* **20**, 25–35.

M. Crouzeix, S. Larsson, S. Piskarev and V. Thomée (1991), 'The stability of rational approximations of analytic semigroups', Technical Report 1991:28, Chalmers University of Technology & the University of Göteborg (Göteborg).

G.E. Forsythe and W.R. Wasow (1960), *Finite Difference Methods for Partial Differential Equations*, John Wiley (New York).

D. Gottlieb and S.A. Orszag (1977), *Numerical Analysis of Spectral Methods*, Soc. Ind. Appl. Math. (Philadelphia).

D.F. Griffiths, I. Christie and A.R. Mitchell (1980), 'Analysis of error growth for explicit difference schemes in conduction-convection problems', *Int. J. Numer. Meth. Engrg* **15**, 1075–1081.

R.D. Grigorieff (1991), 'Time discretization of semigroups by the variable two-step BDF method', in *Numerical Treatment of Differential Equations* (K. Strehmel, ed.), Teubner (Leipzig), 204–216.

B. Gustafsson, H.-O. Kreiss and A. Sundström (1972), 'Stability theory of difference approximations for mixed initial boundary value problems. II', *Math. Comput.* **26**, 649–686.

E. Hairer and G. Wanner (1991), *Solving Ordinary Differential Equations*, Vol. II, Springer (Berlin).

P. Henrici (1974), *Applied and Computational Complex Analysis*, Vol. 1, John Wiley (New York).

R.A. Horn and C.R. Johnson (1990), *Matrix Analysis*, Cambridge University Press (Cambridge).

F. John (1952), 'On integration of parabolic equations by difference methods', *Comm. Pure Appl. Math.* **5**, 155–211.

J.F.B.M. Kraaijevanger (1992), 'Two counterexamples related to the Kreiss matrix theorem', submitted to *BIT*.

J.F.B.M. Kraaijevanger, H.W.J. Lenferink and M.N. Spijker (1987), 'Stepsize restrictions for stability in the numerical solution of ordinary and partial differential equations', *J. Comput. Appl. Math.* **20**, 67–81.

H.-O. Kreiss (1962), 'Über die Stabilitätsdefinition für Differenzengleichungen die partielle Differentialgleichungen approximieren', *BIT* **2**, 153–181.

H.-O. Kreiss (1966), 'Difference approximations for the initial-boundary value problem for hyperbolic differential equations', in *Numerical Solution of Nonlinear Differential Equations* (D. Greenspan, ed.), John Wiley (New York), 141–166.

H.-O. Kreiss (1990), 'Well posed hyperbolic initial boundary value problems and stable difference approximations', in *Proc. Third Int. Conf. on Hyperbolic Problems, Uppsala, Sweden.*

H.-O. Kreiss and L. Wu (1992), 'On the stability definition of difference approximations for the initial boundary value problem', to appear in *Comm. Pure Appl. Math.*

H.J. Landau (1975), 'On Szegő's eigenvalue distribution theorem and non-Hermitian kernels', *J. d'Analyse Math.* **28**, 335–357.

G.I. Laptev (1975), 'Conditions for the uniform well-posedness of the Cauchy problem for systems of equations', *Sov. Math. Dokl.* **16**, 65–69.

H.W.J. Lenferink and M.N. Spijker (1990), 'A generalization of the numerical range of a matrix', *Linear Algebra Appl.* **140**, 251–266.

H.W.J. Lenferink and M.N. Spijker (1991a), 'On a generalization of the resolvent condition in the Kreiss matrix theorem', *Math. Comput.* **57**, 211–220.

H.W.J. Lenferink and M.N. Spijker (1991b), 'On the use of stability regions in the numerical analysis of initial value problems', *Math. Comput.* **57**, 221–237.

R.J. LeVeque and L.N. Trefethen (1984), 'On the resolvent condition in the Kreiss matrix theorem', *BIT* **24**, 584–591.

J.C. López-Marcos and J.M. Sanz-Serna (1988), 'Stability and convergence in numerical analysis III: linear investigation of nonlinear stability', *IMA J. Numer. Anal.* **8**, 71–84.

C. Lubich (1991), 'On the convergence of multistep methods for nonlinear stiff differential equations', *Numer. Math.* **58**, 839–853.

C. Lubich and O. Nevanlinna (1991), 'On resolvent conditions and stability estimates', *BIT* **31**, 293–313.

C.A. McCarthy (1992), 'A strong resolvent condition need not imply power-boundedness', to appear in *J. Math. Anal. Appl.*

C.A. McCarthy and J. Schwartz (1965), 'On the norm of a finite boolean algebra of projections, and applications to theorems of Kreiss and Morton', *Comm. Pure Appl. Math.* **18**, 191–201.

T. Meis and U. Marcowitz (1981), *Numerical Solution of Partial Differential Equations*, Springer (New York).

J. Miller (1968), 'On the resolvent of a linear operator associated with a well-posed Cauchy problem', *Math. Comput.* **22**, 541–548.

J. Miller and G. Strang (1966), 'Matrix theorems for partial differential and difference equations', *Math. Scand.* **18**, 113–123.

K.W. Morton (1964), 'On a matrix theorem due to H.O. Kreiss', *Comm. Pure Appl. Math.* **17**, 375–379.

K.W. Morton (1980), 'Stability of finite difference approximations to a diffusion-convection equation', *Int. J. Num. Meth. Engrg* **15**, 677–683.

O. Nevanlinna (1984), 'Remarks on time discretization of contraction semigroups', Report HTKK–MAT–A225, Helsinki Univ. Techn. (Helsinki).

J.T. Oden and J.N. Reddy (1976), *An Introduction to the Mathematical Theory of Finite Elements*, John Wiley (New York).

S.V. Parter (1962), 'Stability, convergence, and pseudo-stability of finite-difference equations for an over-determined problem', *Numer. Math.* **4**, 277–292.

A. Pazy (1983), *Semigroups of Linear Operators and Applications to Partial Differential Equations*, Springer (New York).

C. Pearcy (1966), 'An elementary proof of the power inequality for the numerical radius', *Michigan Math. J.* **13**, 289–291.

S.C. Reddy and L.N. Trefethen (1990), 'Lax-stability of fully discrete spectral methods via stability regions and pseudo-eigenvalues', *Comput. Meth. Appl. Mech. Engrg* **80**, 147–164.

S.C. Reddy and L.N. Trefethen (1992), 'Stability of the method of lines', *Numer. Math.* **62**, 235–267.

L. Reichel and L.N. Trefethen (1992), 'Eigenvalues and pseudo-eigenvalues of Toeplitz matrices', *Linear Algebra Appl.* **162–164**, 153–185.

R.D. Richtmyer and K.W. Morton (1967), *Difference Methods for Initial-value Problems*, 2nd Ed., John Wiley (New York).

W. Rudin (1973), *Functional Analysis*, McGraw-Hill (New York).

J.C. Smith (1985), 'An inequality for rational functions', *Amer. Math. Monthly* **92**, 740–741.

M.N. Spijker (1972), 'Equivalence theorems for nonlinear finite-difference methods', *Springer Lecture Notes in Mathematics*, Vol. **267**, Springer (Berlin), 233–264.

M.N. Spijker (1985), 'Stepsize restrictions for stability of one-step methods in the numerical solution of initial value problems', *Math. Comput.* **45**, 377–392.

M.N. Spijker (1991), 'On a conjecture by LeVeque and Trefethen related to the Kreiss matrix theorem', *BIT* **31**, 551–555.

W.G. Strang (1960), 'Difference methods for mixed boundary-value problems', *Duke Math. J.* **27**, 221–231.

G. Strang (1964), 'Accurate partial difference methods II. Non–linear problems', *Numer. Math.* **6**, 37–46.

G. Strang and G.J. Fix (1973), *An Analysis of the Finite Element Method*, Prentice-Hall (Englewood Cliffs).

E. Tadmor (1981), 'The equivalence of L_2-stability, the resolvent condition, and strict H-stability', *Linear Algebra Appl.* **41**, 151–159.

V. Thomée (1990), 'Finite difference methods for linear parabolic equations', in *Handbook of Numerical Analysis I* (P.G. Ciarlet and J.L. Lions, eds), North-Holland (Amsterdam), 5–196.

L.N. Trefethen (1988), 'Lax-stability vs. eigenvalue stability of spectral methods', in *Numerical Methods for Fluid Dynamics III* (K.W. Morton and M.J. Baines, eds), Clarendon Press (Oxford), 237–253.

L.N. Trefethen (1992), *Spectra and Pseudospectra*, book in preparation.

J.M. Varah (1979), 'On the separation of two matrices', *SIAM J. Numer. Anal.* **16**, 216–222.

E. Wegert and L.N. Trefethen (1992), 'From the Buffon needle problem to the Kreiss matrix theorem', to appear in *Amer. Math. Monthly*.

Acta Numerica (1993), *pp.* 239–284

Finite element solution of the Navier–Stokes equations

Michel Fortin
Département de mathématiques et de statistique
Université Laval
Québec, Canada
E-mail: mfortin@mat.ulaval.ca

CONTENTS

1. Introduction

Viscous incompressible flows are of considerable interest for applications. Let us mention, for example, the design of hydraulic turbines or rheologically complex flows appearing in many processes involving plastics or molten metals. Their simulation raises a number of difficulties, some of which are likely to remain while others are now resolved. Among the latter are those related to incompressibility which are also present in the simulation of incompressible or nearly incompressible elastic materials. Among the still unresolved are those associated with high Reynolds numbers which are also met in compressible flows. They involve the formation of boundary layers and turbulence, an ever present phenomenon in fluid mechanics, implying that we have to simulate unsteady, highly unstable phenomena.

This article will deal mainly with problems associated with incompressibility effects but will also try to address the other issues. It will not be

an exhaustive presentation and will evidently be somewhat biased by the
prejudices of the author and his ignorance of many areas of an ever growing
literature. The reader might consult the books by Girault and Raviart
(1986) where other mathematical aspects of the problem are treated. The
book by Hughes (1987) is application-oriented and is a good reference for
those interested in the actual implementation of finite element methods.
The reader should also refer to Pironneau (1989) or to Thomasset (1981)
for more information and other aspects of the problem.

2. The finite element method

2.1. Sobolev spaces

Let $L^2(\Omega)$ be the space of square integrable functions. We then define the
Sobolev spaces,

$$H^m(\Omega) = \{v \mid v \in L^2(\Omega),\ D^\alpha v \in L^2(\Omega),\ |\alpha| \le m\} \qquad (2.1)$$

where $D^\alpha v = \partial^{|\alpha|} v / \partial x_1^{\alpha_1} \partial x_2^{\alpha_2} \ldots \partial x_n^{\alpha_n}$, $|\alpha| = \alpha_1 + \alpha_2 + \cdots + \alpha_n$. For our
purpose, the most important of these spaces will be $H^1(\Omega)$ (and some of its
subspaces). We define on $H^m(\Omega)$, the semi-norm

$$|v|_{m,\Omega} = \left(\sum_{|\alpha|=m} \int_\Omega |D^\alpha v(x)|^2 \mathbf{x} \right)^{1/2}. \qquad (2.2)$$

It is then clear that $|v|_{0,\Omega}$ is the usual norm on $L^2(\Omega)$. In general, we shall
use on $H^m(\Omega)$ the standard norm

$$\|v\|_{m,\Omega} = \left(\sum_{|\alpha|\le m} |v|_{m,\Omega}^2 \right)^{1/2}, \qquad (2.3)$$

which on $H_0^1(\Omega)$ reduces to $\|v\|_{1,\Omega} = \left(|v|_{0,\Omega}^2 + |v|_{1,\Omega}^2 \right)^{1/2}$.

2.2. Conforming finite elements

We shall be interested here in *finite element approximations* of $H_0^1(\Omega)$ and
$L^2(\Omega)$. It is not possible to give a complete presentation of the finite element
methods as this would require a book in itself. We refer to Ciarlet (1978),
Ciarlet and Lions (1991), Hughes (1987), Raviart and Thomas (1983) or to
the classical Zienkiewicz (1977) for a general presentation. For more specific
issues and details on many of the topics introduced here, Brezzi and Fortin
(1991) should be a suitable reference. We nevertheless need a minimum of
notation.

    The basic idea of the finite element method is to construct a *partition* T_h of
the domain Ω by subdividing it into triangles or quadrilaterals which will be
called elements. One then builds approximations using polynomial functions

defined element-wise with some continuity requirements at the interfaces between the elements. To approximate $L^2(\Omega)$ no continuity is required while $C^0(\Omega)$-continuity yields correct approximations of $H^1(\Omega)$. Higher-order continuity properties would be required for higher-order Sobolev spaces. To describe finite element approximations more precisely we shall need a few definitions. Let us define on an element $K \in \mathcal{T}_h$ the space of polynomials of degree $\leq k$,

$$P_k(K) = \Big\{ p(x_1, x_2) \mid p(x_1, x_2) = \sum_{i+j \leq k} a_{ij}\, x_1^i\, x_2^j \Big\}. \tag{2.4}$$

The dimension of $P_k(K)$ is $(k+1)(k+2)/2$ for $n = 2$ and for $n = 3$, $(k+1)(k+2)(k+3)/6$. We shall also use, (for $n = 2$)

$$P_{k_1, k_2}(K) = \Big\{ p(x_1, x_2) \mid p(x_1, x_2) = \sum_{\substack{i \leq k_1 \\ j \leq k_2}} a_{ij}\, x_1^i\, x_2^j \Big\} \tag{2.5}$$

the space of polynomials of degree $\leq k_1$ in x_1 and $\leq k_2$ in x_2. In the same way we can define $P_{k_1, k_2, k_3}(K)$ for $n = 3$. The dimension of these spaces is respectively $(k_1 + 1)(k_2 + 1)$ and $(k_1 + 1)(k_2 + 1)(k_3 + 1)$. We then define

$$Q_k(K) = \begin{cases} P_{k,k}(K) & \text{for } n = 2, \\ P_{k,k,k}(K) & \text{for } n = 3. \end{cases} \tag{2.6}$$

The classical finite element approximations are obtained by using polynomials like $P_k(\hat{K})$ or $Q_k(\hat{K})$ on some *reference element* $\hat{K}$ and to carry them over to an arbitrary element K by a change of variable:

$$v_h|_K = \hat{v} \circ F^{-1}, \tag{2.7}$$

where $K = F(\hat{K})$ and $\hat{v}$ is a polynomial function on $\hat{K}$. Continuity is obtained by a suitable choice of the degrees of freedom, that is, interpolation points defining the polynomials. The simplest case is described in the following example.

Example 2.1 (Affine finite elements.) This is the most classical family of finite elements. The reference element is the triangle $\hat{K}$ of Figure 2.1 and we use the affine transformation

$$F(\hat{x}) = x_0 + B\hat{x}, \tag{2.8}$$

where B is a two-by-two matrix.

The element $K = F(\hat{K})$ is an arbitrary triangle and it is not degenerate provided $\det B \neq 0$. We now take $\hat{P} = P_k(\hat{K})$ and choose an appropriate set of degrees of freedom. The standard choices for $k \leq 3$ are presented on Figure 2.2 where the dots represent the degrees of freedom. One notes that this choice of points ensures continuity at the element interfaces. $\square$

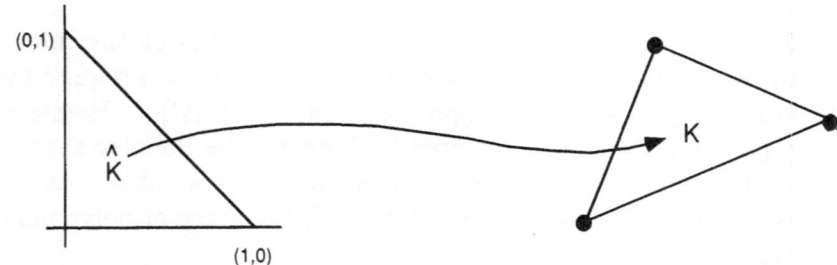

Fig. 2.1. Affine transformation.

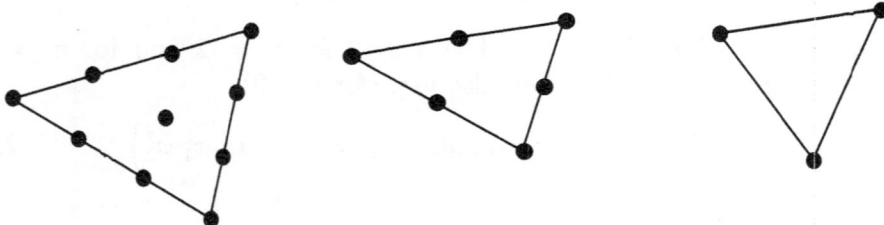

Fig. 2.2. Standard affine elements.

The next example presents a second classical family of finite elements. They are defined on arbitrary quadrilaterals and will not be polynomial functions even though they are obtained by applying a change of variables to a polynomial.

Example 2.2 (Quadrilateral elements.) The reference element concerned is taken to be the square $\hat{K} =]0, 1[\times]0, 1[$. We take $\hat{P} = Q_k(\hat{K})$ and a transformation F with each component in $Q_1(\hat{K})$. We present the standard choice of degrees of freedom for $k = 1$ in Figure 2.3. It must be noted that we need $F \in (Q_1(\hat{K}))^2$ to define a general straight-sided quadrilateral. $\square$

Finally we recall that it is possible to employ curved elements to obtain better approximation properties near the boundary of the domain Ω.

Example 2.3 (Isoparametric elements.) Let us first consider the triangular case. We shall use the same reference element and the same set $\hat{P}$ as in Example 2.1. We now take the transformation $F(\hat{x})$ so that each of

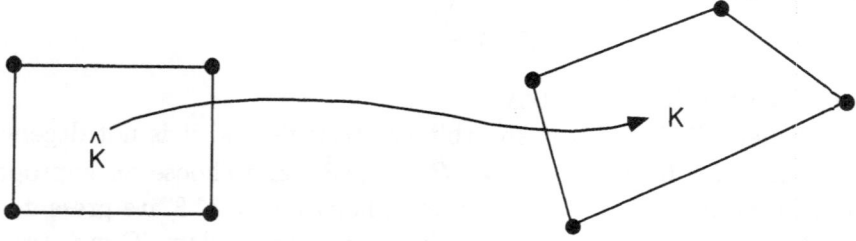

Fig. 2.3. Q_1 isoparametric element.

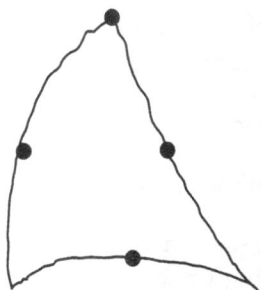

Fig. 2.4. Curved triangular element.

its components F_i belongs to $P_k(\hat{K})$. For $k = 1$ nothing is changed but for $k \geq 2$, the element K now has curved boundaries. We depict the case $k = 2$ in Figure 2.4. It must be noted that the curvature of element boundaries introduces additional terms in the approximation error and the curved elements should be used only when they are really necessary (Ciarlet and Raviart (1972) or Ciarlet (1978)). Similar constructions enable us to define isoparametric quadrilateral elements using $F \in Q_k(K)$. $\square$

We note again that the degrees of freedom have been chosen in order to ensure continuity between elements. We also need some basic results about the accuracy of interpolation by finite element functions. In dealing with $H^1(\Omega)$, one generally employs Lagrange interpolation, that is, the value of the interpolant at the degrees of freedom are computed from the value of the functions at these points, excluding derivatives. There is, however, a difficulty as point values of functions in $H^1(\Omega)$ are not, in general, defined. This can be circumvented by using the technique of Clément (1975) where local averages are employed instead of point values. The details are beyond the scope of this article. We shall only cite a very basic result, assuming r_h to be defined by the usual Lagrange interpolant.

Proposition 2.1 If the mapping F is affine, that is $F(\hat{x}) = x_0 + B\hat{x}$, and if $r_h p_k = p_k$ for any $p_k \in P_k(K)$, we have for $v \in H^s(\Omega)$, $m \leq s$, $1 < s \leq k+1$

$$|v - r_h v|_{m,K} \leq c \, \|B^{-1}\|^m \, \|B\|^s \, |v|_{s,K}. \tag{2.9}$$

As we said above, the condition $s > 1$ is required in order to ensure that point values of the function to be interpolated are well defined and the result can be improved (cf. Brezzi and Fortin (1991)). To obtain global results on Ω, we shall need some assumption to ensure that the partition $\mathcal{T}_h$ is not degenerate, i.e. that the angles of the triangles are bounded away from π.

Let then h_K be the diameter of K, and let us define, for affine elements,

$$\sigma_K = \frac{h_K}{\rho_K} \tag{2.10}$$

where ρ_K is the diameter of the largest inscribed disk (or sphere) in K. We then say that a family of triangulations $(\mathcal{T}_h)_{h\geq 0}$ is regular if

$$\sigma_K < \sigma, \quad \forall K \in \mathcal{T}_h, \; \forall h. \tag{2.11}$$

One may then prove the following result.

Proposition 2.2 If $(\mathcal{T}_h)_{h\geq 0}$ is a regular family of affine partitions, there exists a constant c depending on k and on σ and an interpolation operator Π_1 such that

$$\sum_K h_K^{2m-2}|v - \Pi_1 v|_{m,K}^2 \leq c\|v\|_{1,\Omega}^2 \; m = 0, 1. \; \square \tag{2.12}$$

For more general partitions including general isoparametric elements, the result is qualitatively the same: we have an $\mathcal{O}(h^k)$ approximation provided the family of partitions is regular in a sense which has to be made precise for each type of partition.

Finally, we introduce some notation for the usual spaces of finite element approximations. We thus define

$$\mathcal{L}_k^s = \{v \mid v|_K \in P_k(K), \; v \in H^s(\Omega)\}. \tag{2.13}$$

In the same way we shall write $\mathcal{L}_{[k]}^s$ when $\mathcal{T}_h$ consists of quadrilaterals and the local approximations are built from $Q_k(\hat{K})$ by an appropriate change of variables. We shall also quite often need a class of functions called *bubble functions*. For an element K a bubble function is a function vanishing on ∂K. In particular, we shall denote

$$\begin{cases} B_k = (P_k(K) \cap H_0^1(K)), \\ B_{[k]} = Q_k(K) \cap H_0^1(K)). \end{cases} \tag{2.14}$$

2.3. Scaling argument

In some of the proofs, we shall invoke scaling arguments in order to express the dependence of some quantities on the fineness of the mesh. The standard procedure (cf. Ciarlet (1978)) is to map the quantity to be estimated on a reference element $\hat{K}$ on which it can be computed and then to study the effect of the change of variables which maps $\hat{K}$ to an arbitrary element K in the partition. An interesting variant of this procedure, introduced by Dupont and Scott (1980), consists essentially of separating the two issues of the size and the shape of the element. Indeed, using the change of variables

$$x = h_K \, \hat{x} + b, \tag{2.15}$$

one can easily check the effect of mesh size on a given quantity. In this way
one sees that one has

$$\begin{cases} |v_h|_{1,K} = |\hat{v}_h|_{1,\hat{K}}, \\ |v_h|_{0,K} = h_K \, |\hat{v}_h|_{0,\hat{K}}, \end{cases} \tag{2.16}$$

and many other similar relations. The effect of shape is then treated by an
argument of compactness: a continuous function is bounded on a compact
set. One obtains in this way, for a general transformation,

$$|v_h|_{1,K} = c(k, \theta_0)|\hat{v}_h|_{1,\hat{K}}, \tag{2.17}$$

where k is the degree of polynomials employed and θ_0 is the smallest angle
of the mesh. We refer the reader to Dupont and Scott(1980) or Brezzi and
Fortin(1991) for more details.

3. Presentation of the problem

Let Ω be a domain of $\mathbf{R}^2$ or $\mathbf{R}^3$ and let us denote Γ its boundary. We shall
want to solve in this domain, over a time interval $]0, T[$, the Navier–Stokes
equations of incompressible fluid flow with initial conditions and boundary
conditions. Let ρ be the density of the fluid, $\boldsymbol{u}$ its velocity and p, its pressure.
We thus have to find in Ω, a solution of

$$\rho\left(\frac{\partial \boldsymbol{u}}{\partial t} + \boldsymbol{u}\cdot\mathbf{grad}\,\boldsymbol{u}\right) - 2\mu\,\boldsymbol{A}\boldsymbol{u} + \mathbf{grad}\,p = \rho\boldsymbol{f}, \tag{3.1}$$

$$\mathrm{div}\,\boldsymbol{u} = 0, \tag{3.2}$$

$$\boldsymbol{u}(x, 0) = \boldsymbol{u}_0(x). \tag{3.3}$$

In equation (3.1), we have denoted

$$\boldsymbol{A}\boldsymbol{u} = \begin{cases} \dfrac{\partial^2 u_1}{\partial x_1^2} + \dfrac{1}{2}\dfrac{\partial}{\partial x_2}\left(\dfrac{\partial u_1}{\partial x_2} + \dfrac{\partial u_2}{\partial x_1}\right) \\ \dfrac{\partial^2 u_2}{\partial x_2^2} + \dfrac{1}{2}\dfrac{\partial}{\partial x_1}\left(\dfrac{\partial u_1}{\partial x_2} + \dfrac{\partial u_2}{\partial x_1}\right). \end{cases} \tag{3.4}$$

Taking (3.2) into account, it is easily seen that we have

$$2\boldsymbol{A}\boldsymbol{u} = \triangle\boldsymbol{u}. \tag{3.5}$$

However, the variational formulation and *natural boundary conditions* will,
as we shall see later, be different for these two forms of the equations. We
consider a part Γ_D of Γ on which Dirichlet boundary conditions are given,

$$\boldsymbol{u}|_{\Gamma_D} = 0, \tag{3.6}$$

and a part Γ_N on which Neumann type conditions are specified, that is, in
the present case, a condition on stresses is given. Let $\boldsymbol{n}$ be the outward
unit normal to Γ_n and $\boldsymbol{t}$ the associated tangent vector for a two-dimensional

problem or let t_1, t_2 be tangent vectors for a three-dimensional problem. We then impose, $g = \{g_n, g_{t_1}, g_{t_2}\}$ being given, the boundary conditions on Γ_N

$$-p + 2\mu \frac{\partial u \cdot n}{\partial n} = g_n, \tag{3.7}$$

$$\mu \left(\frac{\partial u \cdot n}{\partial t_i} + \frac{\partial u \cdot t_i}{\partial n} \right) = g_{t_i}, \quad i = 1, 2. \tag{3.8}$$

For two-dimensional problems, we have only one tangent vector and one condition in (3.8) instead of two. We shall, in fact, work with a variational formulation of the Navier–Stokes equations, and for this we shall need to define appropriate function spaces. Let us denote

$$V = (H_D^1(\Omega))^2 = \left\{ v \mid v \in (H^1(\Omega))^2, v|_{\Gamma_D} = 0 \right\} \tag{3.9}$$

$$Q = L^2(\Omega). \tag{3.10}$$

We also define the rate-of-strain tensor $\varepsilon(u)$ by

$$\varepsilon_{ij} = \frac{1}{2} \left(\frac{\partial u_i}{\partial x_j} + \frac{\partial u_j}{\partial x_i} \right). \tag{3.11}$$

Let $\Sigma =]0, T[\times \Omega$ and let us seek a weak solution $u \in L^2(0, T; V)$, $p \in L^2(\Sigma)$ of equations (3.1)–(3.3), that is, let us look for $\{u, p\}$, solution of

$$\begin{cases} \int_\Omega \frac{\partial u}{\partial t} \cdot v \, dx + 2\mu \int_\Omega \varepsilon(u) : \varepsilon(v) \, dx + \int_\Omega u \cdot \text{grad} \, u \cdot v \, dx \\ \qquad\qquad - \int_\Omega f \cdot v \, dx - \int_\Omega p \, \text{div} \, v \, dx = 0, \quad \forall v \in V \qquad (3.12) \\ \int_\Omega q \, \text{div} \, u \, dx = 0, \quad \forall q \in Q, \end{cases}$$

where the meaning of $\partial u / \partial t$ would have to be made precise.

Remark 3.1 It can be easily checked through an integration by parts that the natural boundary conditions associated with this variational formulation are precisely (3.6)–(3.8). This would not be the case had we employed, instead of

$$2\mu \int_\Omega \varepsilon(u) : \varepsilon(v) \, dx,$$

a different bilinear form such as $\mu \int_\Omega \text{grad} \, u : \text{grad} \, v \, dx$ which leads to the same equations inside Ω but with the boundary conditions

$$-p + \mu \frac{\partial u \cdot n}{\partial n} = g_n \text{ on } \Gamma_N, \tag{3.13}$$

$$\mu \frac{\partial u \cdot t}{\partial n} = g_t \text{ on } \Gamma_N. \tag{3.14}$$

It is also possible to obtain as a natural condition

$$\operatorname{rot} \boldsymbol{u}|_\Gamma = g, \tag{3.15}$$

using $\mu \int_\Omega \operatorname{rot} \boldsymbol{u} : \operatorname{rot} \boldsymbol{v} \, dx$ as a bilinear form, which still generates the same differential operator in Ω. $\square$

We refer the reader to Temam (1977) or Lions (1969) for a complete presentation of existence and uniqueness results. One striking point with respect to these equations is the absence of an equation containing $\partial p/\partial t$: thus our system is not of the Cauchy–Kowalevska type. In fact the pressure appears here as a *Lagrange multiplier* associated with the divergence-free condition $\operatorname{div} \boldsymbol{u} = 0$. To understand this we shall, in the next section, consider the simplified steady-state Stokes problem, valid for low-speed or highly viscous flows. For the moment, we shall highlight an additional property of the above equations. They are equations of 'convection–diffusion' type, by which it is meant that they model the mixing of transport phenomena with diffusion. It is well known that in this kind of problem the behaviour of the solution is determined by the relative magnitudes of the convection and the diffusion terms. Diffusion-dominated problems behave like standard parabolic equations while advection-dominated ones, although theoretically parabolic, behave almost as if they were hyperbolic, except in some small regions, 'boundary layers', where diffusion effects reappear with startling consequences. In the case of the Navier–Stokes equations, the ratio of advection to diffusion is expressed by the Reynolds number. It is obtained by non-dimensionalizing the equations and has the form

$$Re = \frac{\rho U d}{\mu}. \tag{3.16}$$

It must be emphasized that a Reynolds number has no absolute meaning: it is a *relative* number. It enables problems in the *same geometry* with similar boundary conditions to be compared. In practice, high Reynolds number problems are difficult and must be handled with care but there is no absolute scale for 'large' or 'small' .

4. The Stokes problem: incompressibility and pressure

4.1. The continuous problem

We shall consider in this section the simplest possible incompressible flow problem, the steady-state Stokes problem, obtained from equations (3.1)–(3.2) by neglecting the time derivative and the inertial terms $\boldsymbol{u} \cdot \operatorname{grad} \boldsymbol{u}$. This *approximation* of the Navier–Stokes equations is valid for very low Reynolds numbers, that is for small velocities or high viscosity. The problem thus becomes:

$$-2\mu\,\boldsymbol{A}\boldsymbol{u} + \operatorname{grad} p = \boldsymbol{f}, \tag{4.1}$$

$$\operatorname{div} \boldsymbol{u} = g, \tag{4.2}$$

$$\boldsymbol{u}|_\Gamma = 0. \tag{4.3}$$

In (4.2) we have introduced a non-zero right-hand side $g \in Q$. This is for the sake of generality and causes no additional difficulty. In most cases, we shall take $g = 0$. We shall describe how the Stokes problem can be seen as a constrained optimization problem and how pressure appears naturally as a Lagrange multiplier. This will enable us to apply the general results of Brezzi (1974), Babuška (1973) or Brezzi and Fortin (1991). This will also help us later in the construction of numerical algorithms for the computation of the pressure. First we define

$$a(\boldsymbol{u}, \boldsymbol{v}) \;=\; 2\mu \int_\Omega \boldsymbol{\varepsilon}(\boldsymbol{u}) : \boldsymbol{\varepsilon}(\boldsymbol{v})\,\mathrm{d}x, \tag{4.4}$$

$$b(\boldsymbol{v}, q) \;=\; -\int_\Omega q \operatorname{div} \boldsymbol{v}\,\mathrm{d}x. \tag{4.5}$$

Clearly, problem (4.1)–(4.3) can be written in the form:

$$\begin{cases} a(\boldsymbol{u}, \boldsymbol{v}) + b(\boldsymbol{v}, p) = (\boldsymbol{f}, \boldsymbol{v}), & \forall \boldsymbol{v} \in V, \\ b(\boldsymbol{u}, q) = (g, q), & \forall q \in Q. \end{cases} \tag{4.6}$$

This problem is nothing but the optimality condition of a saddle-point problem,

$$\inf_{\boldsymbol{v} \in V} \sup_{q \in Q} \mu \int_\Omega |\boldsymbol{\varepsilon}(\boldsymbol{v})|^2\,\mathrm{d}x - \int_\Omega q \operatorname{div} \boldsymbol{v}\,\mathrm{d}x - \int_\Omega \boldsymbol{f}\cdot\boldsymbol{v}\,\mathrm{d}x + \int_\Omega q\,q\,\mathrm{d}x, \tag{4.7}$$

which is equivalent to the contrained minimization problem,

$$\inf_{\operatorname{div}\boldsymbol{v}=g} \mu \int_\Omega |\boldsymbol{\varepsilon}(\boldsymbol{v})|^2\,\mathrm{d}x - \int_\Omega \boldsymbol{f}\cdot\boldsymbol{v}\,\mathrm{d}x. \tag{4.8}$$

In this context, it is clear that the pressure may be seen as the Lagrange multiplier associated with the constraint $\operatorname{div} \boldsymbol{v} = g$. This will also remain true, in a generalized sense, for the full Navier–Stokes problem (3.1)–(3.3). If we now return to problem (4.6), we also see that we are now dealing with a *mixed variational formulation* (Brezzi 1974, Brezzi and Fortin 1991) and we have a general framework in which to study our problem. In general, the existence and uniqueness of the solution of a problem of type (4.6) requires two conditions. The first one is *coercivity* of the bilinear form $a(\cdot, \cdot)$ on V. In the case of the Stokes problem, in the setting defined above, this condition is immediately satisfied and is nothing but Korn's inequality, that is, there exists a constant $\alpha > 0$ such that

$$\int_\Omega \boldsymbol{\varepsilon}(\boldsymbol{v}) : \boldsymbol{\varepsilon}(\boldsymbol{v}) \, \mathrm{d}x \geq \alpha \|\boldsymbol{v}\|^2 \quad \forall \boldsymbol{v} \in V, \tag{4.9}$$

which holds for $V = (H_0^1(\Omega))^n (n = 2, 3)$ but also for more general boundary conditions (see Duvaut and Lions (1972)). The second condition is known as the inf–sup condition, which will be our terminology, but also as the Babuška–Brezzi condition or even the Ladyzhenskaya–Babuška–Brezzi (LBB) condition. It can be written as,

$$\inf_{\boldsymbol{v} \in V} \sup_{q \in Q} \frac{b(\boldsymbol{v}, q)}{\|\boldsymbol{u}\|_V \|q\|_Q} \geq k_0 > 0. \tag{4.10}$$

This looks somewhat abstract and cumbersome. It means in fact that the operator B from V into Q', the dual of V, is surjective. In a more general form it can be written as

$$\inf_{\boldsymbol{v} \in V} \sup_{q \in Q} \frac{b(\boldsymbol{v}, q)}{\|\boldsymbol{u}\|_V \|q\|_{Q/\ker B^t}} \geq k_0 > 0, \tag{4.11}$$

where

$$\ker B^t = \{q|\, b(\boldsymbol{v}, q) = 0, \forall \boldsymbol{v} \in V\} \tag{4.12}$$

and the quotient norm $\|q\|_{Q/\ker B^t}$ is defined by

$$\|q\|_{Q/\ker B^t} = \inf_{q_0 \in \ker B^t} \|q + q_0\|_Q. \tag{4.13}$$

Condition (4.11) then means that the operator B has a closed range in Q' and the p part of the solution is then only defined up to an element of $\ker B^t$. In our case, we have $Q = Q' = L^2(\Omega)$ and the operator B is the divergence operator from V into $L^2(\Omega)$. With $V = (H_0^1(\Omega))^2$, it is not surjective and $\ker B^t = \ker(\mathbf{grad})$ is the subspace of constants. Pressure will then be defined up to a constant. Whenever we have Neumann conditions on part of the boundary, we recover surjectivity and hence uniqueness.

4.2. The dual problem

It is usual, when a Lagrange multiplier is introduced to enforce a constraint, to consider the *dual problem*, that is the problem transformed into this new variable. It is obtained by changing the inf–sup problem (4.7) into a sup–inf problem through reversing the order of operations and eliminating $\boldsymbol{v}$ by performing the minimization in $\boldsymbol{v}$ for a given q. In our case, an easy calculation shows that the dual problem can be written as

$$\sup_q \frac{1}{2} \int_\Omega \boldsymbol{A}^{-1}\mathbf{grad}\, q \cdot \mathbf{grad}\, q \, \mathrm{d}x - \int_\Omega \boldsymbol{A}^{-1}\boldsymbol{f} \cdot \mathbf{grad}\, q \, \mathrm{d}x, \tag{4.14}$$

for which the optimality condition is

$$\operatorname{div} \boldsymbol{A}^{-1} \operatorname{grad} p = \operatorname{div} \boldsymbol{A}^{-1} \boldsymbol{f}. \tag{4.15}$$

As we shall see later, the properties of the discrete dual problem play a crucial role in the analysis of the numerical scheme.

4.3. The discrete problem

We are now in a position to consider discretizations of problem (4.6). To do so, we introduce finite-dimensional subspaces $V_h \subset V$ and $Q_h \subset Q$ and we consider the discrete analogue of (4.6),

$$\begin{cases} a(\boldsymbol{u}_h, \boldsymbol{v}_h) + b(\boldsymbol{v}_h, p_h) = (\boldsymbol{f}, \boldsymbol{v}_h), & \forall \boldsymbol{v}_h \in V_h, \\ b(\boldsymbol{u}_h, q_h) = (g, q_h), & \forall q_h \in Q_h, \end{cases} \tag{4.16}$$

where, as in (4.6) g will be zero in most cases. For such a *conforming approximation*, the general theory of Brezzi (1974) (see Brezzi and Fortin (1991)) applies directly. It relies on the discrete version of conditions (4.9) and (4.11). The first condition is trivial in the present case and follows directly from the inclusion $V_h \subset V$. To consider the second condition, we first identify Q_h and Q_h' just as we identified Q and Q', we let $B_h = \operatorname{div}_h$ be the discrete divergence operator from V_h into Q_h associated with the restriction of the bilinear form $b(\cdot, \cdot)$ to these spaces and let $B_h^t = \operatorname{grad}_h$ be its transpose,

$$\langle \operatorname{div}_h \boldsymbol{u}_h, q_h \rangle = b(\boldsymbol{u}_h, q_h) = \langle \boldsymbol{u}_h, \operatorname{grad}_h q_h \rangle, \quad \boldsymbol{u}_h \in V_h, \ q_h \in Q_h. \tag{4.17}$$

In general, div_h is not the restriction of div to V_h. Indeed, from equation (4.17) we have

$$\operatorname{div}_h \boldsymbol{u}_h = P_{Q_h} \operatorname{div} \boldsymbol{u}_h, \tag{4.18}$$

where P_{Q_h} is the projection operator from Q onto Q_h. As we shall see later, in many actual cases $\operatorname{div}_h \boldsymbol{u}_h$ will be some average of $\operatorname{div} \boldsymbol{u}_h$. This also implies that the kernel of the discrete gradient grad_h,

$$\ker \operatorname{grad}_h = \{ q_h \in Q_h \mid b(\boldsymbol{v}_h, q_h) = 0, \forall \boldsymbol{v}_h \in V_h \}, \tag{4.19}$$

is not necessarily the one-dimensional subspace of constants. Cases will arise in which nonconstant functions have a zero discrete gradient. Such cases will be pathological and will require special care if they are not simply avoided. We can now state the second condition as

$$\inf_{\boldsymbol{v}_h \in V_h} \sup_{q_h \in Q_h} \frac{b(\boldsymbol{v}_h, q_h)}{\|\boldsymbol{v}_h\|_V \|q_h\|_{Q_h/\ker \operatorname{grad}_h}} \geq k_h \geq k_0 > 0. \tag{4.20}$$

The first part ($\geq k_h$) is trivial in a finite dimensional setting. The really important requirement is the existence of a constant k_0 independent of h.

Given coercivity and the discrete inf–sup condition (4.20) we can apply the theory of Brezzi (1974) to obtain the existence and uniqueness of $(\boldsymbol{u}_h, p_h)$ in V_h and $Q_h/\ker \operatorname{\mathbf{grad}}_h$ and we can state

Theorem 4.1   Let $(\boldsymbol{u}, p)$ be the solution of problem (4.6) and $(\boldsymbol{u}_h, p_h)$ be the solution of the discrete problem (4.16). We then have the error estimates:

$$\|\boldsymbol{u} - \boldsymbol{u}_h\|_V \tag{4.21}$$

$$\leq C_1(1/\alpha, 1/k_h)\left\{\inf_{\boldsymbol{v}_h \in V_h} \|\boldsymbol{u} - \boldsymbol{v}_h\|_V + \inf_{q_h \in Q_h} \|p - q_h\|_Q\right\},$$

$$\|p - p_h\|_{Q/\ker(\operatorname{\mathbf{grad}}_h)} \tag{4.22}$$

$$\leq C_2(1/\alpha, 1/k_h^2)\left\{\inf_{\boldsymbol{v}_h \in V_h} \|\boldsymbol{u} - \boldsymbol{v}_h\|_V + \inf_{q_h \in Q_h} \|p - q_h\|_Q\right\}.$$

□

It must be remarked that both constants C_1 and C_2 depend on $1/\alpha$ but that C_2 depends on $1/k_h^2$, which makes the approximation of pressure much more sensitive to a bad behaviour of k_h. In many cases where k_h is not bounded from below but depends on h, it is customary to see acceptable approximate velocities but a disastrous approximate pressure field. We shall develop later another approach to clarify this point. Before doing so, we shall present a criterion for the inf–sup condition and consider some classical examples.

4.4. The inf–sup condition and criteria

The question that now arises is to find some way of checking condition (4.20). Although this is not the only possibility, a quite convenient way is through a criterion introduced in Fortin (1977) which reduces the question to the construction of a suitable interpolation operator. The criterion can be found in a general setting in Brezzi and Fortin (1991). For the present purpose, we consider a special, albeit general enough case. As a starting point, we assume that the continuous inf–sup condition (4.11) holds, which is indeed the case for the problem considered. We then prove:

Lemma 4.1   Suppose that we can build an operator Π_h from V into V_h satisfying

$$b(\Pi_h \boldsymbol{v} - \boldsymbol{v}, q_h) = 0 \qquad \forall q_h \in Q_h, \tag{4.23}$$

$$\|\Pi_h \boldsymbol{v}\|_V \leq c\|\boldsymbol{v}\|_V, \tag{4.24}$$

with a constant c independent of h. Then the discrete inf–sup condition (4.20) holds

Proof. Indeed we have from (4.11), as $Q_h \subset Q$,

$$\sup_{v \in V} \frac{b(v, q_h)}{\|v\|_V} \geq k_0 \|q_h\|_{Q/\ker B^t}. \tag{4.25}$$

But by (4.23) and (4.24), we may write

$$\sup_{v_h \in V_h} \frac{b(v_h, q_h)}{\|v_h\|_V} \geq \sup_{v \in V} \frac{b(\Pi_h v, q_h)}{\|\Pi_h v\|_V} \tag{4.26}$$

$$\geq \sup_{v \in V} \frac{1}{c} \frac{b(v, q_h)}{\|v\|_V} \geq \frac{k_0}{c} \|q_h\|_{Q/\ker B^t},$$

hence the result. $\square$

The use of Lemma 1 requires two things: finding a suitable class of elements, constructing Π_h which satisfies (4.23) and then checking that this operator is uniformly continuous in h, that is (4.24). This last requirement is generally purely technical although establising it could be quite intricate. It is also worth stating here an important fact about the operator Π_h.

Lemma 4.2 If the condition (4.23) holds, then

$$\ker B_h^t \subset \ker B^t, \tag{4.27}$$

that is there are no spurious zero-energy mode.

Proof. This is easily inferred from (4.23). Indeed we must show that any $q_h \in \ker B_h^t$, that satisfies

$$b(v_h, q_h) = 0 \quad \forall v_h \in V_h, \tag{4.28}$$

also satisfies

$$b(v, q_h) = 0 \quad \forall v \in V. \tag{4.29}$$

But $b(v, q_h) = b(\Pi_h v, q_h) = 0$ and the result is immediate. $\square$

Let us come back to the problem of constructing the operator Π_h. This is often done in practice by starting from a standard interpolation operator and by correcting it by some local operations. The following lemma provides a general procedure to do so.

Lemma 4.3 Let us suppose that the finite element approximation has been chosen so that Proposition 2.2 or some analogous result applies with some suitable interpolation operator Π_1, satisfying the continuity requirement,

$$\|\Pi_1 v\|_V \leq c_1 \|v\|_V, \quad \forall v \in V \tag{4.30}$$

We also suppose that there exists a second operator $\Pi_2 \in \mathcal{L}(V, V_h)$ satisfying

$$\|\Pi_2(I - \Pi_1)v\|_V \leq c_2 \|v\|_V, \quad \forall v \in V, \tag{4.31}$$

$$\int_\Omega \operatorname{div}(\boldsymbol{v} - \Pi_2\boldsymbol{v})\, q_h\, \mathrm{d}x \;=\; 0, \quad \forall \boldsymbol{v} \in V,\ \forall q_h \in Q_h, \qquad (4.32)$$

where the constants c_1 and c_2 are independent of h. Then the operator Π_h defined by

$$\Pi_h\boldsymbol{u} = \Pi_1\boldsymbol{u} + \Pi_2(\boldsymbol{u} - \Pi_1\boldsymbol{u}), \quad \boldsymbol{u} \in V, \qquad (4.33)$$

satisfies (4.23) and (4.24).

Proof. It is easy to see that condition (4.23) holds. Indeed

$$
\begin{aligned}
b(\Pi_h w, q_h) &= b(\Pi_2(w - \Pi_1 w), q_h) + b(\Pi_1 w, q_h) \\
&= b(w - \Pi_1 w, q_h) + b(\Pi_1 w, q_h) \\
&= b(w, q_h).
\end{aligned}
\qquad (4.34)
$$

On the other hand,

$$\|\Pi_h w\|_v \le \|\Pi_2(w - \Pi_1 w)\|_V + \|\Pi_1 w\|_V \le (c_1 + c_2)\|w\|_V \qquad (4.35)$$

so that condition (4.24) holds. $\square$

In many cases, Π_1 will be the interpolation operator of Clement (1975) (cf. Proposition 2.2) in $H^1(\Omega)$ for which we have

$$\sum_K h_K^{2r-2}|v - \Pi_1 v|_{r,K}^2 \le c\, \|v\|_{1,\Omega}^2, \quad r = 0, 1. \qquad (4.36)$$

Taking $r = 1$ in (4.36) and using the triangle inequality

$$\|\Pi_1 v\|_V \le \|v - \Pi_1 v\|_V + \|v\|_V \qquad (4.37)$$

yields (4.30).

4.5. The matrix form of the discrete problem

Suppose that we are given a basis $\{\phi_i^V\}_{1\le i\le N}$ of V and a basis $\{\psi_k^Q\}_{1\le k\le M}$ of Q. We can define the matrices

$$\mathcal{A}_{i,j} = a(\phi_i^V, \phi_j^V), \qquad (4.38)$$

$$\mathcal{B}_{i,k} = b(\phi_i^V, \psi_k^Q). \qquad (4.39)$$

Matrix $\mathcal{A}$ is positive definite while $\mathcal{B}$ is a rectangular matrix. We shall also need later the mass matrices

$$\mathcal{M}_{i,j}^V = (\phi_i^V, \phi_j^V)_{0,\Omega}, \qquad (4.40)$$

$$\mathcal{M}_{k,l}^Q = (\psi_k^Q, \psi_l^Q)_{0,\Omega}. \qquad (4.41)$$

We now consider the discrete problem (4.16) and we write

$$
\begin{cases}
\boldsymbol{u}_h = \sum_i \boldsymbol{U}_i \phi_i^V, \\
p_h = \sum_k \boldsymbol{P}_k \psi_k^Q.
\end{cases}
\qquad (4.42)
$$

Using this notation, the discrete problem may be written as

$$\begin{pmatrix} \mathcal{A} & \mathcal{B} \\ \mathcal{B}^t & 0 \end{pmatrix} \begin{pmatrix} U \\ P \end{pmatrix} = \begin{pmatrix} F \\ G \end{pmatrix}. \qquad (4.43)$$

We therefore have to solve a symmetric indefinite system. It is possible, as $\mathcal{A}$ is invertible, to eliminate U from this problem to get a problem in P only:

$$\mathcal{B}\mathcal{A}^{-1}\mathcal{B}^t P = G - \mathcal{B}(\mathcal{A})^{-1}F. \qquad (4.44)$$

We shall come back later to numerical methods adapted to this problem.

4.6. Eigenproblems associated with the discrete inf–sup condition

It should be clear from the earlier analysis that the discrete inf–sup condition is closely related to the behaviour of the dual problem, in particular the discrete dual problem (4.44). Let us indeed go back to (4.20) and let us rewrite it in the notation of the previous subsection. We get

$$\inf_{V \in V_h} \sup_{Q \in Q_h} \frac{\langle \mathcal{B}V, Q \rangle}{\langle \mathcal{A}V, V \rangle^{1/2} \langle \mathcal{M}^Q Q, Q \rangle^{1/2}} = k_h, \qquad (4.45)$$

where we have made the assumption that $\langle \mathcal{A}V, V \rangle^{1/2}$ is employed as a norm on V_h. This involves a Rayleigh's quotient for the *singular value decomposition* of the matrix $\mathcal{B}$ with the norms defined by $\mathcal{A}$ and $\mathcal{M}^Q$ on V_h and Q_h respectively. This can be reduced to solving the generalized eigenvalue problem,

$$\mathcal{B}\mathcal{A}^{-1}\mathcal{B}^t Q_i = \mu_i^2 \mathcal{M}^Q Q_i. \qquad (4.46)$$

The square root of the smallest eigenvalue is nothing but the constant k_h of (4.20) while the square root of the largest one is the norm $\|b\|$ of the bilinear form $b(\cdot,\cdot)$. For more details, we refer to Brezzi and Fortin (1991). This argument shows that all kinds of behaviour is possible: the correct case is when the eigenvalues are bounded away from zero. When some eigenvalues vanish with h, part of the solution will be spoiled. We refer to Malkus (1981) where these eigenvalues have been computed numerically for some elements. A more complete discussion of similar eigenvalue problems and of the condition number of associated systems can be found in Fortin and Pierre (1992).

5. Finite elements for incompressible problems

In this section, we shall present, in a general framework, some classical examples of finite element approximations to the equations of incompressible materials. The problem is *a priori* simple. We are looking for a velocity field in $H_0^1(\Omega))^2$, which implies that all classical constructions hold. In the same way, the pressure which is sought only in $L^2(\Omega)$ can be approximated by a very wide choice of elements. Standard conforming elements (built for $H^1(\Omega)$) will evidently be suitable and we shall say in that case that we have a *continuous pressure approximation*. On the other hand, we are allowed to avoid any continuity of the discrete pressure at interfaces and to use *discontinuous pressure approximations*. As we shall see in the examples, this last case will ensure a better conservation of mass. The difficulty which arises is that our approximations of velocity and pressure cannot be chosen independently but must satisfy a compatibility condition: the discrete inf-sup condition (4.20). Our goal will therefore be to build approximations satisfying this condition while preserving simplicity and efficiency.

5.1. Exact incompressibility

A natural idea when one comes to the problem of approximating divergence-free problems is to try to enforce the constraint strongly, that is, at every point. This can be done quite easily. Indeed, given a choice of a space V_h for the approximate velocities, it would be sufficient to take Q_h so that it contains div V_h to ensure that the divergence of the solution is zero everywhere. Just as many simple ideas, this one leads to a dead-end. What happens, at least for low-degree elements, is that the solution is overconstrained and we have a *locking phenomenon*, that is the only function satisfying the divergence-free constraint is the function identically zero. This is the case in the following simple example.

Example 5.1 (The P_1–P_0 approximation.) We approximate velocity by the simplest finite element: piecewise linear functions on triangles. The divergence is then a subspace of the space of piecewise constants so that using this space for Q_h enforces the divergence-free condition exactly. A simple count, using Euler's relations on a triangulation, however, shows that, on a general mesh, the number of constraints is larger than the number of degrees of freedom and that we have locking. It must, however, be noted that on a composite mesh where triangles are obtained by dividing quadrilaterals by their diagonals (Figure 5.1), a linear dependence appears between the constraints so that non-trivial solutions exist. Nevertheless, the resulting approximation does not satisfy the discrete inf–sup condition. $\square$

Example 5.2 (Quadrilateral elements.) The reader may easily check that the same locking phenomenon will appear on rectangular elements (on a regular grid for instance) if one tries to impose an *exact* divergence-free condition to a bilinear or biquadratic approximation of velocities. $\square$

Example 5.3 (Second-order triangular elements.) The counting procedure also shows that, on a general triangular mesh, piecewise quadratic

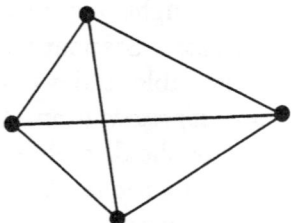

Fig. 5.1. A quadrilateral subdivided by diagonals.

divergence-free elements exist but require far too many triangles to be efficient. However, using the same mesh as in Figure 5.1, it is possible to get a divergence-free approximation satisfying the discrete inf–sup condition; but it is then necessary to filter a spurious mode from the pressure approximation. We refer to Brezzi and Fortin (1991) for details. □

Remark 5.1 The previous example is directly related to the composite approximation of Fraeijs de Veubeke and Sander (1968) for plate problems (see also Ciavaldini and Nédélec (1974)). Those problems indeed require C^1-continuity, and a composite element of degree three can be built on the mesh of Figure 5.1. Taking the curl of this approximation yields a divergence-free function, which is piecewise quadratic and C^0-continuous. The construction is therefore based on the approximation of the *stream function* in $H_0^2(\Omega)$. It must be noted that no similar composite constructions are known for three-dimensional problems. □

Remark 5.2 (Higher order methods.) We would like to recall briefly the statement of a basic result by Scott and Vogelius (1985) which, roughly speaking, says: under minor assumptions on the decomposition $\mathcal{T}_h$ (in triangles) the pair $V_h = (\mathcal{L}_k^1)^2$, $Q_h = \mathcal{L}_{k-1}^1$ satisfies the inf–sup condition for $k \geq 4$. This, in a sense, settles the matter as far as higher order methods are concerned, and leaves only the problem of finding stable lower order approximations. It must, however, be noted that some instabilities might in certain cases remain in the pressure, although they could be filtered out *a posteriori*. In fact the restrictions to which we alluded earlier are that the sides of the triangles should not be collinear as in the special grid of Figure 5.1, which reduces the number of linearly dependent constraints, leaving some of the pressure degrees of freedom unused. This being said, the use of high-order methods is not very popular as they require the delicate manipulation of high-degree polynomials. It has however gained a new popularity.

We have seen that exactly divergence-free methods are delicate, requiring high-order elements or special grids. This leads us to try enforcing the divergence-free condition only approximately, in the hope of obtaining simpler constructions. We have already noted (cf. equation (4.17)) that we have to deal with a discrete divergence operator div_h which is the projection on Q_h of the divergence operator. If Q_h is smaller, or V_h larger, this projection will effectively weaken the divergence-free condition. The effect will be, on one hand, that the verification of the discrete inf–sup condition (4.20) will be easier. On the other hand, the accuracy of the approximation would evidently be impaired by taking Q_h too small so that we shall privilege the enrichment of V_h as a potential cure to our difficulties.

5.2. Simple constructions for approximately divergence-free elements

We shall now introduce a simple and general way of satisfying the inf–sup condition. The basic idea is indeed very simple: the discrete inf–sup condition involves a supremum over $v_h \in V_h$. Making the space V_h larger will make the supremum grow and will intuitively make the condition easier to fulfil. This technique can be further extended to composite elements but for the sake of simplicity, it is worth considering the simpler case first.

The idea of *enriched* elements has been used several times, starting with Crouzeix and Raviart (1973) for discontinuous pressures and Arnold, Brezzi and Douglas (1984) and Arnold, Brezzi and Fortin (1984) for continuous pressures. We present it in the general form given by Brezzi and Pitkäranta (1984) (see also Stenberg (1984)). It consists essentially in stabilizing an element by an enrichment of the velocity field by **bubble functions**, that is functions having their support restricted to one element and vanishing on the boundary of this element. The simplest bubble function is the conforming bubble function, denoted $b_{3,K}$. It is a polynomial function of degree three. If we denote by $\lambda_1, \lambda_2, \lambda_3$, the barycentric coordinates of the triangle we then have $b_{3,K} = \lambda_1 \lambda_2 \lambda_3$. We associate with the finite element discretization $Q_h \subset L^2(\Omega)$ the space

$$M(\mathbf{grad}\, Q_h) = \{\beta \mid \beta|_K = b_{3,K}\mathbf{grad}\, q_h|_K \quad \text{for some } q_h \in Q_h\}. \qquad (5.1)$$

In other words, the restriction of a function $\beta \in M(\mathbf{grad}\, Q_h)$ to an element K is the product of the P_3-bubble functions $b_{3,K}$ and the gradient of a function from $Q_h|_K$.

Remark 5.3 Notice that the space $M(\mathbf{grad}\, Q_h)$ is not defined through some basic space $\widehat{M}$ on the reference element. This can be easily done, if one wants to, in the case of *affine* elements, for all the reasonable choices of Q_h. However this is clearly unnecessary: if we know how to compute q_h on K we also know how to compute $\mathbf{grad}\, q_h$ and there is no need for a reference element. □

We now turn to prove two results, concerning continuous or discontinuous pressures.

Proposition 5.1 (Stability of continuous pressure elements.) We suppose

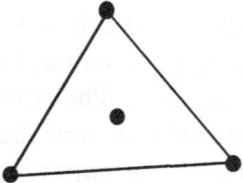 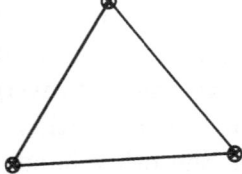

Fig. 5.2. The MINI element.

that there exists $\Pi_1 \in \mathcal{L}(V, V_h)$ satisfying (4.30), that we have $Q_h \subset H^1(\Omega)$ and that $M(\mathbf{grad}\, Q_h)$ is defined as in (5.1). Then the pair (V_h, Q_h) is a stable element, in the sense that it satisfies the inf–sup condition.

Proof. We shall use Lemma 4.3. We already have our operator Π_1 by assumption. We only need to construct Π_2. We define $\Pi_2 : V \rightarrow M(\mathbf{grad}\, Q_h)$, on each element, by requiring

$$\begin{cases} \Pi_2 v|_K \in M(\mathbf{grad}\, Q_h)|_K = b_{3,K} \mathbf{grad}\, Q_h|_K, \\ \int_K (\Pi_2 v - v) \cdot \mathbf{grad}\, q_h \; dx = 0, \quad \forall q_h \in Q_h|_K. \end{cases} \qquad (5.2)$$

Problem (5.2) has obviously a unique solution. It is clear that Π_2 satisfies (4.31) of Lemma 4.3. Finally (4.30) follows by a scaling argument. We thus have the desired result. $\square$

Corollary 5.1 Assume that $Q_h \subset Q$ is any space of continuous piecewise smooth functions. If $(\mathcal{L}_1^1)^2 \oplus M(\mathbf{grad}\, Q_h) \subset V_h$ then the pair (V_h, Q_h) satisfies the inf–sup condition.

Proof. Continuity and piecewise smoothness imply that $Q_h \subset H^1(\Omega)$. The condition $(\mathcal{L}_1^1)^2 \subset V_h$ implies the existence of Π_1 satisfying condition (4.30), and condition $M(\mathbf{grad}\, Q_h) \subset V_h$ is by hypothesis. Hence we can apply Proposition 5.1. $\square$

These results apply, for instance, to the enriched Taylor–Hood element and to the families introduced in Arnold, Brezzi and Fortin (1984).

Example 5.4 (The MINI element.) The first family is defined by

$$V_h = (\mathcal{L}_k^1 \oplus B_{k+2})^2, \quad Q_h = \mathcal{L}_k^1, k \geq 1, \qquad (5.3)$$

where B_{k+2} is defined as in (2.14). The simplest of these elements is the so-called MINI element. It is obtained by taking $k = 1$ in (5.3). This means that a cubic bubble, $(k+2 = 3)$, is added to a simple piecewise linear approximation of velocity while pressure remains piecewise linear. This element is sketched in Figure 5.2. The corresponding **equal interpolation** element, using piecewise linear approximations for both velocity and pressure is not stable in the sense that it does not satisfy the inf–sup condition. This is, in fact, the case for all equal interpolation approximations. $\square$

Fig. 5.3. Enriched Taylor–Hood element.

Example 5.5 The second family is defined by

$$V_h = (\mathcal{L}_k^1 \oplus B_{k+1})^2, \quad Q_h = \mathcal{L}_{k-1}^1, \; k \geq 2. \tag{5.4}$$

In the simplest case, $(k = 2)$, a cubic bubble is added to a piecewise quadratic approximation of velocity while pressure is approximated by piecewise linear functions as in the previous example. This element is sketched in Figure 5.3 Without bubbles, this element is known as the *Taylor–Hood element* and it is already stable. The proof of this will be considered later, as it requires a special technique. □

We turn now to the case of discontinuous pressure elements. Many reasons may lead us to consider such approximations. Probably the most important one is probably the better approximation to the equation of conservation of mass generated by such elements, in comparison with dicontinuous pressure elements. In fact, whenever Q_h contains piecewise constant functions, the divergence-free condition contains, as a particular case, the condition

$$\int_K \operatorname{div} \boldsymbol{v}_h \, dx = 0, \quad \forall K \in \mathcal{T}_h, \tag{5.5}$$

which means that the average divergence is null on every element or, equivalently, that mass is conserved on every element. In the case of continuous pressure approximations, the divergence-free condition is also averaged, but the averages cannot be reduced to a local conservation property. In hard cases this may have important consequences (Fortin and Pelletier (1989)). Discontinuous pressures are also important because they can be combined with a penalty method to eliminate pressure as an unknown, as we shall see in Section 8. Before stating the general result, we shall consider a simple special case which will be the basis for the general setting.

Example 5.6 (The P_2–P_0 and Q_2–Q_0 elements.) These are the basic and simplest stable discontinuous pressure elements. We shall only consider the triangular case in detail as the quadrilateral case can be treated in essentially the same way. The element by itself has no particular property except that pressure is approximated with very low precision. If we refer to estimates (4.22) and (4.23), this implies that we do not achieve the full accuracy expected from second-degree polynomials employed to approximate velocity.

 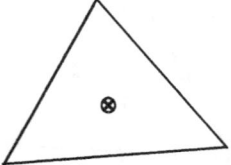

Fig. 5.4. The P_2–P_0 element.

The accuracy of the element is thus not optimal and it is not recommnded in practice.

To check the inf–sup condition, we shall use Lemma 4.1 and build an operator Π_h satisfying the conditions (4.23) and (4.24). To do so we start from the standard interpolation operator Π_1 of Proposition 2.2 and we modify the midside values so that on every side S of K, the resulting new interpolant $\tilde{\Pi}_1 v$ satisfies

$$\int_S (\tilde{\Pi}_1 v_i - v_i)\, \mathrm{d}s = 0 \quad \forall v \in V, i = 1, 2, \tag{5.6}$$

where the v_i are the components of v. Formally, we can work again with Lemma 4.3 and define, on every element K, $\Pi_{2|_K}$ by

$$\begin{cases} \Pi_2 v \in P_2(K), \\ \Pi_2 v_{|_K}(M) = 0, \quad \text{for any vertex } M \text{ of } K, \\[2mm] \displaystyle\int_S \Pi_2 v \ \mathrm{d}s = \int_S v\, \mathrm{d}s, \quad \text{for every side } S \text{ of } K. \end{cases} \tag{5.7}$$

One then defines $\tilde{\Pi}_1$ by

$$\tilde{\Pi}_1 v = \Pi_1 v + \Pi_2 (v - \Pi_1 v). \tag{5.8}$$

It is clear that that $\tilde{\Pi}_1$ satisfies condition (4.23) for we have

$$\int_K \mathrm{div}\,(\Pi_2 v - v)\, \mathrm{d}x = \int_{\partial K} (\Pi_2 v - v)\cdot n\, \mathrm{d}s = 0. \tag{5.9}$$

As to the continuity property, it follows by a scaling argument as

$$|\Pi_2 v|_{1,K} = |\widehat{\Pi_2 v}|_{1,\hat{K}} \le c(2,\theta_0)\|\hat{v}\|_{1,\hat{K}} \tag{5.10}$$
$$\le c(2,\theta_0)(h_K^{-1}|v|_{0,K} + |v|_{1,K}),$$

where $c(2,\theta_0)$ is a constant, depending on the degree of the polynomials employed, which is 2, and on the minimum angle of the mesh as in (2.17). Using this result and the properties of Π_1 yields the result. $\square$

Proposition 5.2 (Stability of discontinuous pressure elements.) Let us suppose that there exists $\tilde{\Pi}_1 \in \mathcal{L}(V, V_h)$ satisfying

$$\int_K \text{div } (v - \tilde{\Pi}_1 v)\, dx = 0 \quad \forall K \in \mathcal{T}_h \tag{5.11}$$

and that $M(\mathbf{grad}\, Q_h) \subset V_h$ as defined in (5.1). Then the pair (V_h, Q_h) is a stable element, in the sense that it satisfies the inf–sup condition.

Proof. We shall proceed by applying Lemma 4.3. We take $\tilde{\Pi}_1$ satisfying (5.11) as operator Π_1. We are not going to define Π_2 on all of V, but only in the subspace

$$V^0 = \left\{ v \mid v \in V, \int_K \text{div } v\, dx = 0, \quad \forall K \in \mathcal{T}_h \right\} \tag{5.12}$$

For every $v \in V^0$ we construct $\Pi_2 v \in M(\mathbf{grad}\, Q_h)$ by requiring that, on each element K,

$$\begin{cases} \Pi_2 v|_K \in M(\mathbf{grad}\, Q_h)|_K = b_{3,K}\mathbf{grad}\, Q_h|_K, \\ \int_K \text{div } (\Pi_2 v - v)\, q_h\, dx = 0, \quad \forall q_h \in Q_h|_K. \end{cases} \tag{5.13}$$

Note that (5.13) is uniquely solvable if $v \in V^0$ since the divergence of a bubble function always has zero mean value (hence the number of nontrivial equations is equal to $\dim(Q_h|_K) - 1$, which is equal to the number of unknowns; the nonsingularity then follows easily). It is clear that Π_2, as given by (5.13), will satisfy (4.31) for all $v \in V^0$. We have to check that

$$\|\Pi_2 v\|_1 \leq c\, \|v\|_1, \tag{5.14}$$

which actually follows again by a scaling argument. It is then easy to see that the operator

$$\Pi_h = \tilde{\Pi}_1 + \Pi_2(I - \tilde{\Pi}_1) \tag{5.15}$$

satisfies the condition of Lemma 4.3 and the inf-sup condition follows. $\square$

Corollary 5.2 (Bi-dimensional triangular case.) Let us assume that $Q_h \subset Q$ is any space of piecewise smooth functions and suppose that

$$(\mathcal{L}_2^1)^2 \oplus M(\mathbf{grad}\, Q_h) \subset V_h.$$

Then the pair (V_h, Q_h) satisfies the inf-sup condition.

Proof. The condition $(\mathcal{L}_2^1)^2 \subset V_h$ implies that we can construct $\tilde{\Pi}_1$ as in Example 5.6. On the other hand we have $M(\mathbf{grad}\, Q_h) \subset V_h$, so that we can apply the previous Proposition 5.2. $\square$

Propositions 4.1, 4.2 and 4.3 require a few comments. They show that almost any element can be stabilized by using bubble functions. For continuous pressure elements this procedure is mainly useful in the case of triangular elements. For discontinuous pressure elements it is possible to stabilize

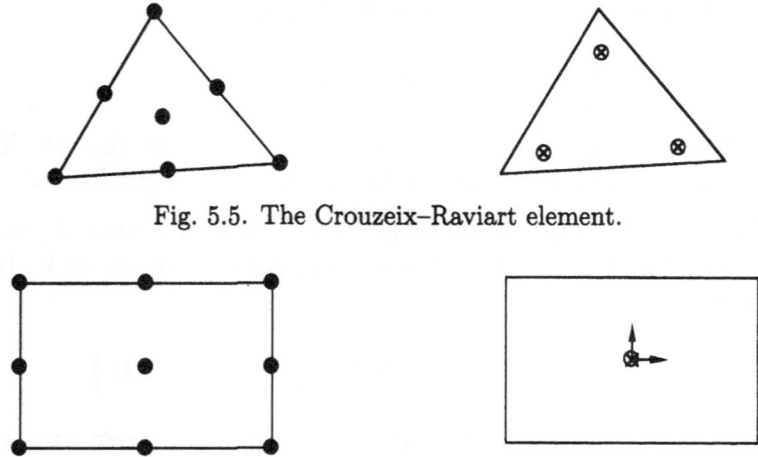

Fig. 5.5. The Crouzeix–Raviart element.

Fig. 5.6. The Q_2–P_1 element.

elements provided that they are already stable for piecewise constant pressure field. Examples of such a procedure can be found in Fortin and Fortin (1985a). Stability with respect to piecewise constant pressure implies that at least one degree of freedom on each side or face of the element is linked to the normal component of the velocity (Bernardi and Raugel (1981) or Fortin (1981)). Let us now consider a few examples of discontinuous pressure elements.

Example 5.7 (The Crouzeix and Raviart element.) We take Q_h to be the space of piecewise linear discontinuous functions. The previous construction then consists in adding cubic bubbles to a piecewise quadratic approximation of the velocity. This element is sketched in Figure 5.5. It provides second-order accuracy and is probably one of the best choices among stable triangular elements. □

This element has a rectangular (or even isoparametric) counterpart which is worth presenting. One interesting fact is that the triangular and rectangular versions are compatible and can be used inside a mixed mesh.

Example 5.8 (The Q_2–P_1 element and generalizations.) Let us consider an approximation of the velocity by a full biquadratic approximation and of the pressure by piecewise *linear* discontinuous functions. It can then be checked that the element is stable using the same kind of argument as in Corollary 5.2. This element, sketched in Figure 5.6, is one of the most popular elements for the approximation of incompressible flows. Although the previous results, as stated, can only be applied to the triangular case, the rectangular case and its isoparametric counterpart can be handled along the same lines. The idea is that once the constant part of the pressure is controlled by integrals on the boundary of the element, one may use *internal*

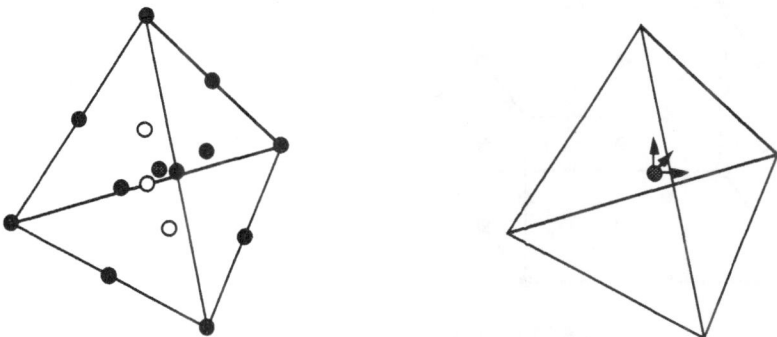

Fig. 5.7. Three-dimensional Crouzeix–Raviart element.

nodes to control the remaining part. In the case of a full biquadratic Q_2 approximation of the velocity, we have two internal degrees of freedom so that a P_1 pressure field can be used, but not a bilinear Q_1, as this would require three internal nodes. Evidently, one could enrich the approximation of the velocity to accommodate any degree of approximation of the pressure (Fortin and Fortin 1985a). It can be easily checked that for $k \geq 3$ a Q_k approximation of the velocity can be combined with a P_{k-1} or a Q_{k-1} approximation of the pressure. The case of $k = 1$ is pathological and will be discussed later. □

Example 5.9 (Three-dimensional tetrahedral discontinuous pressure elements.) The same arguments can be directly translated to the three-dimensional case (cf. Fortin (1981) or Stenberg (1987)). The main difference is that, in order to control the piecewise constant part of the pressure, one needs to use degrees of freedom on the faces of tetrahedra rather than on the edges. The equivalent of the operator $\tilde{\Pi}_1$ requires the integration of fluxes on faces; this requires the use of polynomials of degree greater than or equal to three if we only want to enrich by internal nodes. However, there exists a three-dimensional Crouzeix and Raviart element as sketched in Figure 5.7. It is obtained by enriching a second-degree element (with ten degrees of freedom on vertices and on the edges) by one cubic bubble on each face plus one fourth-degree internal bubble. Moving to higher degree polynomials, one may similarly build enriched elements with any order of accuracy. Finally with polynomials of degree higher than or equal to nine, one may build exactly divergence-free elements, as in the result of Scott and Vogelius (1985) discussed earlier. □

Example 5.10 (Three-dimensional hexahedral discontinuous pressure elements.) It can easily be checked that the three-dimensional Q_2–P_1 element sketched in Figure 5.8 is also a stable element. The Q_2–Q_1 element is not stable. For $k \geq 3$, a Q_k–P_{k-1} approximation is stable and for $k \geq 4$, so will Q_k–Q_{k-1}. □

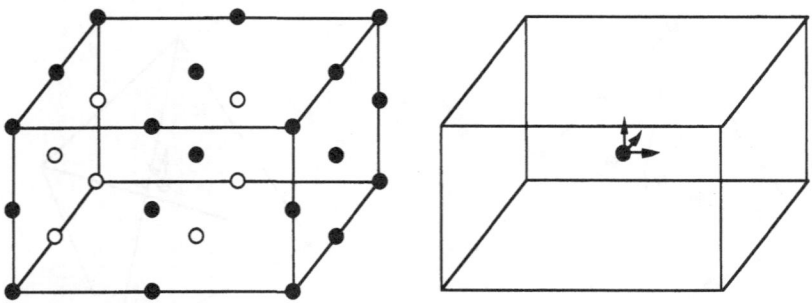

Fig. 5.8. Three-dimensional Q_2-P_1 element.

5.3. Nonconforming elements

We have just seen that it is possible to build approximately divergence-free approximations by enriching the approximation of the velocity. A different way to obtain stable approximations is to employ nonconforming elements, that is elements for which continuity requirements at interfaces have been relaxed. Using nonconforming elements implies that the variational formulation must be modified. In the Stokes problem, for instance, one must define discrete versions of the bilinear forms $a(\cdot, \cdot)$ and $b(\cdot, \cdot)$,

$$a_h(\boldsymbol{u}, \boldsymbol{v}) = 2\mu \sum_K \int_K \boldsymbol{\varepsilon}(\boldsymbol{u}) : \boldsymbol{\varepsilon}(\boldsymbol{v}) \, dx, \tag{5.16}$$

$$b_h(\boldsymbol{v}, q) = \sum_k \int_K q \, \mathrm{div}\, \boldsymbol{v} \, dx. \tag{5.17}$$

These discrete forms are defined even for functions which are discontinuous at interfaces, as only derivatives inside elements are involved. The discrete problem can now be written

$$\begin{cases} a_h(\boldsymbol{u}_h, \boldsymbol{v}_h) + b_h(\boldsymbol{v}_h, p_h) = (\boldsymbol{f}, \boldsymbol{v}_h), & \forall \boldsymbol{v}_h \in V_h, \\ b_h(\boldsymbol{u}_h, q_h) = 0, & \forall q_h \in Q_h. \end{cases} \tag{5.18}$$

It is then possible to perform an error analysis of the problem. We refer to Brezzi and Fortin (1991) or to the original work of Crouzeix and Raviart (1973) for precise results the development of which is beyond the scope of this article. Let us simply say that nonconformity introduces additional consistency terms in the error analysis. These terms have to be properly bounded and the key for this is the *generalized patch-test*: 'for a nonconforming approximation of degree k to be optimal with respect to error estimates, the moments $\int_S v_h p_{k-1} \, ds$, must be continuous at any interface S, for any polynomial p_{k-1} of degree $k-1$'. The simplest of these elements is described in the following example. It was introduced in Crouzeix and Raviart (1973).

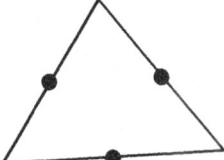 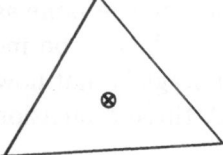

Fig. 5.9. The nonconforming P_1 element.

Example 5.11 (The nonconforming P_1 element.) We consider an approximation of the velocity by functions which are piecewise linear but are continuous only at midside points at element interfaces. This implies that $\int_S v \, ds$ is continuous as the midpoint rule is exact for polynomials of degree one and the patch–test is therefore satisfied. The pressure is piecewise constant and the element is sketched in Figure 5.9. This is the simplest first-order accurate element for incompressible problems. As pressure is discontinuous, one has local conservation of mass. The three-dimensional analogue is readily built, using values at the barycentre of the faces as degrees of freedom. □

It is also possible to construct higher order nonconforming elements. This is easily done for odd degree polynomials. One can find, for instance, a third-order nonconforming element in the paper of Crouzeix and Raviart (1973) in which a polynomial of degree three, enriched by bubbles of degree four, is employed. Continuity is then required at three Gauss–Legendre points on each element side. This implies that the element passes the correct patch test and the values at those Gauss–Legendre points can be used, with the addition of some internal nodes, as degrees of freedom. For even degree polynomials, a pathology arises and a different way must be found, as described in the next example.

Example 5.12 (The Fortin–Soulié nonconforming element.) It is easy to see that, in the two-dimensional case, the construction of a nonconforming element of degree two (or more generally of even degree), leads to unexpected difficulties. To satisfy the patch test and obtain the correct accuracy, one should ensure continuity at the two Gauss–Legendre points on the sides of elements. The trouble is that these six points cannot be used as degrees of freedom for a polynomial of degree two as one would like to do following the previous example: there exists a *nonconforming bubble* which vanishes at all six Gauss–Legendre points. It is expressed, in barycentric coordinates, as

$$b_{nc}(\lambda_1, \lambda_2, \lambda_3) = 2 - 3(\lambda_1^2 + \lambda_2^2 + \lambda_3^2). \tag{5.19}$$

The way around this difficulty is to construct second-order nonconforming methods in the same way as one built the element of Example 5.5: by enriching a standard conforming element of degree two by the nonconforming bubble (5.19). We refer to Fortin and Soulié (1983) for details. The degrees of freedom are the same as in a Crouzeix–Raviart element and only the bubble function has to be modified in the code, essentially a one line change. The advantage is that now only polynomials of degree two have to be manipulated. A three-dimensional version has also been derived in Fortin (1985). □

Finally, let us note that it is possible to build quadrilateral nonconforming elements along the same lines, that is by enriching a standard element by a function satisfying the patch test. A Q_1 nonconforming element can, for example, be obtained by adding to the standard conforming element a function of the form $\phi(x,y) = xy$ on $]-1,1[\times]-1,1[$ (Fortin and Soulié, 1983). It is also possible to add a function of the form $x^2 - y - y^2$ to a P_1 approximation (Rannacher and Turěk, 1992).

5.4. Taylor–Hood elements and generalizations

There exists another class of stable elements which is not covered by the previous analysis and which are worth a presentation. This class contains the Taylor–Hood element and its generalizations (Hood and Taylor, 1973; Bercovier and Pironneau, 1977; Brezzi and Falk, 1991). They essentially consist of taking, for triangular elements

$$V_h = \mathcal{L}_k^1, \quad Q_h = \mathcal{L}_{k-1}^1, \tag{5.20}$$

that is continuous pressure elements with the pressure one degree lower than the velocity. This yields the right order of accuracy as one only approximates pressure in $L^2(\Omega)$. The corresponding quadrilateral elements are also widely employed and the three-dimensional counterpart is quite popular. Because it contains an important idea, Verfürth's trick, we rapidly sketch the proof of stability for the original Taylor–Hood elemencorresponding to $k = 2$ in (5.20). The proof proceeds in two steps, the first being very general.

Lemma 5.1 Let Ω be a bounded domain of $\mathbf{R}^n$ with Lipschitz continuous boundary. Let $V_h \subset (H_0^1(\Omega))^2 = V$ and $Q_h \subset H^1(\Omega)$. Suppose that there exists a linear operator Π_h^0 from V into V_h and a constant c, independent of h, such that

$$\|v_h - v\|_{r,\Omega} \le c \sum_K \left(h_K^{2-2r} \|v\|_{1,K}^2 \right)^{1/2}, \quad \forall v \in V, r = 0,1. \tag{5.21}$$

Then there exist two positive constants c_1 and c_2 such that for every $q_h \in Q_h$

$$\sup_{v_h \in V_h} \frac{\int_\Omega q_h \operatorname{div} v_h \, dx}{\|v_h\|_1} \ge c_1 \|q_h\|_{0/\mathbf{R}} - c_2 \sum_k \left(h_k^2 \|\operatorname{\mathbf{grad}} q_h\|_{0,K}^2 \right)^{1/2}. \tag{5.22}$$

We refer to Brezzi and Fortin(1991) for the proof. Let us remark again that this is general and holds for any continuous pressure approximation. Now let us return to the special case of the quadratic–linear approximation of Taylor and Hood.

Lemma 5.2 (Stability of the Taylor–Hood element.) Let $V_h = (\mathcal{L}_2^1)^2 \cap (H_0^1(\Omega))^2$ and $Q_h = \mathcal{L}_1^1$. Then, if any element of $\mathcal{T}_h$ has no more than one edge on the boundary, there exists a positive constant c_3 such that for every $q_h \in Q_h$,

$$\sup_{v_h \in V_h} \frac{\int_\Omega q_h \operatorname{div} v_h \, dx}{\|v_h\|_1} \geq \left(\sum_K h_K^2 |q_h|_{1,K}^2 \right)^{1/2}. \tag{5.23}$$

Proof. We shall prove the result by constructing a suitable $\bar{v}_h$. Let $q_h \in Q_h$ be given and let K be an element of $\mathcal{T}_h$. We define $\bar{v}_h$ on K by

$$\begin{cases} \bar{v}_h = 0 & \text{at the vertices of } K, \\ \bar{v}_h = -t_e(\operatorname{grad} q_h \cdot t_e)|e|^2, \end{cases} \tag{5.24}$$

at the midpoint of every edge e of K, denoting by $|e|$ the length of e and by t_e the unit tangent vector to e, with some chosen orientation. One easily checks that

$$\|\bar{v}_h\|_{1,K} \leq ch_K|q_h|_{1,K}. \tag{5.25}$$

Now, we use a quadrature formula, which is exact for any polynomial of degree 2,

$$\int_K p_2(x) \, dx = \frac{\operatorname{meas}(K)}{3} \sum_M p_2(M), \tag{5.26}$$

where the sum is taken over the midpoints M of the edges of K. We then have, with the choice (5.24),

$$\begin{aligned} \int_\Omega q_h \operatorname{div} \bar{v}_h \, dx &= -\int_\Omega \operatorname{grad} q_h \cdot \bar{v}_h \, dx \\ &= -\sum_K \int_K \operatorname{grad} q_h \cdot \bar{v}_h \, dx \\ &= -\sum_K \sum_M (\operatorname{grad} q_h \cdot \bar{v}_h)(M) \frac{\operatorname{meas}(K)}{3} \\ &= \sum_K \sum_M |\operatorname{grad} q_h \cdot t_e|^2 |e|^2 \frac{\operatorname{meas}(K)}{3} \\ &\geq C \sum_k h_k^2 \|\operatorname{grad} q_h\|_{0,K}^2, \end{aligned} \tag{5.27}$$

where in the last inequality we have implicitly used a nondegeneracy condition $|e| \geq \sigma h_K$ and the hypothesis that two sides of K are internal so that $\bar{v}_h$ is defined from $\operatorname{grad} q_h \cdot t_e$ in at least two directions on every triangle.

From (5.25) and (5.27) we get

$$\frac{\int_\Omega q_h \operatorname{div} \bar{\boldsymbol{v}}_h \, dx}{\|\bar{\boldsymbol{v}}_h\|_1} \geq c \frac{\sum_k h_K^2 \|\operatorname{\mathbf{grad}} q_h\|_{0,K}^2}{\left(\sum_k h_K^2 \|\operatorname{\mathbf{grad}} q_h\|_{0,K}^2\right)^{1/2}}, \tag{5.28}$$

which is the desired result. $\square$

We can now prove

Proposition 5.3 (Stability of the Taylor–Hood element.) The pair $V_h = (\mathcal{L}_2^1)^2 \cap (H_0^1(\Omega))^2$ and $Q_h = \mathcal{L}_1^1$ is a stable element for the Stokes problem, that is it satisfies the discrete inf–sup condition.

Proof. We multiply the inequality (5.22) by c_3 and (5.23) by c_2 and we add them to get

$$(c_3 + c_2) \sup_{\boldsymbol{v}_h \in V_h} \frac{\int_\Omega q_h \operatorname{div} \boldsymbol{v}_h \, dx}{\|\boldsymbol{v}_h\|_1} \geq c_1 c_3 \|q_h\|_{0/\mathrm{R}}, \tag{5.29}$$

which is the desired condition. $\square$

This idea of combining a 'bad inequality' -like (5.22) and a 'good inequality in a bad norm' -like (5.23) is due to Verfürth (1984). It can be applied to other situations, for example to the study of stabilized methods presented in the next sections.

6. The Q_1–P_0 element (or 'what might go wrong')

We have introduced, in the previous sections, the discrete inf–sup condition (4.20). It is important in practice to know what should be expected if this condition is *not satisfied*. It is clear that the trouble will arise in the dual problem, that is, with the pressure. The most classical of difficulties is the appearance of a spurious zero-energy mode in the dual problem. All functions in $\ker(\operatorname{\mathbf{grad}}_h)$ are zero-energy modes in the dual problem. Those which are nonconstant are known as *chequerboard modes* because of the first discovered case:

Example 6.1 (The Q_1–P_0 element and the chequerboard model.) We consider a Q_1–P_0 approximation, that is we approximate velocity by bilinear elements and pressure by piecewise constants. Moreover, we restrict ourselves to a regular and rectangular mesh. Then, if we colour the rectangles like the squares of a chequerboard, there exists a spurious zero-energy mode taking value 0 on white squares and value 1 on black squares (Figure 6.1).

Fig. 6.1. The chequerboard mode.

This mode is defined up to a multiplicative constant and often manifests itself by huge values. In particular, a small displacement of a node by a value of ϵ transforms the zero eigenvalue into an $\mathcal{O}(\epsilon)$ eigenvalue, making an $\mathcal{O}(1/\epsilon)$ chequerboard mode to appear. $\square$

Other examples of zero-energy modes are met in equal-interpolation approximations, that is approximations in which pressure and velocities are approximated by polynomials of the same degree. Most of the time, but not always, they are strongly mesh-dependent and are present only on special regular meshes. The exactly divergence-free element of Example 5.1 on the crossgrid mesh of Figure 5.1 also suffers from exactly the same chequerboard mode as the Q_1–P_0 element.

This, however, is not the only way in which things can go wrong. Another way is that some nonzero eigenvalues become vanishingly small when h decreases, implying that the constant in condition (4.20) is not bounded from below and goes to zero with h. The result is at best a loss in the order of convergence or, worse still, a total loss of convergence. Again, the Q_1–P_0 element provides us with the simplest example. If we consider a regular rectangular mesh and compute the eigenvalues of the dual problem (Malkus, 1981), we see that a large number of them become smaller as h decreases. They can be associated with eigenvectors consisting of a restriction of the chequerboard mode described above to a 2×2 patch of elements. In all cases, a sign of instability first appears in the pressure. It is only in very severe cases that velocities are polluted in a visible way. Derivatives of velocities are however likely to suffer so that computing the vorticity is a good indicator of trouble. To make things still better, it is possible to build special meshes on which the Q_1–P_0 approximation is stable. One of them is presented in the next figure and was introduced by Letallec and Ruas (1986). It is also possible to show that on a regular mesh, formed of 2×2 patches of elements, things are not so bad as would appear from previous considerations: velocity converges at the right order and pressure can be filtered by projecting it on a proper subspace. A proof can be found in Brezzi and Fortin (1991).

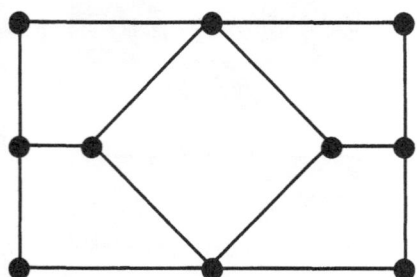

Fig. 6.2. A special mesh for Q_1–P_0.

7. Stabilization techniques

Up to now, we have obtained stable finite element pairs for the approximation of the velocity and the pressure by a clever choice of polynomial spaces.
There is, however, another possiblity which has received much attention in
recent years: stabilization can be achieved by modifying the variational formulation of the problem. The idea was introduced by Brooks and Hughes
(1982) for the stabilization of finite element methods for first-order advection problems. It was later extended in Hughes, Franca and Balestra (1986),
Hughes and Franca (1987) and Franca and Hughes (1988) to the Stokes problem, improving on the idea of Brezzi and Pitkäranta (1984) that we shall
develop later. Our emphasis will be on the variant of Douglas and Wang
(1989) which we consider to be most suitable for the Stokes problem or,
more generally, for mixed problems. But let us first consider the formulation of Brezzi and Pitkäranta (1984) which is simple and contains all the
basic ideas.

Example 7.1 (The stabilization of Brezzi and Pitkäranta.) The principle is very simple and consists of considering a perturbation of the Stokes
problem, that is to modify the problem (4.6) into

$$\begin{cases} a(\boldsymbol{u}_\epsilon, \boldsymbol{v}) + b(\boldsymbol{v}, p_\epsilon) = (\boldsymbol{f}, \boldsymbol{v}), & \forall \boldsymbol{v} \in V, \\ b(\boldsymbol{u}_\epsilon, q) = \epsilon \int_\Omega \mathbf{grad}\, p_\epsilon \cdot \mathbf{grad}\, q\, dx, & \forall q \in Q. \end{cases} \tag{7.1}$$

This is the variational formulation of the problem,

$$-2\mu\, \boldsymbol{A}\boldsymbol{u}_\epsilon + \mathbf{grad}\, p_\epsilon = \boldsymbol{f}, \tag{7.2}$$

$$\mathrm{div}\, \boldsymbol{u}_\epsilon + \epsilon \triangle p_\epsilon = 0, \tag{7.3}$$

$$\boldsymbol{u}_\epsilon|_\Gamma = 0, \quad \left.\frac{\partial p_\epsilon}{\partial n}\right|_\Gamma = 0. \tag{7.4}$$

We see from (7.4) that a parasitic Neumann boundary condition has been
introduced for the pressure. In practice, this will imply a boundary layer

effect and pressure values will be polluted near the boundary. Nevertheless, one can guess how this model stabilizes an unstable finite element method: chequerboard modes are highly oscillatory and they will be removed by the smoothing effect of the Laplace operator.

The proof that the solution obtained from the method of Example 7.1 is stable will be derived in two steps. First we shall try to obtain a bound on the difference between the solution of the perturbed problem and the solution of the standard Stokes problem. It can be proved (Brezzi and Fortin, 1991) that one has the following estimate.

Proposition 7.1 Let (u, p) tbe the solution of Problem (4.6) and (u_ϵ, p_ϵ) be the solution of Problem (7.1). Then we have

$$\|u - u_\epsilon\|_1 + \|p - p_\epsilon\|_0 \le c\sqrt{\epsilon}\|p\|_1. \tag{7.5}$$

$\square$

We refer to Brezzi and Fortin (1991) for a proof.

This result is not optimal and one can get an $\mathcal{O}(\epsilon^{\frac{3}{2}-\delta})$ estimate if p is smooth enough. However (7.5) is sufficient for our present purpose. Indeed, taking $\epsilon = O(h^2)$ will make the error in (7.5) of the same order as the error in a standard approximation by piecewise linear functions. Therefore, we can discretize Problem (7.1) with the simplest possible elements, such as a P_1–P_1 or a Q_1–Q_1 approximation and obtain results converging with the correct asymptotic accuracy. This gain is, however, not as complete as one would like. The choice of ϵ is critical: if it is too small, pressure oscillation remains while if it is too large, boundary layer effects will spoil the solution. What we would ultimately like to find would be a more robust formulation. A first step toward this is to employ the *Galerkin-least-squares* formulation as in Hughes and Franca (1987). To understand it better, we return to the Lagrangian of Problem (4.7) which we change tentatively to

$$\inf_{v \in V} \sup_{q \in Q} \mu \int_\Omega |\varepsilon(v)|^2 \, dx - \int_\Omega q \operatorname{div} v \, dx \tag{7.6}$$

$$- \int_\Omega f \cdot v \, dx - \epsilon \int_\Omega |Av + \operatorname{grad} q - f|^2 \, dx.$$

Note that we have added a squared term with a negative sign. This is because we want to stabilize the pressure which is the dual variable in the saddle-point problem. As in the Galerkin-least-squares method, this squared term corresponds to one of the equations in the strong form, namely (7.2). We could have added the square of the second equation to improve the coercivity properties of the problem with respect to u. In the present case, this is of no use as the bilinear form $a(\cdot, \cdot)$ is already fully coercive. Examples of cases where this modification would be useful can be found in Brezzi, Fortin and Marini (1992).

Formulation (7.6) needs a few comments. The first one is that it is in fact ill defined: indeed, for $v \in V$, Av is not square-integrable. We should therefore move to a more regular space, with the side effect of a more difficult approximation, or weaken the formulation as we shall do later. The second comment is that something will go wrong with the coercivity with respect to u as we now have a bilinear form

$$a(u, v) = \mu \int_\Omega \varepsilon(u) : \varepsilon(v) \, dx - \epsilon \int_\Omega Av \cdot Av \, dx. \tag{7.7}$$

The negative sign impairs the coercivity of $a(\cdot, \cdot)$. Only for discrete problems can this be cured, by taking ϵ small (e.g., $\mathcal{O}(h^2)$), and by using the equivalence of norms on a finite-dimensional space, more precisely an *inverse inequality* of the form

$$\|Av_h\|_0 \le \frac{C}{h} \|\varepsilon(v_h)\|_0. \tag{7.8}$$

Let us return to the first point. In order to be able to employ a standard finite element approximation, we shall write the least-squares terms in the form

$$\epsilon \sum_K \int_K |Av + \mathbf{grad}\, q - f|^2 \, dx. \tag{7.9}$$

This is now well defined on the space

$$W = \{(v, q) | Av + \mathbf{grad}\, q_{|K} \in L^2(K), \forall K \in \mathcal{T}_h\}. \tag{7.10}$$

Standard finite element discretizations of $H^1(\Omega) \times L^2(\Omega)$ are also contained in W as the restriction to an element is a regular polynomial function. This modification does not, however, cure the problem of coercivity. The answer to this second issue is a formulation introduced by Douglas and Wang (1989) where the variational problem (4.6) is modified into

$$a(u_\epsilon, v) + b(v, p_\epsilon)$$
$$+ \epsilon \sum_K \int_K (Au_\epsilon + \mathbf{grad}\, p_\epsilon - f) \cdot Av \, dx = (f, v), \quad \forall v \in V,$$
$$b(u_\epsilon, q) - \epsilon \sum_K \int_K (Au_\epsilon + \mathbf{grad}\, p_\epsilon - f) \cdot \mathbf{grad}\, q \, dx = 0, \quad \forall q \in Q. \tag{7.11}$$

This differs by one sign change from what would be obtained by the optimality conditions of Problem (7.6). This sign change is nevertheless crucial: choosing $v = u_\epsilon$ and $q = p_\epsilon$ in (7.11) and substracting the two equations one gets

$$\mu \int_\Omega |\varepsilon(u)_\epsilon|^2 \, dx + \epsilon \sum_K \int_K |Au_\epsilon + \mathbf{grad}\, p_\epsilon|^2 \, dx \le c(\|f\|^2). \tag{7.12}$$

We have therefore obtained stability in the space W defined in (7.10) for any value of ϵ. It might appear that we have not really stabilized $\mathbf{grad}\, p$ but only $A\boldsymbol{u} + \mathbf{grad}\, p$. However, as $\boldsymbol{u}$ is generally smooth, the stabilization effort really bears on p. Formulations of this type have been used with success with many finite element formulations. They allow equal-interpolation approximations of low order which would otherwise be forbidden in standard variational formulations. Everything is not as nice as it would seem from the above discussion: the solution still suffers from a parasitic boundary condition on p and a boundary layer effect. The source of trouble is that the term $A\boldsymbol{u}$ in $A\boldsymbol{u} + \mathbf{grad}\, p - \boldsymbol{f}$ is not computed accurately in a standard finite element approximation. It will normally be approximated at one order lower than the other terms. The limiting case is the piecewise linear one where $A\boldsymbol{u}|_K$ is always identically zero. This will oblige us to take again $\epsilon = \mathcal{O}(h^2)$ to recover the corrrect error estimate. Moreover, this lack of accuracy in one term spoils the solution in a visible way near the boundary. Many techniques have been advocated to remove this boundary layer effect (e.g., Brezzi and Douglas (1988)). The most popular one consists in substracting boundary effects by adding a correcting term to the formulation. For example, one might modify Problem (7.11) into

$$
\left\{
\begin{aligned}
&a(\boldsymbol{u}_\epsilon, \boldsymbol{v}) + b(\boldsymbol{v}, p_\epsilon) \\
&\quad + \epsilon \sum_K \int_K (A\boldsymbol{u}_\epsilon + \mathbf{grad}\, p_\epsilon - \boldsymbol{f}) \cdot A\boldsymbol{v}\, dx = (\boldsymbol{f}, \boldsymbol{v}), \quad \forall \boldsymbol{v} \in V, \\
&b(\boldsymbol{u}_\epsilon, q) - \epsilon \sum_K \int_K (A\boldsymbol{u}_\epsilon + \mathbf{grad}\, p_\epsilon - \boldsymbol{f}) \cdot \mathbf{grad}\, q\, dx \\
&\quad - \int_{\partial\Omega} ((A\boldsymbol{u}_\epsilon + \mathbf{grad}\, p_\epsilon - \boldsymbol{f}) \cdot \boldsymbol{n}\, q\, ds = 0, \quad \forall q \in Q.
\end{aligned}
\right.
$$

$$(7.13)$$

Numerical results obtained through such modifications are good (cf., e.g., Leborgne (1992)). However, coercivity properties are lost and getting a solution from the discretized problem becomes delicate. The correct way of eliminating boundary layer effects is still an open problem. To conclude, stabilized formulations are an important new idea in the approximation of incompressility, an idea which is likely to see new developments in future years.

8. Numerical methods for the discretized problems

Given a stable approximation, we now have the practical task of effectively computing the approximate solution. We shall deal with two different issues, namely the treatment of the incompressibility condition and the treatment of the full nonlinear Navier–Stokes problem.

8.1. Penalty methods

The numerical methods which we want to introduce are based mainly on techniques derived from penalty methods. We shall therefore describe these briefly. We use the steady-state Stokes problem as a prototype but the idea applies to any incompressible problem. A penalty method is then a perturbation of our original problem (4.6) into

$$
\begin{cases}
a(\boldsymbol{u}_\epsilon, \boldsymbol{v}) + b(\boldsymbol{v}, p_\epsilon) = (\boldsymbol{f}, \boldsymbol{v}), & \forall \boldsymbol{v} \in V, \\
b(\boldsymbol{u}_\epsilon, q) - \epsilon \int_\Omega p_\epsilon q \, \mathrm{d}x = 0, & \forall q \in Q.
\end{cases}
\tag{8.1}
$$

It can be shown (Bercovier (1978) or Brezzi and Fortin (1991)) that the error induced by this (regular) perturbation is $\mathcal{O}(\epsilon)$. Let us now consider the matrix form of the discrete problem already presented in Section(4.5). The problem becomes

$$
\begin{pmatrix} \mathcal{A} & \mathcal{B} \\ \mathcal{B}^t & -\epsilon \mathcal{M}^Q \end{pmatrix} \begin{pmatrix} U \\ P \end{pmatrix} = \begin{pmatrix} F \\ 0 \end{pmatrix}.
\tag{8.2}
$$

But the matrix $\mathcal{M}^Q$ is invertible and it is possible to eliminate pressure from these equations to obtain

$$
\mathcal{A}U + \frac{1}{\epsilon} \mathcal{B}^t (\mathcal{M}^Q)^{-1} \mathcal{B}U = F.
\tag{8.3}
$$

Once U has been obtained by solving (8.3), one can calculate the pressure by

$$
P = \frac{1}{\epsilon} (\mathcal{M}^Q)^{-1} \mathcal{B}U.
\tag{8.4}
$$

This procedure is in fact usable only if the matrix $\mathcal{M}^Q$ is easily invertible (Bercovier, Engelman and Gresho (1982)). For discontinuous pressure approximations described in Section 5, we can invert $\mathcal{M}^Q$ element by element and the numerical implementation is direct. It must be said that this simplification also has some disadvantages: the system (8.3) is ill conditioned for ϵ small. Care must be taken if one wishes to get an accurate solution, and the convergence of iterative methods, such as a conjugate-gradient method is jeopardized. For continuous pressure approximations, $(\mathcal{M}^Q)^{-1}$ is a full matrix and the reduced problem is not tractable. The perturbed problem (8.2) is nevertheless employed as it cures the singularity (p is defined up to an additive constant) of the original problem in the case of pure Dirichlet conditions on $\boldsymbol{u}$.

8.2. The augmented Lagrangian method

We briefly describe here how a simple iterative procedure, called the augmented Lagrangian method, can be employed to remove penalty errors and to efficiently compute a solution of the original problem (4.16). Our presentation will be sketchy by necessity. We refer to Fortin and Glowinski (1983) for a precise analysis of the method. This technique is also closely related to the artificial compressibility method introduced by Chorin (1968) and widely used under different names.

Suppose that we choose P_0, an arbitrary initial guess for the pressure. We then compute, P_n being known, P_{n+1} from the relation,

$$\begin{pmatrix} \mathcal{A} & \mathcal{B} \\ \mathcal{B}^t & -\epsilon\mathcal{M}^Q \end{pmatrix} \begin{pmatrix} U_{n+1} \\ P_{n+1} \end{pmatrix} = \begin{pmatrix} F \\ -\epsilon\mathcal{M}^Q P_n \end{pmatrix}. \tag{8.5}$$

If $\mathcal{M}^Q$ is easily invertible, one can write this in the decoupled form,

$$\mathcal{A}U_n + \frac{1}{\epsilon}\mathcal{B}^t(\mathcal{M}^Q)^{-1}\mathcal{B}U_n = F - \mathcal{B}^t P_n, \tag{8.6}$$

$$\mathcal{M}^Q P_{n+1} = \mathcal{M}^Q P_n + \frac{1}{\epsilon}\mathcal{B}U_n. \tag{8.7}$$

This is a special case of a more general algorithm, Uzawa's algorithm, for the numerical solution of saddle-point problems. Convergence is easily proved for any positive value of ϵ. Taking ϵ small, (say 10^{-6}) makes the algorithm convergent to machine precision in two or three iterations. In fact taking ϵ small makes the dual problem in p very well conditioned (cf. Fortin and Pierre (1992)) so that this iteration, which is in fact a gradient method for the dual problem, converges very rapidly. The price we pay is that Problem (8.6) in U is ill conditioned. When an iterative solution is needed, as is often the case in three-dimensional problems, a balance should be kept between the convergence of the iteration for solving Problem (8.6) and the convergence of the outer iteration in (8.7). Methods of this type have also been used as preconditioners for conjugate-gradient methods (e.g., Fortin (1989)).

8.3. Nonlinear problems

When the Navier–Stokes problem is considered, we have to solve a large nonlinear system. The most popular method is Newton's method which reduces this solution to a sequence of linear incompressible problems. The augmented Lagrangian method can then be used to solve these linear problems. Under some restrictions on the choice of ϵ it can also be incorporated to Newton's iteration (Fortin and Fortin, 1985b). The most efficient solution method is, however, to employ a conjugate-gradient-like iteration such as the GMRES method of Saad and Schultz (1986) with a suitable preconditioning. One then needs only to compute products of some vectors and the Jacobian matrix and this can be approximated by differences, avoiding the actual computation of the Jacobian. A very good description of this technique can be found in Shakib, Hughes and Zdeněk (1989) for compressible

problems and can be transposed directly to incompressible problems. For
incompressible problems, some approximate augmented Lagrangian method
can be used as a preconditioner (Fortin, 1989).

9. Time-dependent problems

Our original problem was time-dependent and we now return to this aspect.
The standard procedure for the discretization of a time-dependent problem
is to first consider a discretization in space, reducing it to a large system of
ordinary differential equations and then to employ some numerical scheme
for this system. The choice of scheme can then be made from a vast collection
of ODE solvers.

9.1. Time discretization, projection methods

One important point in the choice of a time discretization is that the system
is not of Cauchy–Kovalevska type as there is no derivative in time of the
pressure in the equations. In fact, in this respect, the problem is related to
the so-called algebraic–differential systems (Petzold, 1983). It can be seen
that the pressure part is elliptic. Indeed, taking the divergence of equation
(3.1), we obtain the *Poisson pressure equation*

$$- \triangle p = \operatorname{div} (\boldsymbol{u} \cdot \operatorname{\mathbf{grad}} \boldsymbol{u}) - \operatorname{div} \boldsymbol{f}, \tag{9.1}$$

in which we have used (3.2) to eliminate a number of terms. This equation
(9.1) holds at all times. It has been widely employed in the construction
of time-stepping procedures, but difficulties arise from the absence of pressure boundary conditions. There is, in reality, no rigorous way to obtain
such conditions apart from some iterative procedure or the construction of
an integral equation on the boundary of the domain like in Glowinski and
Pironneau (1979) (see also Gresho and Sani (1987)). This being said, it is
possible to include the solution of a Neumann problem in p into a fractional
step method, such as the projection method introduced in Chorin (1968) and
developed in Fortin, Peyret and Temam (1971). This scheme, in its simplest
form, would consists of an advection step followed by a projection on the
subspace of divergence-free functions. However, it is not immediately clear
in which space should the projection take place. The two obvious choices
are $L^2(\Omega)$ or $H_0^1(\Omega)$. Let us consider in some detail these two cases.

Lemma 9.1 (The $L^2(\Omega)$ projection.) Let $\boldsymbol{u}$ be given in $(L^2(\Omega))^n$. Then
$\boldsymbol{u}$ can be written as

$$\boldsymbol{u} = \boldsymbol{u}_0 + \operatorname{\mathbf{grad}} p_0 + \operatorname{\mathbf{grad}} p_1, \tag{9.2}$$

with $p_0 \in H_0^1(\Omega)$, $p_1 \in \mathcal{H}(\Omega)$ and $\boldsymbol{u}_0 \in H_0(\Omega)$, where

$$\left\{ \begin{array}{l} \mathcal{H}(\Omega) = \{q \,|\, q \in H^1(\Omega), \triangle p = 0\}, \\ H_0(\Omega) = \{\boldsymbol{v} \,|\, \boldsymbol{v} \in (L^2(\Omega))^n, \operatorname{div} \boldsymbol{v} = 0, \boldsymbol{v} \cdot \boldsymbol{n}_{|\partial\Omega} = 0\}. \end{array} \right. \tag{9.3}$$

Proof. The idea is essentially to solve a Dirichlet problem,

$$- \triangle p_0 = \operatorname{div} \boldsymbol{u}, \qquad (9.4)$$

to compute $\boldsymbol{u}_1 = \boldsymbol{u} - \mathbf{grad}\, p_0$, to solve a Neumann problem

$$- \triangle p_1 = 0, \quad \frac{\partial p_1}{\partial \boldsymbol{n}} = \boldsymbol{u}_1 \cdot \boldsymbol{n} \text{ on } \partial\Omega, \qquad (9.5)$$

and then finally obtain $\boldsymbol{u}_0 = \boldsymbol{u}_1 - \mathbf{grad}\, p_1$. $\square$

A few remarks are needed about the boundary conditions employed. First, the condition $\boldsymbol{u}_0 \cdot \boldsymbol{n}$ has to be justified for it does not make sense, *a priori*, to write a boundary condition for a function in $(L^2(\Omega))^n$. We refer to Temam (1977) for this justification. Second, the two problems (9.4)–(9.5) may be combined into one:

$$- \triangle p = \operatorname{div} \boldsymbol{u}, \quad \frac{\partial p}{\partial \boldsymbol{n}} = \boldsymbol{u} \cdot \boldsymbol{n} \text{ on } \partial\Omega, \qquad (9.6)$$

provided $\boldsymbol{u}$ slightly more regular, namely if $\operatorname{div} \boldsymbol{u} \in L^2(\Omega)$. Let us now see how one can use this approach to construct a fractional step method.

Example 9.1 (The L^2 projection scheme.) This scheme will be a fractional step method and many variants are possible, depending on the implicit or explicit character of the first step. We shall consider here the implicit variant, which we feel is more reliable but other cases can be easily formulated. Moreover, we shall not explicitly introduce a space discretization and we shall, formally, write the scheme without any such discretization. Let then $\boldsymbol{u}^n$ and p^n be known at time step n. We shall compute the solution at the next time step $n + 1$ in two substeps. First, we solve, denoting by δt the time step,

$$\frac{\boldsymbol{u}^{n+\frac{1}{2}} - \boldsymbol{u}^n}{\delta t} + \boldsymbol{u}^{n+\frac{1}{2}} \cdot \mathbf{grad}\, \boldsymbol{u}^{n+\frac{1}{2}} - 2\mu\, \boldsymbol{A}\boldsymbol{u}^{n+\frac{1}{2}} + \mathbf{grad}\, p^n = \boldsymbol{f}. \qquad (9.7)$$

This is a nonlinear problem which can be solved either by a Newton method or an approximate Newton method. No incompressibility condition is imposed on $\boldsymbol{u}^{n+\frac{1}{2}}$ and the next step intends to correct this deficiency by projecting it on the divergence-free subspace $H_0(\Omega)$. This amounts to solving a Neumann problem:

$$- \triangle \delta p = \operatorname{div} \boldsymbol{u}^{n+\frac{1}{2}}, \quad \frac{\partial \delta p}{\partial \boldsymbol{n}} = \boldsymbol{u}^{n+\frac{1}{2}} \cdot \boldsymbol{n} \text{ on } \partial\Omega \qquad (9.8)$$

and then to compute

$$\begin{cases} \boldsymbol{u}^{n+1} = \boldsymbol{u}^{n+\frac{1}{2}} - \mathbf{grad}\, \delta p, \\ p^{n+1} = p^n + \delta p. \end{cases} \qquad (9.9)$$

This is simple, but something is wrong: u^{n+1} does not satisfy the correct boundary conditions because the projection step only requires $u^{n+1} \cdot n$ to be null and leaves the tangential condition $u^{n+1} \cdot t$ undetermined. In practice, this problem is bypassed, in a discretized setting, by inserting the correct values at boundary nodes after the projection step. This is a new projection step, in some nonexplicit topology. The result is a scheme which is essentially first order in δt. $\square$

To do improve this, we would like to be able to project in $(H_0^1(\Omega))^n$-norm. But this is essentially equivalent to solving a Stokes problem.

Lemma 9.2 (The $(H_0^1(\Omega))^n$-projection.) Let u be given in $(H_0^1(\Omega))^n$. Then u can be written as

$$u = u_0 + u_p, \tag{9.10}$$

where $u_0 \in V_0(\Omega)$ with

$$V_0(\Omega) = \{v \mid v \in (H_0^1(\Omega))^n, \operatorname{div} v = 0\}, \tag{9.11}$$

and is the solution of the problem

$$\begin{cases} A u_0 + \operatorname{\mathbf{grad}} m = A u, \\ \operatorname{div} u_0 = 0, \end{cases} \tag{9.12}$$

where m is analogous to a pressure and serves to enforce the divergence-free condition.

Proof. The problem is to find u_0 as the solution of the constrained minimization problem,

$$\inf_{u_0 \in V_0} \|\varepsilon(u_0) - \varepsilon(u)\|_{1,\Omega}^2. \tag{9.13}$$

Introducing the Lagrange multiplier m and writing the optimality conditions of the Lagrangian obtained, one gets (9.12). $\square$

Using this result we are naturally led to a new projection scheme.

Example 9.2 (The H_0^1-projection scheme.) Let u^n and p^n be known at time step n. We shall compute the solution at the next time step $n + 1$ in two substeps. First, we solve, denoting by δt the time step,

$$\frac{u^{n+\frac{1}{2}} - u^n}{\delta t} + u^{n+\frac{1}{2}} \cdot \operatorname{\mathbf{grad}} u^{n+\frac{1}{2}} - 2\mu \, A u^{n+\frac{1}{2}} + \operatorname{\mathbf{grad}} p^n = f. \tag{9.14}$$

We then project $u^{n+\frac{1}{2}}$ by solving

$$\begin{cases} A u^{n+1} + \operatorname{\mathbf{grad}} \delta p = A u^{n+\frac{1}{2}}, \\ \operatorname{div} u_{n+1} = 0, \\ p^{n+1} = p^n + \delta p. \end{cases} \tag{9.15}$$

This is only one of possible variants and second-order methods can also be built using the θ-scheme such as in Bristeau, Glowinski and Périaux (198?) or other techniques (Bell, Colella and Glaz, 1989). The price we pay for this better handling of boundary conditions in the projection step is that the Stokes Problem (9.15) defining u^{n+1} is harder to solve than the Neumann problem (9.15). Using this method means that one should dispose of an efficient and simple Stokes solver. $\square$

Finally, a brute force method, that is a fully implicit scheme, can also be employed.

Example 9.3 (The fully implicit scheme.) Let u^n and p^n be known at time step n. We compute the solution at the next time step $n+1$ by solving:

$$\begin{cases} \dfrac{u^{n+1} - u^n}{\delta t} + u^{n+1} \cdot \mathbf{grad}\, u^{n+1} - 2\mu\, Au^{n+1} + \mathbf{grad}\, p^{n+1} = f, \\[2mm] \operatorname{div} u^{n+1} = 0. \end{cases} \tag{9.16}$$

Now u^{n+1} is the solution of a nonlinear incompressible problem. One possible way to solve this is by the penalty method already discussed. An equivalent way of introducing it is through the 'artificial compressibility method' of Chorin (1968) which is usually written as a perturbation of the above scheme:

$$\begin{cases} \dfrac{u^{n+1} - u^n}{\delta t} + u^{n+1} \cdot \mathbf{grad}\, u^{n+1} - 2\mu\, Au^{n+1} + \mathbf{grad}\, p^{n+1} = f, \\[2mm] \epsilon \dfrac{p^{n+1} - p^n}{\delta t} - \operatorname{div} u^{n+1} = 0. \end{cases} \tag{9.17}$$

Using the second equation, the first one may be written as

$$\frac{u^{n+1} - u^n}{\delta t} + u^{n+1} \cdot \mathbf{grad}\, u^{n+1} \tag{9.18}$$

$$-2\mu\, Au^{n+1} + \frac{\delta t}{\epsilon} \mathbf{grad}\operatorname{div} u^{n+1} + \mathbf{grad}\, p^n = f,$$

which is nothing but a penalty method for the solution of (9.16). One also sees that ϵ should be small with respect to δt which may give rise to severe ill-conditioning. An iterative variant based on the augmented Lagrangian method is therefore much more preferable. $\square$

Remark 9.1 In the implicit scheme (9.16) we have used an implicit Euler's scheme which is a stiffly-stable implicit method for ordinary differential equations (cf. Crouzeix and Mignot (1984). This strong stability property is highly desirable for large systems. However, it is now quite well established that to detect bifurcations to unsteady solutions in nonlinear Navier–Stokes problems correctly, a second-order scheme is essential (Fortin, Fortin and Gervais, 1991). A reasonable solution is through Gear's method which is a two-step implicit stiffly-stable scheme. It requires knowledge of u^n and u^{n-1} to compute u^{n+1}.

$$\begin{cases} \dfrac{\frac{3}{2}u^{n+1} - 2u^n + \frac{1}{2}u^{n-1}}{\delta t} + u^{n+1}\cdot\mathbf{grad}\, u^{n+1} \\ \qquad\qquad -2\mu\, Au^{n+1} + \mathbf{grad}\, p^{n+1} = f, \\ \mathrm{div}\, u^{n+1} = 0. \end{cases} \tag{9.19}$$

This scheme has been successfully employed for the computation of Hopf bifurcations. ☐

Remark 9.2 An interesting variant for a totally implicit scheme consists in using a method of characteristics for the discretization of advection terms (cf. Pironneau (1989)). The simplest way to do so can be summarized in the following algorithm.

$$\text{For any vertex } V \text{ of coordinates } x \text{ compute } x_* = x - \delta t \tag{9.20}$$

$$\text{Compute } u_*(x, t_n) = u(x_*, t_n). \tag{9.21}$$

To compute u^{n+1} one then solves

$$\begin{cases} \dfrac{u^{n+1} - u_*}{\delta t} - 2\mu\, Au^{n+1} + \mathbf{grad}\, p^{n+1} = f, \\ \\ \mathrm{div}\, u^{n+1} = 0. \end{cases} \tag{9.22}$$

The problem to solve in u^{n+1} is then a linear problem which can be solved by any suitable Stokes solver. More sophisticated versions of this idea are currently employed in industrial codes.

10. Conclusion

The possible issues to be considered in the numerical solution of the Navier–Stokes equations are so numerous that only a small fraction of them has been addressed here. Some, such as solution algorithms, have only been sketched. Finally, questions related to *a posteriori* error estimations and adaptivity have been completely ignored. The main difficulty remaining in the field is certainly the treatment of flows at high Reynolds number. Boundary layers imply delicate questions of mesh adaptation. Turbulence models, which try to represent the macroscopic effects of the small scales of the flow, are also an important issue. With respect to the treatment of incompressibilty which was our main topic, three-dimensional problems remain a challenge in both the construction of accurate elements and in the design of efficient solution methods. We hope that this article shall be useful as a guide into the rapidly changing world of computational fluid dynamics.

REFERENCES

D.N. Arnold, F. Brezzi and J. Douglas (1984), 'PEERS: a new mixed finite element for plane elasticity', *Japan J. Appl. Math.* **1**, 347–367.

D.N. Arnold, F. Brezzi and M. Fortin (1984), 'A stable finite element for the Stokes equations', *Calcolo* **21**, 337–344.

I.Babuška (1973), 'The finite element method with lagrangian multipliers', *Numer. Math.* **20**, 179–192.

J.B. Bell, P. Colella and H.M.Glaz (1989), 'A second-order projection method for the incompressible Navier–Stokes equations', *J. Comput. Phys.* **85**, 257–283.

M. Bercovier (1978), 'Perturbation of a mixed variational problem, applications to mixed finite element methods', *R.A.I.R.O. Anal. Numer.* **12**, 211–236.

M. Bercovier, M. Engelman and P. Gresho (1982), 'Consistent and reduced integration penalty methods for incompressible media using several old and new elements', *Int. J. Numer. Meth. in Fluids* **2**, 25–42.

M. Bercovier and O.A. Pironneau (1977), 'Error estimates for finite element method solution of the Stokes problem in the primitive variables', *Numer. Math.* **33**, 211–224.

C. Bernardi and G. Raugel (1981), 'Méthodes d'éléments finis mixtes pour les équations de Stokes et de Navier–Stokes dans un polygone non convexe', *Calcolo* **18**, 255–291.

F. Brezzi (1974), 'On the existence, uniqueness and approximation of saddle point problems arising from lagrangian multipliers', *R.A.I.R.O. Anal. Numer.* **8**, 129–151.

F. Brezzi and J. Douglas (1988), 'Stabilized mixed methods for the Stokes problem', *Numer. Math.* **53**, 225–235.

F. Brezzi and R.S. Falk (1991), 'Stability of a higher-order Hood–Taylor method', *SIAM J. Numer. Anal.* **28**.

F. Brezzi and M. Fortin (1991), *Mixed and Hybrid Finite Element Methods*, Springer (New-York).

F. Brezzi and J. Pitkäranta (1984), 'On the stabilization of finite element approximations of the Stokes equations', in *Efficient Solutions of Elliptic Systems, Notes on Numerical Fluid Mechanics*, Vol 10, (W. Hackbush, ed.), Braunschweig Wiesbaden (Vieweg).

F. Brezzi, M. Fortin and L.D. Marini (1992), 'Mixed finite element methods with continuous stresses', to appear.

M.O. Bristeau, R. Glowinski and J. Périaux (1987), 'Numerical methods for the navier–Stokes equations, applications to the simulation of compressible and incompressible viscous flows', *Finite Elements in Physics*, (R. Grüber, ed.) Computer Physics, North–Holland (Amsterdam).

A. Brooks and T.J.R. Hughes (1982), 'Streamline upwind/Petrov–Galerkin formulation for convection dominated flows with particular emphasis on the incompressible Navier-Stokes equations', *Comput. Meth. Appl. Mech. Eng.* **32**, 199–259.

P.G. Ciarlet (1978), *The Finite Element Method for Elliptic Problems*, North-Holland (Amsterdam).

P.G. Ciarlet and J.L. Lions (1991), *Handbook of Numerical Analysis, Volume II, Finite Element Methods*, North-Holland (Amsterdam).

P.G. Ciarlet and P.A. Raviart (1972), 'Interpolation theory over curved elements with applications to finite element methods', *Comput. Meth. Appl. Mech. Eng.* **1**, 217–249.

J.F. Ciavaldini and J.C. Nédélec (1974), 'Sur l'élément de Fraeijs de Veubeke et Sander', *R.A.I.R.O. Anal. Numer.* **8**, 29–45.

A.J. Chorin (1968), 'Numerical soluton of the Navier–Stokes equations', *Math. Comput.* **22**, 745–762.

P. Clément (1975), 'Approximation by finite element functions using local regularization', *R.A.I.R.O. Anal. Numer.* **9**, 77–84.

M. Crouzeix and A.L. Mignot (1984), *Analyse Numérique des Équations Différentielles*, Masson (Paris).

M. Crouzeix and P.A. Raviart (1973), 'Conforming and non-conforming finite element methods for solving the stationary Stokes equations', *R.A.I.R.O. Anal. Numer.* **7**, 33–76.

J. Douglas and J. Wang (1989), 'An absolutely stabilized finite element method for the Stokes problem', *Math. Comput.* **52**, 495–508.

T. Dupont and L.R. Scott (1980), 'Polynomial approximation of functions in Sobolev spaces', *Math. Comput.* **34**, 441–463.

G. Duvaut and J.L. Lions (1972), *Les Inéquations en Mécanique et en Physique*, Dunod (Paris).

M. Fortin (1977), 'An analysis of the convergence of mixed finite element methods', *R.A.I.R.O. Anal. Numer.* **11**, 341–354.

M. Fortin (1981), 'Old and new finite elements for incompressible flows', *Int. J. Numer. Meth. in Fluids* **1**, 347–364.

M. Fortin (1985), 'A three-dimensional quadratic non-conforming element', *Numer. Math.* **46**, 269–279.

M. Fortin (1989), 'Some iterative methods for incompressible flow problems ', *Comput. Phys. Commun.* **53**, 393–399.

A. Fortin and M. Fortin (1985), 'Newer and newer elements for incompressible flow', *Finite Elements in Fluids 6*, (R.H. Gallagher, G.F. Carey, J.T. Oden and O.C. Zienkiewicz, eds), John Wiley (Chichester).

A. Fortin and M. Fortin (1985a), 'A generalization of Uzawa's algorithm for the solution of the Navier–Stokes equations', *Comm. Appl. Numer. Methods* **1**, 205–208.

M. Fortin and R. Glowinski (1983), *Augmented Lagrangian Methods*, North-Holland (Amsterdam).

A. Fortin and D. Pelletier (1989), 'Are FEM solutions of incompressible flows really incompressible?', *Int. J. Numer. Meth. Fluids* **9**, 99–112.

M. Fortin and R. Pierre (1992), 'Stability analysis of discrete generalized Stokes problems', *Numer. Meth. Part. Diff. Eqns* **8**, 303–323.

M. Fortin and M. Soulié (1983), 'A non-conforming piecewise quadratic finite element on triangles', *Int. J. Numer. Meth. Eng.* **19**, 505–520.

A. Fortin, M. Fortin and J.J. Gervais (1991), 'Complex transition to chaotic flow in a periodic array of cylinders', *Theor. Comput. Fluid Dynam.* **3**, 79–93.

M. Fortin, R. Peyret and R. Temam (1971), 'Résolution numérique des équations de Navier–Stokes pour un fluide visqueux incompressible', *J. Mécanique* **10**, 3, 357–339.

B. Fraeijs de Veubeke and G. Sander (1968), 'An equilibrium model for plate bending', *Int. J. Solids and Structures* **4**, 447–468.

L.P. Franca and T.J.R. Hughes (1988), 'Two classes of finite element methods', *Comput. Meth. Appl. Mech. Eng.* **69**, 89–129.

V. Girault and P.A. Raviart (1986), *Finite Element Methods for Navier–Stokes Equations, Theory and Algorithms*, Springer (Berlin).

R. Glowinski (1984), *Numerical Methods for Nonlinear Variational Problems*, Springer (Berlin).

R. Glowinski and O. Pironneau (1979), 'Numerical methods for the first biharmonic equation and for the two-dimensional Stokes problem', *SIAM Rev.* **17**, 167–212.

P.M. Gresho and R.L. Sani (1987), 'On pressure boundary conditions for the incompressible Navier–Stokes equations', *Int. J. Numer. Meth. Fluids* **7**, 1111–1145

P. Hood and C. Taylor (1973), 'Numerical solution of the Navier–Stokes equations using the finite element technique', *Comput. Fluids* **1**, 1–28.

T.J.R. Hughes (1987), *The Finite Element Method: Linear Static and Dynamic Finite Element Analysis*, Prentice-Hall (Englewood Cliffs, NJ).

T.J.R. Hughes and L.P. Franca (1987), 'A new finite element formulation for computational fluid dynamics: VII. The Stokes problem with various well-posed boundary conditions, symmetric formulations that converge for all velocity-pressure spaces', *Comput. Meth. Appl. Mech. Eng.* **65**, 85–96.

T.J.R. Hughes, L.P. Franca and M. Balestra (1986), 'A new finite element formulation of computational fluid dynamics: a stable Petrov–Galerkin formulation of the Stokes problem accomodating equal-order interpolations', *Comput. Meth. Appl. Mech. Eng.* **59**, 85–99.

G. Leborgne (1992), Thèse, Ecole Polytechnique de Paris.

P. Letallec and V. Ruas (1986), 'On the convergence of the bilinear velocity-constant pressure finite method in viscous flow', *Comput. Meth. Appl. Mech. Eng.* **54**, 235–243.

J.L. Lions (1969). *Quelques Méthodes de Résolution des Problèmes aux Limites non Linéaires*, Dunod (Paris).

D.S. Malkus (1981), 'Eigenproblems associated with the discrete LBB-condition for incompressible finite elements', *Int. J. Eng. Sci.* **19**, 1299–1310.

L. Petzold (1983), 'Automatic selection of methods for solving stiff and nonstiff systems of ordinary differential equations', *SIAM J. Sci. Stat. Comput.* **4**, 136–148.

O.A. Pironneau (1989), *Finite Element Methods for Fluids*, John Wiley (Chichester).

R. Rannacher and S. Turěk (1992), 'A simple nonconforming quadrilateral Stokes element', *Numer. Meth. for Part. Diff. Eqns* **8**, 97–111.

P.A. Raviart and J.M. Thomas (1983), *Introduction à l'Analyse Numérique des Équations aux Dérivées Partielles*, Masson (Paris).

Y. Saad and M.H. Schultz (1986), 'A generalized minimum residual method for solving nonsymmetric linear systems, *SIAM J. Sci. Stat. Comput.* **7**, 856–869.

L.R. Scott and M. Vogelius (1985), 'Norm estimates for a maximal right inverse of the divergence operator in spaces of piecewise polynomials', *Math. Modelling Numer. Anal.* **9**, 11–43.

F. Shakib, T.J.R. Hughes and J. Zdeněk (1989), 'A multi–element group preconditioning GMRES algorithm for nonsymmetric systems arising in finite element analysis', *Comput. Meth. Appl. Mech. Eng.* **75**, 415–456.

R. Stenberg (1984), 'Analysis of mixed finite element methods for the Stokes problem: a unified approach', *Math. Comput.* **42**, 9–23.

R. Stenberg (1987), 'On some three-dimensional finite elements for incompressible media', *Comput. Meth. Appl. Mech. Eng.* **63**, 261–269.

R. Temam (1977), *Navier–Stokes Equations*, North-Holland (Amsterdam).

F. Thomasset (1981), *Implementation of Finite Element Methods for Navier–Stokes Equations, Springer Series in Computer Physics*, Springer (Berlin).

R. Verfürth (1984), 'Error estimates for a mixed finite element approximation of the Stokes equation', *R.A.I.R.O. Anal. Numer.* **18**, 175–182.

O.C. Zienkiewicz (1977), *The Finite Element Method*, McGraw–Hill (London).

Acta Numerica (1993), *pp.* 285–326

Old and new convergence proofs for multigrid methods

Harry Yserentant

Mathematisches Institut

Universität Tübingen

D-7400 Tübingen, Germany

E-mail: harry@tue-num2.mathematik.uni-tuebingen.de

Multigrid methods are the fastest known methods for the solution of the large systems of equations arising from the discretization of partial differential equations. For self-adjoint and coercive linear elliptic boundary value problems (with Laplace's equation and the equations of linear elasticity as two typical examples), the convergence theory reached a mature, if not its final state. The present article reviews old and new developments for this type of equation and describes the recent advances.

CONTENTS

1. Introduction

The discretization of partial differential equations leads to very large systems of equations. For two-dimensional problems, several ten thousand unknowns are not unusual, and in three space dimensions, more than one million unknowns can be reached very easily. The direct solution of systems of this size is prohibitively expensive, both with respect to the amount of storage and to the computational work. Therefore iterative methods like the Gauß–Seidel or the Jacobi iteration have been used from the beginning of the numerical treatment of partial differential equations.

An important step was Young's successive over-relaxation method (1950) which is much faster than the closely related Gauß–Seidel iteration. Nevertheless, this method shares with direct elimination methods the disadvantage that the amount of work does not remain proportional to the number of unknowns; the computer time needed to solve a problem grows more rapidly than the size of the problem. The standard reference on iterative methods is Varga (1962). For a recent treatment, see Hackbusch (1991).

Multigrid methods were the first to overcome this complexity barrier. Multigrid methods are composed of simple basic iterations. Probably the first working multigrid method was developed and analysed by Fedorenko (1964) for the Laplace equation on the unit square. Bachvalov (1966) considered the theoretically much more complex case of variable coefficients. Although the basic idea of combining discretizations on different grids in an iterative scheme appears to be very natural, the potential of this idea was not recognized before the middle of the 1970s. At this time, the multigrid idea began to spread.

The report of Hackbusch (1976) and the paper of Brandt (1977) were the historical breakthrough. The first big multigrid conference in 1981 in Köln was a culmination point of the development; the conference proceedings edited by Hackbusch and Trottenberg (1982) are still a basic reference. With Hackbusch's 1985 monograph, the first stage in multigrid theory came to an end.

Today, multigrid methods are used in nearly every field where partial differential equations are solved by numerical methods. They are applied in computational fluid dynamics as well as in semiconductor simulations. The bibliographies in McCormick (1987) and Wesseling (1992) each contain several hundred references.

The field of multigrid methods has became too large to review in a single article. Therefore, in this paper, we restrict our attention to the class of problems which is best understood, namely to self-adjoint and coercive linear elliptic boundary value problems. For mathematicians, the typical equation in this class is the Laplace equation. People, who are more oriented to real life, would probably think of the partial differential equations of structural mechanics.

Hackbusch (1982) and Braess and Hackbusch (1983) gave the first really satisfying convergence proofs for multigrid methods applied to this class of problem. The main problem with these proofs, and with all other convergence proofs appearing up until the beginning of the 1990s, is that they are based on regularity properties of the boundary value problem which are rarely satisfied in practice. In addition, the underlying finite element or finite difference meshes have to be quasi-uniform, i.e. all discretization cells have to be of approximately the same size. Although these assumptions are common in the theory of finite element methods, they are unrealistic.

These problems led Bank and Dupont (1980) and Axelsson and Gustafsson (1983) to the development of the two-level hierarchical basis methods. Yserentant (1986b) and Bank, Dupont and Yserentant (1988) extended this idea to the multilevel case. These methods have a simpler structure than the usual multigrid methods and do not depend, by their construction, on the restrictive assumptions mentioned earlier. Hierarchical basis methods have been shown to be very efficient in adaptive finite element codes; see Bank (1990) and Deuflhard, Leinen and Yserentant (1989).

Another development of the 1980s were the domain decomposition methods with the Schwarz alternating method as an early example; see Chan *et al.* (1989), for example. Recently these independent fields merged in a joint abstract theory which is flexible enough to treat many, at first sight completely different iteration schemes. The basic references are Bramble, Pasciak, Wang and Xu (1991a, 1991b), Bramble and Pasciak (1991), Dryja and Widlund (1991) and, especially in regard to terminology, Xu (1992b). This unified theory is one of the main topics of the present review article.

Fast iterative methods for the systems of equations

$$Au = f \tag{1.1}$$

resulting from the discretization of self-adjoint, the coercive linear elliptic boundary value problems are not only of interest in their own field but also of interest elsewhere.

For example, such methods can be utilized for the efficient solution of saddle point problems

$$\begin{pmatrix} A & B^T \\ B & 0 \end{pmatrix} \begin{pmatrix} u \\ v \end{pmatrix} = \begin{pmatrix} f \\ g \end{pmatrix} \tag{1.2}$$

as they arise from the discretization of the Stokes equation. Such approaches are described and analysed in the papers of Bramble and Pasciak (1988) and Bank, Welfert and Yserentant (1990).

Fast iterative methods for the equation (1.1) can also be used to construct comparably fast methods for the solution of systems

$$(A + M) u = f \tag{1.3}$$

arising from the discretization of boundary value problems with lower order terms (here represented by M) making the system indefinite and/or unsymmetric. Helmholtz type and convection–diffusion equations fall into this class. Methods of this type are described in Yserentant (1986c), Vassilevski (1992) and Xu (1992a). The mathematical background of these papers is an observation concerning the finite element discretization of perturbed elliptic boundary value problems which has been made by Schatz (1974). The analysis of multigrid methods, which can be directly applied to such boundary value problems, is also based on such perturbation arguments; see

Bank (1981), for example. A completely different approach can be found in
Yserentant (1988).

If M is not self-adjoint and becomes the dominating part in (1.3), i.e.
for convection-dominated problems, for example, the construction of appropriate fast solvers becomes more difficult and is still in its infancy. Until
the present day, most iterative methods for such problems were constructed
on a more or less heuristic basis. Multigrid methods based on incomplete
factorizations turned out to be very efficient. For certain model problems,
a rigorous analysis is possible; see Hackbusch (1985), Wittum (1989b) or
Stevenson (1992). With his frequency decomposition multigrid methods,
Hackbusch (1989a,b) presented a very promising approach. These methods
are also well suited to boundary value problems with strong anisotropies.
It should be mentioned that conjugate-gradient-like algorithms play an important role in the solution of nonsymmetric linear algebraic equations like
(1.3). For a survey of recent developments, see Freund, Golub and Nachtigal
(1992).

Last, but not least, fast iterative solvers for standard symmetric, positive
definite finite element equations can be applied to related nonlinear boundary value problems via approximate Newton techniques; see Bank and Rose
(1982) and Deuflhard (1992). Often, such methods based on inner–outer
iterations present an alternative to nonlinear multigrid methods which treat
the boundary value problem directly. Information about nonlinear multigrid methods can be found in Hackbusch's 1985 book. A very elaborate
convergence analysis is given in Hackbusch and Reusken (1989).

The rest of this paper is organized as follows. In Section 2, we introduce
a very general class of approximate subspace correction methods for the
solution of abstract linear equations $Au = f$ with self-adjoint and positive
definite operators A replacing the usual matrices. The classical multigrid
or, as we often prefer to say, multilevel methods as well as many domain
decomposition methods are such subspace correction methods. How multigrid methods can be interpreted in this sense, will be discussed in detail.
In addition to the classical multigrid algorithms, we present the hierarchical basis methods, which are extremely well suited to adaptively generated,
nonuniform finite element meshes.

In Section 3, a first convergence proof for two-grid methods is given. This
proof follows the lines given in Bank and Dupont (1981). The two-grid
convergence result can be utilized to prove the convergence of the so called
multigrid W-cycle. All early convergence proofs for multigrid methods followed this strategy.

A more sophisticated multigrid convergence proof based on the ideas of
Braess and Hackbusch (1983) will be presented in Section 4. Contrary to
the two grid–multigrid analysis, this convergence proof also applies to the
multigrid V-cycle which is simpler than the W-cycle.

In Section 5, a general convergence theory for the recursively defined, multiplicative subspace correction methods introduced in Section 2 is developed. In Section 6, this abstract theory is applied to multigrid methods for the solution of finite element equations and to the hierarchical basis multigrid method of Bank *et al.* (1988). In addition to a regularity dependent convergence result, which is closely related to the result of Wittum (1989a), a first regularity free convergence estimate is derived.

Utilizing the results presented in Section 7, one can show that multigrid methods reach an optimal complexity which is independent of the regularity properties of the boundary value problem. These results are mainly due to Oswald (1990, 1991) and to the forthcoming paper of Dahmen and Kunoth (1991). Many of the tools employed in these papers were taken from the classical approximation theory and the theory of function spaces. A self-contained presentation for the case of second-order problems can be found in Bornemann and Yserentant (1992).

Finally, in Section 8, we devote our attention to additive multilevel methods which are a special case of the additive subspace correction methods already introduced in Section 2. The most prominent examples of these methods are the hierarchical basis solver (Yserentant, 1986b) and the recent multilevel nodal basis method of Bramble, Pasciak and Xu (1990) and Xu (1989). Our presentation follows Xu (1992b) and Yserentant (1990). These methods are more flexible and simpler than the usual recursively defined multilevel methods and fit very well to nonuniformly refined grids. In addition, they present advantages for implementation on parallel computers.

The present survey article is strongly influenced by the recent work of Bramble, Pasciak, Wang and Xu. Although not explicitly stated at every place, often we follow their argumentation very closely, especially in Sections 2, 5 and 8. The merits of these authors are herewith explicitly acknowledged.

Special thanks also to Randy Bank and Wolfgang Hackbusch who laid the foundations of multigrid convergence theory. They have supported me in many respects.

2. Subspace correction and multilevel methods

We begin this section with a very abstract formulation of a discrete elliptic boundary value problem. Let S be a finite dimensional space. We assume that S is equipped with an inner product $a(u, v)$ inducing the norm

$$\|u\| = a(u, u)^{1/2} \tag{2.1}$$

and a second inner product (u, v) inducing the norm

$$\|u\|_0 = (u, u)^{1/2}. \tag{2.2}$$

We introduce a symmetric and positive definite operator $A : \mathcal{S} \to \mathcal{S}$ by the relation

$$(Au, v) = a(u, v), \quad v \in \mathcal{S}, \tag{2.3}$$

where symmetric and positive definite here is always understood to be symmetric and positive definite with respect to the inner product (u, v). Our aim is the construction and analysis of a general class of fast solvers for the abstract linear equation

$$Au = f. \tag{2.4}$$

This equation is equivalent to the problem of finding a $u \in \mathcal{S}$ satisfying the relation

$$a(u, v) = (f, v) \tag{2.5}$$

for all elements $v \in \mathcal{S}$. We remark that the inner product (2.2) does not enter the final form of the algorithms as they are implemented on the computer, and that the constants in our central abstract convergence theorems will be invariant under a change in this inner product.

In the applications that we have in mind, (2.1) is the norm induced by the elliptic boundary value problem under consideration whereas (2.2) is chosen to be a L_2-like inner product. To give an example, let $\Omega \subseteq \mathbb{R}^2$ be a bounded polygonal domain. As a model problem, we consider the differential equation

$$- \sum_{i,j=1}^{2} D_j(a_{ij} D_i u) = f \tag{2.6}$$

on Ω with homogeneous boundary conditions $u = 0$ on the boundary of Ω. The weak formulation of this boundary value problem is to find a function $u \in H_0^1(\Omega)$ satisfying the relation

$$a(u, v) = \int_{\Omega} fv \, dx \tag{2.7}$$

for all $v \in H_0^1(\Omega)$ where, in this example, the bilinear form $a(u, v)$ is given by the integral expression

$$a(u, v) = \int_{\Omega} \sum_{i,j=1}^{2} a_{ij} D_i u D_j v \, dx. \tag{2.8}$$

We assume that the a_{ij} are continuously differentiable functions, that

$$a_{ij} = a_{ji}, \tag{2.9}$$

and that there are positive constants M and δ with

$$\delta \sum_{i=1}^{2} \xi_i^2 \leq \sum_{i,j=1}^{2} a_{ij}(x)\xi_i\xi_j \leq M \sum_{i=1}^{2} \xi_i^2 \tag{2.10}$$

for all $x \in \Omega$ and all $\xi_1, \xi_2 \in \mathbb{R}$. These conditions guarantee that (2.8) defines an inner product on $H_0^1(\Omega)$ which is equivalent to the usual inner product on this space.

By a triangulation $\mathcal{T}$ of Ω, we mean a set of triangles such that the intersection of two such triangles is either empty or consists of a common edge or a common vertex. Here we start with an intentionally coarse initial triangulation $\mathcal{T}_0$ of Ω. The triangulation $\mathcal{T}_0$ is refined several times, giving a family of nested triangulations $\mathcal{T}_0, \mathcal{T}_1, \mathcal{T}_2, \ldots$. For ease of presentation, we consider only uniformly refined families of triangulations in this article, at least from a rigorous point of view. Thus a triangle of $\mathcal{T}_{k+1}$ is generated by subdividing of triangle of $\mathcal{T}_k$ into four congruent subtriangles.

Nevertheless one should keep in mind that nonuniformly refined meshes are absolutely necessary to approximate solutions with singularities arising from corners, cracks, interfaces or nonlinearities. On an informal basis, we will discuss whether and in which way the presented results can be generalized to such sequences of grids.

For triangular grids, the most successful nonuniform refinement scheme is due to Bank and Weiser (1985). It is also described in Bank et al. (1988). The scheme is based on the regular subdivision of triangles as described earlier and on carefully chosen additional bisections of triangles. Refinement schemes, which are based exclusively on the bisection of triangles, are discussed in Bänsch (1991) and Rivara (1984). The nonuniform refinement of tetrahedral meshes in three space dimensions is a harder challenge. In Bänsch (1991), the bisection of tetrahedra is utilized. The refinement strategy of Bank and Weiser can also be generalized to three dimensions.

Corresponding to the triangulations $\mathcal{T}_k$ we have finite element spaces $\mathcal{S}_k$. In our example, $\mathcal{S}_k$ consists of all functions which are continuous on Ω and linear on the triangles in $\mathcal{T}_k$ and which vanish on the boundary of Ω. By construction, $\mathcal{S}_k$ is a subspace of $\mathcal{S}_l$ for $k \leq l$. The extension of the presented results to higher order spaces is more or less obvious.

For the rest of this paper, we fix a final level j and the corresponding finite element space $\mathcal{S} = \mathcal{S}_j$. The discrete boundary value problem corresponding to the abstract linear problem (2.4), (2.5) is to find a function $u \in \mathcal{S}$ satisfying the relation

$$a(u, v) = \int_\Omega fv \, \mathrm{d}x \qquad (2.11)$$

for all functions $v \in \mathcal{S}$.

As mentioned earlier, the inner product (2.2) is usually a L_2-like inner product with an appropriate weight function. Our choice is

$$(u, v) = \sum_{T \in \mathcal{T}_0} \frac{1}{\mathrm{area}(T)} \int_T uv \, \mathrm{d}x. \qquad (2.12)$$

The task of the weights here is to make our estimates independent of the size of the triangles in the initial triangulation. In the three-dimensional case, these factors have to be replaced by other factors behaving like $1/\text{diam}(T)^2$.

After this illustrating example, which will accompany the whole paper, we return to the general theory. Let $\mathcal{W}_0, \mathcal{W}_1, \ldots, \mathcal{W}_J$ be subspaces of $\mathcal{S}$. We assume that every $u \in \mathcal{S}$ can be written as

$$u = w_0 + w_1 + \ldots + w_J, \quad w_l \in \mathcal{W}_l. \tag{2.13}$$

We neither assume that this representation is unique, nor that the spaces $\mathcal{W}_l$ are nested.

We need two kinds of orthogonal projections onto the spaces $\mathcal{W}_l$. The projections $Q_l : \mathcal{S} \to \mathcal{W}_l$ are defined by

$$(Q_l u, w_l) = (u, w_l), \quad w_l \in \mathcal{W}_l, \tag{2.14}$$

and the projections $P_l : \mathcal{S} \to \mathcal{W}_l$ by

$$a(P_l u, w_l) = a(u, w_l), \quad w_l \in \mathcal{W}_l. \tag{2.15}$$

If $u \in \mathcal{S}$ is the solution of (2.4), and (2.5), respectively, $P_l u \in \mathcal{W}_l$ is the Ritz approximation of this solution in $\mathcal{W}_l$.

The basic building block of the iterative methods considered here are the *subspace corrections*

$$\tilde{u} \leftarrow \tilde{u} + P_l(u - \tilde{u}) \tag{2.16}$$

with respect to the spaces $\mathcal{W}_l$. The subspace correction (2.16) makes the error $u - \tilde{u}$ between the exact solution and the new approximation a-orthogonal to the space $\mathcal{W}_l$.

To express these subspace corrections in terms of the right-hand side f and the approximations $\tilde{u}$, we introduce the Ritz approximations $A_l : \mathcal{W}_l \to \mathcal{W}_l$ of the operator A with respect to the spaces $\mathcal{W}_l$. They are defined by

$$(A_l u, v) = (Au, v), \quad u, v \in \mathcal{W}_l, \tag{2.17}$$

or equivalently by

$$(A_l u, v) = a(u, v), \quad u, v \in \mathcal{W}_l. \tag{2.18}$$

The operators A, A_l, P_l and Q_l are connected by the relation

$$A_l P_l = Q_l A. \tag{2.19}$$

(2.19) easily follows from

$$(A_l P_l u, w_l) = a(P_l u, w_l) = a(u, w_l) = (Au, w_l) = (Q_l A u, w_l).$$

By (2.19), the Ritz approximation $P_l u$ of the solution u of (2.4) satisfies the equation

$$A_l P_l u = Q_l f, \tag{2.20}$$

and the subspace correction (2.16) can be written as

$$\tilde{u} \leftarrow \tilde{u} + A_l^{-1}Q_l(f - A\tilde{u}). \tag{2.21}$$

It requires the computation of the Ritz approximation

$$P_l(u - \tilde{u}) = A_l^{-1}Q_l(f - A\tilde{u}) \tag{2.22}$$

of the error $u - \tilde{u}$.

The problem is that, for sufficiently large and complicated subspaces $\mathcal{W}_l$, the computation of this approximate defect is far too expensive to lead to a reasonable method. Therefore we replace the subspace corrections (2.21) by *approximate subspace corrections*

$$\tilde{u} \leftarrow \tilde{u} + B_l^{-1}Q_l(f - A\tilde{u}) \tag{2.23}$$

with symmetric and positive definite operators $B_l : \mathcal{W}_l \to \mathcal{W}_l$. The operators B_l should have the property that the correction term

$$d_l = B_l^{-1}Q_l(f - A\tilde{u}) \tag{2.24}$$

can easily be computed as the solution of the linear system

$$(B_l d_l, w_l) = (f - A\tilde{u}, w_l), \quad w_l \in \mathcal{W}_l. \tag{2.25}$$

Here we should remark that the computation of the right-hand side of (2.25)

$$(f - A\tilde{u}, w_l) = (f, w_l) - a(\tilde{u}, w_l) \tag{2.26}$$

does not require an explicit knowledge of the abstract operator A but only of the bilinear form $a(u, v)$ and of the linear functional representing the right-hand side of the equation.

A common situation is that the computation of the correction term in (2.23) consists of, say, m steps of a given convergent iterative procedure

$$\tilde{w}_l \leftarrow \tilde{w}_l + \hat{B}_l^{-1}(r_l - A_l\tilde{w}_l) \tag{2.27}$$

for the solution of the equation

$$A_l w_l = r_l, \quad r_l = Q_l(f - A\tilde{u}), \tag{2.28}$$

with a symmetric positive definite operator $\hat{B}_l : \mathcal{W}_l \to \mathcal{W}_l$, that means of m Jacobi steps, for example, where one starts with $w_l = 0$. Then the operator B_l is given by

$$B_l^{-1} = (I - (I - \hat{B}_l^{-1}A_l)^m)A_l^{-1}, \tag{2.29}$$

and is automatically symmetric and positive definite.

If one combines the single subspaces corrections

$$\tilde{u} \leftarrow \tilde{u} + B_l^{-1}Q_l(f - A\tilde{u}) \tag{2.30}$$

sequentially in the order $l = 0, 1, \ldots, J$, one obtains the *multiplicative subspace correction method* corresponding to the subspaces $\mathcal{W}_0, \ldots, \mathcal{W}_J$ of $\mathcal{S}$.

These method generalize the classical Gauß–Seidel iteration where the subspaces are one-dimensional and are spanned by basis functions.

Sufficient for the convergence of this composed method is that the iterations

$$w_l \leftarrow w_l + B_l^{-1}(f_l - A_l w_l) \tag{2.31}$$

for the solution of the equations $A_l w_l = f_l$ on $\mathcal{W}_l$ converge for arbitrarily chosen right-hand sides f_l and the initial approximation $w_l = 0$; this follows from Theorem 5.1 and the finite dimension of $\mathcal{S}$. If one assumes that

$$(A_l w_l, w_l) \leq \omega\,(B_l w_l, w_l), \quad w_l \in \mathcal{W}_l, \tag{2.32}$$

this condition is equivalent to

$$0 < \omega < 2. \tag{2.33}$$

We remark that the condition (2.33) is automatically satisfied, if the B_l themselves represent multiplicative subspace correction methods with exact subspace solvers, for example Gauß–Seidel iterations for the approximate solutions of the linear systems involving the operator A_l. For this particular choice, we have $\omega = 1$.

Classical multigrid methods for the solution of finite element equations like (2.11) fall into the category of such multiplicative subspace correction methods. The *multigrid V-cycle* is a multiplicative subspace correction method with the coarse grid spaces $\mathcal{S}_l$ as subspaces $\mathcal{W}_l$.

The V-cycle is usually defined by recursion on the number j of refinement levels. For the initial level 0, when only one grid is present, the equations are solved exactly. For two or more levels, one proceeds as follows.

Beginning with an approximation $u_0 = \tilde{u}$ of the finite element equation $A_j u = f$, first a *coarse grid correction* is performed. For the two-level case, one computes the approximate defect $d = P_{j-1}(u - \tilde{u}) \in \mathcal{S}_{j-1}$ as the solution of the level $j - 1$ equation

$$A_{j-1} d = Q_{j-1}(f - A_j u) \tag{2.34}$$

and sets

$$u_1 = u_0 + d. \tag{2.35}$$

Then further intermediate approximations $u_2, \ldots, u_{m+1} \in \mathcal{S}_j$ are determined by m so called *smoothing steps*

$$u_{i+1} = u_i + \hat{B}_j^{-1}(f - A_j u_i). \tag{2.36}$$

One ends with $\tilde{u} = u_{m+1}$ as the new approximation for the solution of the equation $A_j u = f$. For more than two levels, the coarse grid equation (2.34) is approximately solved by a call of the method for the level $j - 1$.

If the coarse level equations (2.34) are not solved by *one* but by *two* calls of the method for the preceding level, one speaks of a *W-cycle multigrid method*. Other cycling strategies are possible but will not be discussed here.

For the multigrid V-cycle, the number $j+1$ of levels and the number $J+1$ of subspace corrections coincide. Compared with the V-cycle, additional subspace corrections are added in the W-cycle. The number of subspace corrections exceeds the number of levels, although each of the subspaces $\mathcal{W}_l$, $l = 0, \ldots, J$, is one of the spaces $\mathcal{S}_k$, $k = 0, \ldots, j$.

The reason for the extremely fast convergence of multigrid methods in comparison to the underlying smoothers is that these iterations are very selective for the different components of the error. Fast oscillating components (with respect to the given level) are strongly reduced whereas the remaining components are nearly unaffected. The error is *smoothed*, as the term 'smoothing step' indicates. As the smooth components of the error are already small because of the preceding coarse grid correction, the composed method can be very efficient.

For simple model problems (constant coefficients, square grids, periodic boundary conditions, etc.) the interaction of the smoothing steps and the coarse level corrections can be quantitatively studied using a *Fourier* or *local mode analysis*. We refer to Hackbusch's 1985 book, to Stüben and Trottenberg (1982) and to Brandt (1982).

For any good iterative solver for the solution of finite element equations, the amount of work per iteration step should be proportional to the number of unknowns. Next we check whether this condition is satisfied for the multigrid methods introduced earlier.

Without regarding the algorithmic realization in detail (recall only (2.26)), we get the recursion formula

$$W_j = p\,W_{j-1} + Cn_j \tag{2.37}$$

for the work W_j necessary to perform one step of the multigrid method for the solution of a equation in $\mathcal{S}_j$. n_j denotes the dimension of $\mathcal{S}_j$. $p = 1$ corresponds to the V-cycle and $p = 2$ to the W-cycle. Here we have assumed that the amount of work per cycle, except for the approximate solution of the coarse level equation (2.34), behaves like Cn_j, which is the case for all reasonable smoothers. The recursion formula (2.37) yields

$$W_j = p^j W_0 + C \sum_{k=1}^{j} p^{j-k} n_k. \tag{2.38}$$

If one disregards the work for the solution of the equations on the level 0, a simple analysis shows that the operation count W_j for the single multigrid cycle behaves like $\mathcal{O}(n_j)$ if and only if the dimensions n_k are related by

$$n_k \le cq^{j-k} n_j, \quad k = 1, \ldots, j, \tag{2.39}$$

where $q < 1$ for the V-cycle and $q < 1/2$ for the W-cycle. This means that the dimensions of the spaces $\mathcal{S}_k$ have to increase geometrically. These conditions are satisfied for our model problem (where the dimension essentially grows by the factor 4 from one level to the next), but they can cause problems for adaptively generated, nonuniformly refined meshes.

An interesting modification of the classical multigrid methods, especially as it concerns the application to such adaptively generated, highly nonuniform meshes, is the *hierarchical basis multigrid method* introduced by Bank et al. (1988).

Compared with classical multigrid methods, it works with smaller spaces $\mathcal{W}_l$. Nevertheless, for two-dimensional problems, it reaches a similar efficiency as those of classical multigrid methods. Its structure fits very well to nonuniformly refined meshes and allows the use of simple data structures. Under some mild restrictions, the W-cycle version also works in three space dimensions as can be shown along the lines given in Bank and Dupont (1980), Braess (1981) and Axelsson and Vassilevski (1989).

To describe this method, we have to realize that a function in the finite element space $\mathcal{S}_k$ is uniquely determined by its values at the nodes $x \in \mathcal{N}_k$ which are the vertices of the triangles in the triangulation $\mathcal{T}_k$ which do not lie on the boundary of Ω. Therefore we can define an interpolation operator $\mathcal{I}_k : \mathcal{S} \to \mathcal{S}_k$ by

$$(\mathcal{I}_k u)(x) = u(x), \quad x \in \mathcal{N}_k. \tag{2.40}$$

Utilizing these interpolation operators, we can define the subspace

$$\mathcal{W}_k = \{\mathcal{I}_k u - \mathcal{I}_{k-1} u \,|\, u \in \mathcal{S}\} \tag{2.41}$$

of $\mathcal{S}_k$ as the image of $\mathcal{S}$ (or of $\mathcal{S}_k$) under the operator $\mathcal{I}_k - \mathcal{I}_{k-1}$. The functions in this space $\mathcal{W}_k$ vanish at the nodes $x \in \mathcal{N}_{k-1}$. Therefore they are given by their values at the nodes $x \in \mathcal{N}_k \setminus \mathcal{N}_{k-1}$.

In the hierarchical basis methods, the spaces $\mathcal{S}_k$ are replaced by the spaces (2.41). As with classical multigrid methods, approximate solvers B_k of a very simple structure can be used. For a survey, we refer to Yserentant (1992).

In comparison with the recursion

$$W_j = W_{j-1} + C n_j \tag{2.42}$$

for the work necessary to perform one multigrid V-cycle, the corresponding recursion formula

$$W_j = W_{j-1} + C(n_j - n_{j-1}) \tag{2.43}$$

for the hierarchical basis multigrid method has a different quality. It yields the operation count

$$W_j \leq W_0 + C n_j \tag{2.44}$$

independently of any assumption on the distribution of the unknowns among the levels.

For nonuniformly refined families of grids, the only alternative to the hierarchical basis multigrid method are multigrid methods in which the spaces (2.41) are enriched by those basis functions of S_k which are associated with the nodes in $\mathcal{N}_{k-1}$ having neighbours in $\mathcal{N}_k \setminus \mathcal{N}_{k-1}$. From a computational point of view, this corresponds to *local smoothing procedures*. The theoretical understanding of such methods began with Bramble *et al.* (1991b).

The recursively defined multiplicative subspace correction methods can be seen as generalizations of the Gauß–Seidel method. The corresponding Jacobi-type iterations have recently been the focus of much interest. Because of their simpler structure, these *additive subspace correction methods*

$$\tilde{u} \leftarrow \tilde{u} + \sum_{l=0}^{J} B_l^{-1} Q_l (f - A\tilde{u}) \tag{2.45}$$

offer many advantages as preconditioners for the conjugate gradient method.

With subspaces $\mathcal{W}_l$ as in the hierarchical basis multigrid method and Jacobi-type methods as approximate solvers B_l, the additive subspace correction method for the solution of the finite element equations (2.11) becomes the hierarchical basis solver; see Yserentant (1986b, 1990, 1992). For the choice $\mathcal{W}_l = S_l$, one obtains the multilevel nodal basis preconditioner of Bramble *et al.* (1990) and Xu (1989); see also Yserentant (1990).

3. An analysis for the two-level case

The first general convergence proofs for classical, recursively defined multigrid methods for finite element equations like (2.11) stem from the end of the 1970s. They are mainly the work of Wolfgang Hackbusch and of Randolph E. Bank and Todd Dupont; see Hackbusch (1981, 1985) and Bank and Dupont (1981).

It is Hackbusch's merit to have identified and clearly separated the two main building blocks which lay the foundation to all standard convergence proofs and which became the basis of a countless number of articles appearing until the present day. These properties are the *smoothing property*, which essentially describes the necessary relations between the approximate subspace solvers and the finite element equations, and the *approximation property*, which describes the interaction of the different levels. Both properties will be discussed in this section.

In this and the next section, we assume that the eigenvalues of the error propagation operators

$$I - \hat{B}_k^{-1} A_k \tag{3.1}$$

of the smoothing iterations

$$w_k \leftarrow w_k + \widehat{B}_k^{-1}(f_k - A_k w_k) \tag{3.2}$$

are nonnegative. This condition is equivalent to

$$(w_k, A_k w_k) \leq (w_k, \widehat{B}_k w_k), \quad w_k \in \mathcal{S}_k, \tag{3.3}$$

and guarantees the convergence of the iteration (3.2). The condition (3.3) is stronger than (2.33). (3.3) is our version of the smoothing property.

(3.3) holds, if the operators $\widehat{B}_k$ are properly scaled, i.e. if the iteration (3.2) is sufficiently damped. If (3.2) represents a symmetric (block) Gauß–Seidel iteration, (3.3) is automatically satisfied.

Our second assumption concerns the spaces

$$\mathcal{V}_k = \{P_k u - P_{k-1} u \,|\, u \in \mathcal{S}\} \subseteq \mathcal{S}_k, \tag{3.4}$$

i.e. the a-orthogonal complements of the spaces $\mathcal{S}_{k-1}$ in $\mathcal{S}_k$. We assume that there exists a constant K with

$$(v_k, \widehat{B}_k v_k) \leq K(v_k, A_k v_k), \quad v_k \in \mathcal{V}_k. \tag{3.5}$$

(3.5) is the counterpart to (3.3). As the norms induced by the $\widehat{B}_k$ are generally much stronger than the energy norm induced by the A_k, (3.5) can be interpreted as an approximation property of the functions in $\mathcal{S}_k$ by functions from the subspace $\mathcal{S}_{k-1}$.

Without any doubt, (3.5) is a much more critical assumption than (3.3). In the finite element case, and especially for our model problem, (3.5) is essentially equivalent to the Aubin–Nitsche Lemma; see Ciarlet (1978), for example. Assume that, for $k = 1, \ldots, j$ and all $u_k \in \mathcal{S}_k$,

$$c_1 4^k \|u_k\|_0^2 \leq (u_k, \widehat{B}_k u_k) \leq c_2 4^k \|u_k\|_0^2. \tag{3.6}$$

Because of the scaling of the L_2-like norm (2.12) by the areas of the triangles in the initial triangulation and the fact that the diameter of the triangles shrinks by the factor 2 from one level to the next, smoothers like the Jacobi iteration or the symmetric point Gauß–Seidel iteration have this property. With (3.6), the condition (3.5) is equivalent to the estimate

$$\|P_k u - P_{k-1} u\|_0 \leq c\, 2^{-k} \|P_k u - P_{k-1} u\| \tag{3.7}$$

for the functions $u \in \mathcal{S}$, or, with an infinite sequence of spaces $\mathcal{S}_k$, even equivalent to the estimate

$$\|u - P_k u\|_0 \leq \tfrac{1}{3} c\, 2^{-k} \|u - P_k u\| \tag{3.8}$$

for the functions u in the continuous solution space $H_0^1(\Omega)$. (3.8) and (3.7), respectively, imply (3.5) if the upper estimate in (3.6) holds.

It is well known that (3.8) holds only for H^2-regular problems, i.e. if the solution u of the continuous problem (2.6) belongs to H^2 for right-hand sides $f \in L_2$ and satisfies an estimate

$$\|u\|_{H^2} \leq \tilde{c}\|f\|_{L_2}. \tag{3.9}$$

This holds only if the domain Ω has a C^2-boundary or if Ω is convex. For domains Ω with re-entrant corners, (3.9) is wrong.

This fact restricts the applicability of the classical multigrid convergence theory, although, using differently weighted L_2-norms (with weights depending on the interior angles of the domain) and properly refined triangulations, the algebraic estimates derived in this and the next section can also be applied to certain problems on domains with re-entrant corners; see Yserentant (1986a, 1983) and S. Zhang (1990).

The basic result of this section is an estimate for the convergence rate of the algorithm described by (2.34), (2.35) and (2.36), i.e. for the case in which the coarse grid correction is exactly determined and only two levels are present.

Theorem 3.1 If $u_0 \in S$ is the initial approximation for the solution $u \in S$ of the finite element equation $Au = f$, the new approximation $u_{m+1} \in S$ (after a full two-grid cycle (2.34), (2.35), (2.36)) satisfies the estimate

$$\|u - u_{m+1}\|^2 \leq K\gamma(m)\|u - u_0\|^2, \tag{3.10}$$

where the generic constant $\gamma(m)$ is given by

$$\gamma(m) = \frac{1}{2m+1}\left(1 - \frac{1}{2m+1}\right)^{2m} \tag{3.11}$$

and K is the constant from the approximation property (3.5).

Proof. The main ingredient of the proof is a biorthogonal basis $\psi_1, \ldots, \psi_n$ of S with

$$(\psi_i, \widehat{B}\psi_l) = \delta_{il}, \quad (\psi_i, A\psi_l) = \lambda_i\delta_{il}, \tag{3.12}$$

where we suppress the subscript j for a while; the existence of such a basis follows from basic facts of linear algebra. Then, for $u = \sum_{i=1}^n a_i\psi_i \in S$, the norm (2.1) is given by

$$\|u\|^2 = \sum_{i=1}^n \lambda_i a_i^2. \tag{3.13}$$

If we introduce the discrete norm

$$\|u\| = (u, \widehat{B}u)^{1/2}, \quad u \in S, \tag{3.14}$$

this norm has the representation

$$\|u\|^2 = \sum_{i=1}^n a_i^2. \tag{3.15}$$

The proof of Theorem 3.1 is based on the eigenfunction expansion

$$Gv = \sum_{i=1}^{n}(1 - \lambda_i)a_i\psi_i \tag{3.16}$$

for $v = \sum_{i=1}^{n} a_i\psi_i$ of the error propagation operator

$$G = I - \hat{B}^{-1}A \tag{3.17}$$

for the smoothing process. Introducing the errors

$$e_k = u - u_k, \quad k = 0,\dots,m+1, \tag{3.18}$$

between the exact solution $u \in \mathcal{S}$ and the intermediate approximations $u_0, u_1, \dots, u_{m+1}$ of u, the errors $e_2, \dots, e_m$ can be expressed as

$$e_{k+1} = G^k e_1, \quad k = 0,\dots,m, \tag{3.19}$$

in terms of the error e_1 after the coarse grid correction. If $e_1 = \sum_{i=1}^{n} a_i\psi_i$, one finds

$$\|e_{m+1}\|^2 = \|\sum_{i=1}^{n}(1 - \lambda_i)^m a_i\psi_i\|^2 = \sum_{i=1}^{n}\lambda_i(1 - \lambda_i)^{2m} a_i^2. \tag{3.20}$$

The smoothing property (3.3) is equivalent to the bound

$$\lambda_i \leq 1, \quad i = 1,\dots,n, \tag{3.21}$$

for the eigenvalues λ_i. Therefore

$$\max_{i=1,\dots,n} \lambda_i(1 - \lambda_i)^{2m} \leq \max_{0 \leq \lambda \leq 1} \lambda(1 - \lambda)^{2m} = \gamma(m) \tag{3.22}$$

and

$$\|e_{m+1}\|^2 \leq \gamma(m)\sum_{i=1}^{n} a_i^2 = \gamma(m)\|e_1\|^2. \tag{3.23}$$

As, by the approximation property (3.5),

$$\|e_1\|^2 = \|e_0 - P_{j-1}e_0\|^2 \leq K\|e_0 - P_{j-1}e_0\|^2 \leq K\|e_0\|^2, \tag{3.24}$$

the proposition

$$\|e_{m+1}\|^2 \leq K\gamma(m)\|e_0\|^2 \tag{3.25}$$

of Theorem 3.1 follows. $\square$

The theorem states that the two-grid method converges and that its convergence rate $K\gamma(m)$ becomes even arbitrarily small as soon as the number m of smoothing steps is sufficiently large. As the constant K in the approximation property (3.5) does not depend on j, the convergence rate is independent of the grid size.

This fact can be utilized to prove the convergence of the W-cycle by a relatively simple recursion argument, which is the idea behind all early multigrid convergence proofs. A detailed discussion can be found in Hackbusch's 1985 book or in Stüben and Trottenberg (1982).

Although the two-grid–multigrid analysis is very suggestive and has a broad range of application, the results obtained in this way do not completely satisfy for the given case of self-adjoint, coercive elliptic boundary value problems. Experience says that, applied to problems of this class, not only the W-cycle but also the much simpler and cheaper V-cycle converges very fast. In addition, *one* smoothing step per level turns out to be sufficient.

4. A convergence proof for the V-cycle

The much more sophisticated convergence analysis of Braess and Hackbusch (1983) supports these observations theoretically. In this section, we derive their result in an algebraic language as in Yserentant (1983). Closely related estimates are proven in Bank and Douglas (1985). The assumptions in this section are the same as in the previous section.

Following the original work and contrary to the definition given in Section 2, in this section we assume that the order of the coarse grid correction (2.34), (2.35) and of the smoothing steps (2.36) is reversed. This is not an essential change because it is easy to see that the convergence rates of both versions are equal. The order that we chose in Section 2 seems to be more natural from the point of view of subspace correction methods. This version will be analysed in Section 6.

Theorem 4.1 If $u \in S$ denotes the exact solution of the equation to be solved and if $u_0 \in S$ is the given initial approximation of u, the new approximation $u_{m+1} \in S$, obtained by a multigrid V-cycle or W-cycle, satisfies the estimate

$$\|u - u_{m+1}\|^2 \leq \frac{c}{c + 2m} \|u - u_0\|^2, \qquad (4.1)$$

where $c = K^2$ and K is the constant from the approximation property (3.5).

As in the two-level proof of the previous section, the proof is based on the eigenfunction expansion (3.16) of the error propagation operator G for the smoothing process which is given by (3.17).

As the eigenvalues λ_i are not greater than 1, we have $1 - \lambda_i \geq 0$ for all i and can define the powers G^α, $\alpha \geq 0$, of G by

$$G^\alpha v = \sum_{i=1}^{n} (1 - \lambda_i)^\alpha a_i \psi_i, \qquad (4.2)$$

where $v = \sum_{i=1}^{n} a_i \psi_i$ is the eigenfunction expansion of the function $v \in S$.

The main ingredient of the proof is the functional

$$\rho(v) = \begin{cases} \|G^{1/2}v\|^2/\|v\|^2 & , & v \neq 0 \\ 0 & , & v = 0 \end{cases} \tag{4.3}$$

for the elements $v \in S$. Note that always $\rho(v) \leq 1$. $\rho(v)$ can be seen as a measure for the smoothness of v. If $\rho(v)$ is small compared with 1, the smoothed element Gv has a small norm compared with v.

Our first lemma describes the success of the coarse grid correction in terms of this kind of smoothness of the error.

Lemma 4.2   For all functions $v \in S$, $v - P_{j-1}v$ satisfies the estimate

$$\|v - P_{j-1}v\|^2 \leq \min\{1, K(1 - \rho(v))\} \|v\|^2. \tag{4.4}$$

Proof.   Let $v \in S$ and $v - P_{j-1}v$ have the eigenfunction expansions

$$v = \sum_{i=1}^{n} a_i \psi_i, \quad v - P_{j-1}v = \sum_{i=1}^{n} b_i \psi_i.$$

Then

$$\|v - P_{j-1}v\|^2 = a(v - P_{j-1}v, v) = \sum_{i=1}^{n} \lambda_i a_i b_i.$$

With the Schwarz inequality, this yields

$$\|v - P_{j-1}v\|^2 \leq \left(\sum_{i=1}^{n} b_i^2\right)^{1/2} \left(\sum_{i=1}^{n} \lambda_i^2 a_i^2\right)^{1/2}$$

$$= \left(\sum_{i=1}^{n} b_i^2\right)^{1/2} \left\{\sum_{i=1}^{n} \lambda_i a_i^2 - \sum_{i=1}^{n} \lambda_i(1 - \lambda_i)a_i^2\right\}^{1/2}$$

$$= \|v - P_{j-1}v\| \{ \|v\|^2 - \|G^{1/2}v\|^2 \}^{1/2}$$

$$= \|v - P_{j-1}v\| \{1 - \rho(v)\}^{1/2} \|v\|.$$

Inserting (3.5), which means

$$\|v - P_{j-1}v\|^2 \leq K\|v - P_{j-1}v\|^2,$$

one obtains

$$\|v - P_{j-1}v\|^2 \leq K(1 - \rho(v)) \|v\|^2.$$

This proves the proposition. $\square$

Together with the next lemma describing the effect of the smoothing iterations, Lemma 4.2 forms the backbone of the proof of Theorem 4.1.

Lemma 4.3   For all functions $v \in S$, $G^k v$ satisfies the estimate

$$\|G^k v\| \leq \rho(G^k v)^k \|v\|. \tag{4.5}$$

Proof. Let $\mu_i = 1 - \lambda_i$. Because of $\mu_i \geq 0$ and utilizing Hölder's inequality, one obtains, for all $v = \sum_{i=1}^{n} a_i \psi_i$,

$$\|G^k v\|^2 = \sum_{i=1}^{n} \lambda_i (\mu_i^k a_i)^2$$

$$= \sum_{i=1}^{n} (\lambda_i \mu_i^{2k+1} a_i^2)^{\frac{2k}{2k+1}} (\lambda_i a_i^2)^{\frac{1}{2k+1}}$$

$$\leq \left(\sum_{i=1}^{n} \lambda_i \mu_i^{2k+1} a_i^2 \right)^{\frac{2k}{2k+1}} \left(\sum_{i=1}^{n} \lambda_i a_i^2 \right)^{\frac{1}{2k+1}}$$

$$= \|G^{k+1/2} v\|^{\frac{4k}{2k+1}} \|v\|^{\frac{2}{2k+1}}.$$

This estimate is equivalent to

$$\|G^k v\| \|G^k v\|^{2k} \leq \|G^{1/2}(G^k v)\|^{2k} \|v\|.$$

This is the proposition. $\square$

Now we can prove the theorem. Denoting by $d = P_{j-1} e_m$ the exact coarse grid correction and by $\tilde{d} \in S_{j-1}$ the approximate coarse grid correction computed by p steps of the method for the level $j - 1$, one obtains, utilizing

$$e_{m+1} = (e_m - P_{j-1} e_m) + (d - \tilde{d}), \qquad (4.6)$$

the relation

$$\|e_{m+1}\|^2 = \|e_m - P_{j-1} e_m\|^2 + \|d - \tilde{d}\|^2. \qquad (4.7)$$

With an upper bound δ_{j-1} for the convergence rate of the method on the preceding level $j - 1$,

$$\|e_{m+1}\|^2 \leq \|e_m - P_{j-1} e_m\|^2 + \delta_{j-1}^{2p} \|P_{j-1} e_m\|^2 \qquad (4.8)$$

follows. This equation can be rewritten as

$$\|e_{m+1}\|^2 \leq (1 - \delta_{j-1}^{2p}) \|e_m - P_{j-1} e_m\|^2 + \delta_{j-1}^{2p} \|e_m\|^2. \qquad (4.9)$$

Lemma 4.2 yields

$$\|e_m - P_{j-1} e_m\|^2 \leq \min\{1, K(1 - \rho(e_m))\} \|e_m\|^2, \qquad (4.10)$$

and Lemma 4.3

$$\|e_m\|^2 \leq \rho(e_m)^{2m} \|e_0\|^2. \qquad (4.11)$$

If we insert these relations, we have proven the estimate

$$\|u - u_{m+1}\|^2 \leq \delta_j^2 \|u - u_0\|^2, \qquad (4.12)$$

where δ_j is given by the relatively complicated expression

$$\delta_j^2 = \max_{0 \leq \rho \leq 1} \rho^{2m} [(1 - \delta_{j-1}^{2p}) \min\{1, K(1 - \rho)\} + \delta_{j-1}^{2p}]. \qquad (4.13)$$

Together with

$$\delta_0 = 0, \tag{4.14}$$

this recursion leads to an estimate for the convergence rates δ_j.

For a fixed $\rho \in [0, 1]$, the function

$$\varepsilon \to \rho^{2m}[(1 - \varepsilon)\min\{1, K(1 - \rho)\} + \varepsilon] \tag{4.15}$$

increases monotonically. Introducing the abbreviation $c(m) = c/(c + 2m)$, the assumption

$$\delta_{j-1}^2 \leq c(m) \tag{4.16}$$

leads therefore to

$$\delta_j^2 \leq \max_{0 \leq \rho \leq 1} R(\rho), \tag{4.17}$$

where the function $R(\rho)$ is given by

$$R(\rho) = \rho^{2m}[(1 - c(m))\min\{1, K(1 - \rho)\} + c(m)]. \tag{4.18}$$

As necessarily $K \geq 1$, $R(\rho)$ is monotonically increasing on the interval $[0, 1]$. This proves the estimate

$$\delta_j^2 \leq R(1) = c(m) \tag{4.19}$$

for the convergence rate of the multigrid method.

Later, in Wittum (1989a), Theorem 4.1 has been generalized to the case that the error propagation operator (3.1) can have negative eigenvalues. Both the analysis of Braess and Hackbusch and that of Wittum do not take the internal structure of the smoothers into account. A result refined in this respect has been proven by Stevenson (1992). Reusken (1992) examined the convergence of multigrid methods with respect to the maximum norm. He shows that, up to a logarithmic factor, one can obtain the same convergence estimates as for the energy norm studied here. Another convergence proof, based on projection arguments and norm estimates instead of eigenfunction expansions, can be found in Mandel, McCormick and Bank (1987).

5. General multiplicative methods

A main drawback of all these approaches is their strong dependence on the regularity properties of the boundary value problem and of the considered family of grids which is reflected in the assumption (3.5). This fact makes it extremely difficult to apply these theories in a rigorous sense to problems with singularities caused by re-entrant corners, jumps in the boundary conditions, by interfaces, and so on.

Recently Bramble et al. (1991a, 1991b) developed an alternative convergence theory which overcomes these difficulties to a large extent.

This theory can be formulated in the abstract framework of the multiplicative subspace correction methods introduced in Section 2. We remark that the case of nonoverlapping subspaces $\mathcal{W}_k$ (which covers the hierarchical basis multigrid method) has implicitly been treated in Bank *et al.* (1988).

With some slight modifications as presented in Bramble and Pasciak (1991), Xu (1992b), or in the present paper, the theory shows that the convergence rate of multigrid methods is uniformly bounded independently of any regularity of the boundary value under consideration. It does not show that the convergence rate tends to zero if one increases the number of smoothing steps per level.

In this section, we develop the abstract theory for the multiplicative subspace correction methods introduced in Section 2. The application to our model problem and to other elliptic boundary value problems will be discussed in the next section.

The theory is based on the decomposition of the space $\mathcal{S}$ into a direct sum

$$\mathcal{S} = \mathcal{V}_0 \oplus \mathcal{V}_1 \oplus \ldots \oplus \mathcal{V}_J \tag{5.1}$$

of subspaces $\mathcal{V}_k \subseteq \mathcal{W}_k$. These subspaces $\mathcal{V}_k$ are only a tool for the theoretical analysis, they do not enter the practical computation. Often, this fact gives a lot of freedom in the choice of these subspaces and makes the convergence theory very flexible.

Two assumptions have to be fulfilled to apply the theory. The first assumption concerns the *stability of the decomposition*. We require that there exists a constant K_1 such that, for all $v_k \in \mathcal{V}_k$,

$$\sum_{k=0}^{J}(B_k v_k, v_k) \leq K_1 \| \sum_{k=0}^{J} v_k \|^2. \tag{5.2}$$

The second assumption is a *Cauchy–Schwarz type inequality*. We assume that there exist constants $\gamma_{kl} = \gamma_{lk}$ with

$$a(w_k, v_l) \leq \gamma_{kl} (B_k w_k, w_k)^{1/2}(B_l v_l, v_l)^{1/2} \tag{5.3}$$

for $k \leq l$, all $w_k \in \mathcal{W}_k$, and all $v_l \in \mathcal{V}_l$ such that

$$\sum_{k,l=0}^{J} \gamma_{kl} x_k y_l \leq K_2(\sum_{k=0}^{J} x_k^2)^{1/2}(\sum_{l=0}^{J} y_l^2)^{1/2} \tag{5.4}$$

holds for all $x_k, y_l \in \mathbb{R}$. That means, we require that the spectral radius of the matrix (γ_{kl}) is bounded by a constant K_2.

In addition, we assume that the constant ω in (2.32) satisfies the condition

$$\omega < 2 \tag{5.5}$$

which is equivalent to the convergence of the basic iterations (2.31).

(5.3) includes the Cauchy–Schwarz type inequality

$$a(v_k, v_l) \leq \gamma_{kl} (B_k v_k, v_k)^{1/2} (B_l v_l, v_l)^{1/2} \tag{5.6}$$

for $v_k \in \mathcal{V}_k, v_l \in \mathcal{V}_l$, and $k, l = 0, \ldots, J$. (5.6) and (5.4) imply that, corresponding to (5.2),

$$\| \sum_{k=0}^{J} v_k \|^2 \leq K_2 \sum_{k=0}^{J} (B_k v_k, v_k) \tag{5.7}$$

for all $v_k \in \mathcal{V}_k$. Therefore the expression

$$\| \sum_{k=0}^{J} v_k \|^2 = \sum_{k=0}^{J} (B_k v_k, v_k) \tag{5.8}$$

defines a norm on $\mathcal{S}$ which is, up to the constants K_1 and K_2, equivalent to the norm (2.1) induced by the abstract boundary value problem itself.

With the orthogonal projections P_k onto the spaces $\mathcal{W}_k$, the exact subspace corrections (2.16) are

$$\tilde{u} \leftarrow \tilde{u} + P_k(u - \tilde{u}). \tag{5.9}$$

If we define the operators

$$T_k := B_k^{-1} A_k P_k = B_k^{-1} Q_k A, \tag{5.10}$$

the approximate subspace corrections (2.23) are correspondingly given by

$$\tilde{u} \leftarrow \tilde{u} + T_k(u - \tilde{u}). \tag{5.11}$$

After the substep (2.23), the new error is

$$\tilde{u} - u \leftarrow (I - T_k)(\tilde{u} - u). \tag{5.12}$$

Thus the convergence rate of the multiplicative subspace correction method with respect to the norm (2.1) is the induced norm of the operator

$$E = (I - T_J) \ldots (I - T_0). \tag{5.13}$$

Theorem 5.1 Every cycle of the abstract multiplicative subspace correction method introduced in Section 2 reduces the norm (2.1) of the error at least by the factor $\|E\|$ where

$$\|E\|^2 \leq 1 - \frac{2 - \omega}{K_1(1 + K_2)^2}. \tag{5.14}$$

This factor depends only on the constant K_1 from the stability assumption (5.2), on the constant K_2 from (5.4), and on the constant $\omega < 2$ from equation (2.32).

There are several, in principle, very closely related versions of this theorem in the papers of Bramble, Pasciak, Wang and Xu. The present version bears most resemblance to that of Xu (1992b).

The proof of the theorem is somewhat technical. It is not so easy to detect the idea hidden behind it except that the terms considered are cleverly arranged and split up. The following two lemmas are the main tools:

Lemma 5.2 For all $v_k \in \mathcal{V}_k$ and all $u_k \in \mathcal{S}$ (!),

$$\sum_{k=0}^{J} a(v_k, u_k) \leq \sqrt{K_1} \, \Big\| \sum_{k=0}^{J} v_k \Big\| \, \Big(\sum_{k=0}^{J} a(T_k u_k, u_k) \Big)^{1/2}. \tag{5.15}$$

Proof. One has

$$\sum_{k=0}^{J} a(v_k, u_k) = \sum_{k=0}^{J} (B_k^{1/2} v_k, B_k^{-1/2} A_k P_k u_k)$$

$$\leq \Big(\sum_{k=0}^{J} \|B_k^{1/2} v_k\|_0^2 \Big)^{1/2} \Big(\sum_{k=0}^{J} \|B_k^{-1/2} A_k P_k u_k\|_0^2 \Big)^{1/2}$$

$$= \Big(\sum_{k=0}^{J} (B_k v_k, v_k) \Big)^{1/2} \Big(\sum_{k=0}^{J} (B_k^{-1} A_k P_k u_k, A_k P_k u_k) \Big)^{1/2}$$

$$= \Big(\sum_{k=0}^{J} (B_k v_k, v_k) \Big)^{1/2} \Big(\sum_{k=0}^{J} a(T_k u_k, u_k) \Big)^{1/2}.$$

With the stability assumption (5.2), the proposition follows. $\square$

Lemma 5.3 For all $u \in \mathcal{S}$,

$$\|T_k u\|^2 \leq \omega \, a(T_k u, u). \tag{5.16}$$

Proof. Because

$$\|T_k u\|^2 = (T_k u, A_k T_k u) \leq \omega \, (T_k u, B_k T_k u)$$
$$= \omega \, (T_k u, B_k B_k^{-1} A_k P_k u) = \omega \, a(T_k u, P_k u) = \omega \, a(T_k u, u),$$

the proposition is a simple consequence of (2.32). $\square$

Now we are ready to prove Theorem 5.1. Proposition (5.14) is equivalent to the estimate

$$(2 - \omega)\|v\|^2 \leq K_1(1 + K_2)^2(\|v\|^2 - \|Ev\|^2) \tag{5.17}$$

for all $v \in \mathcal{S}$. With $E_{-1} = I$ and

$$E_k = (I - T_k) \ldots (I - T_0), \quad k = 1, \ldots J,$$

one obtains

$$\|E_{k-1}v\|^2 - \|E_k v\|^2 = 2\, a(T_k E_{k-1}v, E_{k-1}v) - \|T_k E_{k-1}v\|^2.$$

With Lemma 5.3,

$$\|E_{k-1}v\|^2 - \|E_k v\|^2 \geq (2-\omega)\, a(T_k E_{k-1}v, E_{k-1}v)$$

follows. Because of $E_J = E$, summation yields

$$\|v\|^2 - \|Ev\|^2 \geq (2-\omega) \sum_{k=0}^{J} a(T_k E_{k-1}v, E_{k-1}v).$$

Because $\omega < 2$, (5.17) therefore follows from

$$\|v\|^2 \leq K_1(1+K_2)^2 \sum_{k=0}^{J} a(T_k E_{k-1}v, E_{k-1}v). \tag{5.18}$$

For the proof of (5.18), let

$$v = \sum_{l=0}^{J} v_l, \quad v_l \in \mathcal{V}_l.$$

Then

$$\|v\|^2 = \sum_{l=0}^{J} a(E_{l-1}v, v_l) + \sum_{l=1}^{J} a((I - E_{l-1})v, v_l). \tag{5.19}$$

By Lemma 5.2,

$$\sum_{l=0}^{J} a(E_{l-1}v, v_l) \leq \sqrt{K_1}\, \|v\| \left(\sum_{k=0}^{J} a(T_k E_{k-1}v, E_{k-1}v) \right)^{1/2}. \tag{5.20}$$

For the second term on the right-hand side of (5.19), because

$$I - E_{l-1} = \sum_{k=0}^{l-1} T_k E_{k-1},$$

and utilizing (5.3), one gets the estimate

$$
\sum_{l=1}^{J} a((I - E_{l-1})v, v_l) = \sum_{l=1}^{J} \sum_{k=0}^{l-1} a(T_k E_{k-1}v, v_l)
$$

$$
\leq \sum_{l=1}^{J} \sum_{k=0}^{l-1} \gamma_{kl} \left(B_k T_k E_{k-1}v, T_k E_{k-1}v\right)^{1/2} (B_l v_l, v_l)^{1/2}
$$

$$
\leq \sum_{l=0}^{J} \sum_{k=0}^{J} \gamma_{kl} \left(B_k T_k E_{k-1}v, T_k E_{k-1}v\right)^{1/2} (B_l v_l, v_l)^{1/2} \tag{5.21}
$$

$$
\leq K_2 \left(\sum_{k=0}^{J} (B_k T_k E_{k-1}v, T_k E_{k-1}v) \right)^{1/2} \left(\sum_{l=0}^{J} (B_l v_l, v_l) \right)^{1/2}
$$

$$
= K_2 \left(\sum_{k=0}^{J} (B_k v_k, v_k) \right)^{1/2} \left(\sum_{k=0}^{J} a(T_k E_{k-1}v, E_{k-1}v) \right)^{1/2}.
$$

By the assumption (5.2),

$$
\sum_{l=1}^{J} a((I - E_{l-1})v, v_l) \leq K_2 \sqrt{K_1} \, \|v\| \left(\sum_{k=0}^{J} a(T_k E_{k-1}v, E_{k-1}v) \right)^{1/2} \tag{5.22}
$$

follows. Combined with (5.20), one obtains (5.18). This finishes the proof of Theorem 5.1.

Sometimes (for multigrid methods with simple smoothers, for example) it is possible to prove the stronger estimate

$$
a(w_k, w_l') \leq \gamma_{kl} \left(B_k w_k, w_k\right)^{1/2} (B_l w_l', w_l')^{1/2} \tag{5.23}
$$

for $w_k \in \mathcal{W}_k, w_l' \in \mathcal{W}_l, k, l = 0, \ldots, J$, which implies (5.3). In this case, the estimate in Theorem 5.1 holds *independently of the order in which the subspace corrections are performed.*

There are situations in which the proof of the Cauchy–Schwarz type inequality (5.3) causes problems, especially if the coefficients functions of the differential operator under consideration are not smooth or even not differentiable. In such cases, it is often still possible to prove the norm estimate (5.7), provided that the energy norm (2.1) behaves like the energy norm induced by a boundary value problem for which one can prove estimates like (5.3), (5.4). The norm estimate (5.7) is sufficient to derive an estimate for the norm of the error propagation operator (5.13) which does not deteriorate too rapidly in terms of the number of subspaces $\mathcal{W}_k$.

Theorem 5.4 Assuming only (5.7) instead of (5.3) and (5.4), the norm of the error propagation operator (5.13) satisfies the estimate

$$
\|E\|^2 \leq 1 - \frac{2 - \omega}{K_1 (1 + \sqrt{\omega K_2 J})^2}. \tag{5.24}
$$

Proof. Instead of (5.21), one obtains

$$\sum_{l=1}^{J} a((I - E_{l-1})v, v_l) = \sum_{l=1}^{J} \sum_{k=0}^{l-1} a(T_k E_{k-1}v, v_l)$$

$$= \sum_{k=0}^{J-1} \sum_{l=k+1}^{J} a(T_k E_{k-1}v, v_l)$$

$$\leq \left(\sum_{k=0}^{J-1} \|T_k E_{k-1}v\|^2 \right)^{1/2} \left(\sum_{k=0}^{J-1} \| \sum_{l=k+1}^{J} v_l \|^2 \right)^{1/2}.$$

For the first factor on the right-hand side, one gets, by Lemma 5.3,

$$\sum_{k=0}^{J-1} \|T_k E_{k-1}v\|^2 \leq \omega \sum_{k=0}^{J} a(T_k E_{k-1}v, E_{k-1}v).$$

Using (5.7), the second factor can be estimated as follows.

$$\sum_{k=0}^{J-1} \| \sum_{l=k+1}^{J} v_l \|^2 \leq K_2 \sum_{k=0}^{J-1} \sum_{l=k+1}^{J} (B_l v_l, v_l)$$

$$\leq K_2 J \sum_{l=0}^{J} (B_l v_l, v_l) \leq K_1 K_2 J \|v\|^2.$$

This yields the estimate

$$\sum_{l=1}^{J} a((I - E_{l-1})v, v_l) \leq \sqrt{\omega K_1 K_2 J} \|v\| \left(\sum_{k=0}^{J} a(T_k E_{k-1}v, E_{k-1}v) \right)^{1/2}$$

which replaces (5.22). $\square$

6. The application to multilevel algorithms

In this section we apply the abstract theory presented in the last section to the model problem of Section 2. We prove convergence results for the multigrid methods introduced there.

We begin with the classical multigrid method where the subspaces $\mathcal{W}_k$ are the finite element spaces $\mathcal{S}_k$. In the notation of the previous section, the error propagation operator of the V-cycle is

$$E_V = E_V^{(j)}, \quad E_V^{(k)} = (I - T_k) \dots (I - T_0). \tag{6.1}$$

Because $A_0 = B_0$, i.e. $T_0 = P_0$, the $E_V^{(k)}$ satisfy the recursion

$$E_V^{(0)} = I - P_0, \quad E_V^{(k+1)} = (I - T_{k+1})E_V^{(k)}. \tag{6.2}$$

The corresponding recursion for the W-cycle version is

$$E_W^{(0)} = I - P_0, \quad E_W^{(k+1)} = (I - T_{k+1})E_W^{(k)} E_W^{(k)}. \tag{6.3}$$

It follows by induction that

$$E_W^{(k)} = E_V^{(k)} R_W^{(k)}, \quad \|R_W^{(k)}\| \leq 1. \tag{6.4}$$

Therefore one gets, for the energy norm of the error propagation operator $E_W = E_W^{(j)}$ of the W-cycle,

$$\|E_W\| \leq \|E_V\|. \tag{6.5}$$

Moreover, if the order of the coarse grid corrections and the smoothing steps is again reversed, the W-cycle always reduces the energy norm of the error by at least the same factor as the V-cycle.

For the analysis of the V-cycle multigrid method, we assume that, for $k \geq 1$, the operators $B_k : S_k \to S_k$, the smoothers, satisfy the estimate

$$c_1 \|u_k\|^2 \leq (u_k, B_k u_k) \leq c_2 4^k \|u_k\|_0^2 \tag{6.6}$$

for all $u_k \in S_k$. This condition is less restrictive than (3.6) and also covers certain symmetric block Gauß–Seidel schemes, for example.

The crucial point for the application of Theorem 5.1 is the choice of the spaces $V_k \subseteq S_k$. Recall that these subspaces do not enter into the computational process.

The most obvious choice is the a-orthogonal decomposition of S, i.e. the decomposition of S into $V_0 = S_0$ and

$$V_k = \{P_k u - P_{k-1} u \,|\, u \in S\} \subseteq S_k \tag{6.7}$$

for $k = 1, \ldots, j$. As it has already been discussed in Section 3, for H^2-regular problems,

$$4^k \|v_k\|_0^2 \leq C \|v_k\|^2, \quad v_k \in V_k \tag{6.8}$$

holds. Because

$$\|P_0 u\|^2 + \sum_{k=1}^{j} \|P_k u - P_{k-1} u\|^2 = \|u\|^2 \tag{6.9}$$

for the functions $u \in S$, (6.8) is equivalent to

$$\|v_0\|^2 + \sum_{k=1}^{j} 4^k \|v_k\|_0^2 \leq C \|\sum_{k=0}^{j} v_k\|^2, \quad v_k \in V_k. \tag{6.10}$$

With assumption (6.6), this yields (5.2), i.e.

$$\sum_{k=0}^{j} (B_k v_k, v_k) \leq K_1 \|\sum_{k=0}^{j} v_k\|^2. \tag{6.11}$$

The Cauchy–Schwarz type inequality (5.3) is trivial because

$$a(w_k, v_l) = 0, \quad w_k \in W_k, \ v_l \in V_l \ (k < l).$$

Thus we have shown that, for H^2-regular problems, every V-cycle (and every W-cycle) reduces the energy norm of the error at least by a factor

$$1 - \mathcal{O}(1)$$

which is uniformly less than 1 regardless of the number of refinement levels. This has already been proven in Wittum (1989a) and, with the restriction that the smoothing iterations are sufficiently damped, in Braess and Hackbusch (1983) and Bank and Douglas (1985); see Section 4. Therefore this estimate is surely not the most spectacular application of the abstract theory developed in the last section.

But before we discuss other choices of the spaces $\mathcal{V}_k$ leading to improved convergence estimates, we turn to the hierarchical basis multigrid method also described in Section 2.

For the hierarchical basis multigrid method, the finite element space S is already the direct sum of the spaces $\mathcal{W}_0 = S_0$ and

$$\mathcal{W}_k = \{\mathcal{I}_k u - \mathcal{I}_{k-1} u \,|\, u \in S\}, \quad k = 1, \ldots, j, \tag{6.12}$$

introduced in Section 2. Therefore the only possible choice for the subspaces $\mathcal{V}_k$ here are the spaces $\mathcal{W}_k$ itself.

It has been shown by Yserentant (1986b) that the decomposition of S into these spaces $\mathcal{V}_k$ is stable in the sense that, for all $v_k \in \mathcal{V}_k$,

$$\|v_0\|^2 + \sum_{k=1}^{j} 4^k \|v_k\|_0^2 \le C_1 (j+1)^2 \left\| \sum_{k=0}^{j} v_k \right\|^2. \tag{6.13}$$

If we assume that the level 0 equations are again solved exactly, i.e. that $B_0 = A_0$, and that, for $k \ge 1$, the operators $B_k : \mathcal{W}_k \to \mathcal{W}_k$ satisfy an estimate

$$c_1 4^k \|w_k\|_0^2 \le (w_k, B_k w_k) \le c_2 4^k \|w_k\|_0^2 \tag{6.14}$$

for all $w_k \in \mathcal{W}_k$, a stability condition like (5.2), namely

$$\sum_{k=0}^{j} (B_k v_k, v_k) \le K_1^* (j+1)^2 \left\| \sum_{k=0}^{j} v_k \right\|^2. \tag{6.15}$$

follows. The constant $K_1 = K_1^*(j+1)^2$ depends here on the number j of refinement levels.

The proof of (6.13) is based on the estimate

$$\|\mathcal{I}_k u\|^2 \le C(j - k + 1) \|u\|^2, \quad u \in S, \tag{6.16}$$

for the energy norm of the interpolation operators $\mathcal{I}_k : S \to S_k$. On one hand, this is a very robust estimate which is not affected by arbitrarily large jumps in the coefficient functions across the boundaries of the triangles in the initial triangulation. Unfortunately, on the other hand, it is dimension dependent. For three space dimensions, the logarithmic factor has to be replaced by a factor which grows exponentially in the number $j - k$ of the remaining refinement levels. Details can be found in Yserentant (1986b, 1992) and Bank *et al.* (1988).

The Cauchy–Schwarz type inequality (5.3) follows from (6.14) and Lemma 6.1 which has essentially been proven in Yserentant (1986b). Related results can be found in Xu (1992b), Bramble and Pasciak (1991), and in X. Zhang (1991).

Lemma 6.1 There is a constant C, depending only on the constants in (2.10) describing the ellipticity of the boundary value problem, on the variation of the coefficient functions, and on the shape regularity of the triangles, such that, for $k \le l$ and all functions $u \in \mathcal{S}_k$ and $v \in \mathcal{S}_l$,

$$a(u, v) \le C\,(\frac{1}{\sqrt{2}})^{l-k}\,\|u\|\,2^l\|v\|_0. \tag{6.17}$$

Proof. For $l > k + 1$, we fix a triangle $T \in \mathcal{T}_k$ and prove the local estimate

$$a(u, v)|_T \le C\,(\frac{1}{\sqrt{2}})^{l-k}\,|u|_{1;T}\,2^l\|v\|_{0;T}. \tag{6.18}$$

This estimate implies the global estimate (6.17). The basic idea is to split v into the function $v_0 \in \mathcal{S}_l$ given by

$$v_0(x) = \begin{cases} v(x) & , \quad x \in \mathcal{N}_l \cap \partial T \\ 0 & , \quad x \in \mathcal{N}_l \setminus \partial T \end{cases}$$

and into $v_1 = v - v_0$. Then the inner product $a(u, v)|_T$ can be written as

$$a(u, v)|_T = a(u, v_0)|_T + a(u, v_1)|_T.$$

The essential point is that v_1 vanishes on the boundary of T. Therefore we obtain, by partial integration and the product rule,

$$a(u, v_1)|_T = -\sum_{i,j=1}^{2} \int_T D_j a_{ij} D_i u\, v_1\, dx - \sum_{i,j=1}^{2} \int_T a_{ij} D_j D_i u\, v_1\, dx. \tag{6.19}$$

As u is linear on T, the second term on the right-hand side of the equation (6.19) vanishes. Assuming $T \subseteq T' \in \mathcal{T}_0$ and

$$|(D_j a_{ij})(x)| \le M_1 \mathrm{diam}(T')^{-1}, \quad x \in T',$$

the first term on the right-hand side of equation (6.19) and therefore $a(u, v_1)$ can be estimated to be

$$a(u, v_1)|_T \le c_1 |u|_{1;T}\|v_1\|_{0;T}. \tag{6.20}$$

The function v_0 vanishes outside a boundary strip S of T with

$$\frac{\mathrm{area}(S)}{\mathrm{area}(T)} = 1 - (1 - 3(\tfrac{1}{2})^{l-k})^2 \le 6(\tfrac{1}{2})^{l-k}.$$

Therefore

$$a(u, v_0)|_T \leq M|u|_{1;S}|v_0|_{1;S}.$$

As the restriction of u to T is linear,

$$|u|_{1,2;S}^2 = \text{area}(S)|u|_{1,\infty;T}^2 = \frac{\text{area}(S)}{\text{area}(T)}|u|_{1,2;T}^2.$$

Utilizing the inverse inequality

$$|v_0|_{1;S} \leq c_2 2^l \|v_0\|_{0;S},$$

the inner product $a(u, v_0)|_T$ can be estimated to be

$$a(u, v_0)|_T \leq c_3 (\frac{1}{\sqrt{2}})^{l-k}|u|_{1;T} 2^l \|v_0\|_{0;S}. \tag{6.21}$$

We remark that the factor 2^l enters because we normalized the L_2-like norm $\|\cdot\|_{0;T}$ according to (2.12) by the areas of the triangles in the initial triangulation. Because

$$\|v_0\|_{0;S} \leq c_4 \|v\|_{0;T}, \quad \|v_1\|_{0;T} \leq c_4 \|v\|_{0;T}$$

one obtains the proposition combining (6.20) and (6.21).

For $l = k, k + 1$, the proposition follows from the usual Cauchy–Schwarz inequality and the inverse estimate given earlier. $\square$

As result, every step of the hierarchical basis multigrid method reduces the energy norm of the error by at least a factor behaving like

$$1 - \mathcal{O}(1/j^2).$$

Thus Theorem 5.1 leads to an alternative proof of the main convergence theorem in Bank *et al.* (1988) for the special case that the coefficient functions of the differential operator are continuously differentiable. Note that the diameter of the triangles shrinks by the factor 2^{-j} in the transition from level 0 to level j. Therefore j grows logarithmically in the gridsize, which means very slowly. If the subspace corrections are repeated in the reversed order after every cycle, one gets a symmetrized iterative procedure which can be accelerated by the conjugate gradient method. Usually the hierarchical basis multigrid method is applied in this form, so that every step reduces the error in fact by a factor behaving like

$$1 - \mathcal{O}(1/j).$$

We remark that the fact, that the considered finite element functions are piecewise linear, is not essential for the Cauchy–Schwarz type inequality (6.17). With an additional factor 2^k on the right-hand side of (6.20) arising from the the second term on the right-hand side of equation (6.19) and a new

constant in (6.21), the proof of Lemma 6.1 transfers to the case of higher
order polynomials of fixed degree.

The subspace decomposition needed in the construction and analysis of
the hierarchical basis multigrid method can also be used to derive an alternative convergence result for the usual multigrid method. How, is described
in the next section by the example of L_2-like decompositions. One finds
that every multigrid cycle reduces the energy norm of the error at least by
a factor behaving like

$$1 - \mathcal{O}(1/j^2),$$

as for the hierarchical basis multigrid method. For a symmetrized version
accelerated by the conjugate gradient method, one again obtains a better
reduction factor

$$1 - \mathcal{O}(1/j).$$

Contrary to the asymptotically better estimates derived earlier, these estimates (as well as the estimates for the hierarchical basis multigrid method)
do not depend on the regularity of the boundary value problem and are even
independent of jumps of the coefficient functions across the boundaries of
the triangles in the initial triangulation. On the other hand, contrary to the
estimate earlier, they are restricted to two space dimensions.

Without any essential change, the analysis of the hierarchical basis multigrid method can be transferred to the case of nonuniformly refined grids.
Utilizing the same splitting of $\mathcal{S}$, one can also analyse multigrid methods
which are based on local smoothing procedures.

7. L_2-like subspace decompositions

The best, in a certain sense, subspace decomposition is the orthogonal decomposition of $\mathcal{S}$ into $\mathcal{V}_0 = \mathcal{S}_0$ and the orthogonal complements

$$\mathcal{V}_k = \{Q_k u - Q_{k-1} u \,|\, u \in \mathcal{S}\} \subseteq \mathcal{S}_k \tag{7.1}$$

for $k = 1, \ldots, j$. This decomposition has been used for the first time in
the analysis of multigrid methods in Bramble *et al.* (1991b) and, for the
analysis of closely related additive multilevel methods as discussed in the
next section, in Bramble *et al.* (1990) and Xu (1989).

The stability (5.2) of this decomposition can be essentially derived from
the error estimate

$$\|u - Q_k u\|_0 \leq C_1 2^{-k} \|u\|, \tag{7.2}$$

which holds for all functions in $H^1(\Omega)$. For H^2-regular problems, this error
estimate follows from the Aubin–Nitsche Lemma which is also the basis for
the classical proofs in Sections 3 and 4. Here we have less regular boundary
value problems in mind. An elementary proof of (7.2), which does not rely on

such regularity assumptions and which is based on local quasi-interpolants, can be found in Yserentant (1990), for example. Utilizing (7.2), one can show very easily that the discrete norm

$$\|u\|^2 = \|Q_0 u\|^2 + \sum_{k=1}^{j} 4^k \|Q_k u - Q_{k-1} u\|_0^2 \tag{7.3}$$

satisfies the estimate

$$\|u\|^2 \leq K_1^*(j+1)\|u\|^2, \tag{7.4}$$

which contains an additional logarithmic factor compared with (6.10).

Utilizing the equivalence of certain Besov and Sobolev spaces, Oswald (1990, 1991, 1992) and Dahmen and Kunoth (1991) recently developed a very general framework to compare norms like (7.3) with Sobolev norms. Especially, they could improve (7.4) to

$$\|u\|^2 \leq K_1 \|u\|^2 \tag{7.5}$$

with a constant K_1 neither depending on the number of refinement levels nor on regularity properties of the boundary value problem. In Bornemann and Yserentant (1992), a more specialized, but relatively elementary proof of (7.5) is given. The influence of boundary conditions and nonuniform refinements is discussed very carefully in this article.

Results, which are related to (7.5), have been proven in Bramble and Pasciak (1991), Xu (1992b), and X. Zhang (1992). These articles are based on the regularity theory of elliptic equations, although the degree of regularity finally enters only in the size of the constants.

Supposing again the property (6.6) of the smoothers, (7.5) yields the first basic assumption (5.2). In addition, with (7.5) (or also with (7.2)) one obtains

$$4^k \|v_k\|_0^2 \leq C \|v_k\|^2, \quad v_k \in \mathcal{V}_k, \tag{7.6}$$

so that, on $\mathcal{V}_k$, the energy norm $\|\cdot\|$ induced by the boundary value problem under consideration is equivalent to the scaled L_2-like norm $2^k \|\cdot\|_0$.

For the proof of the second basic assumption (5.3), (5.4) of the general theory in Section 5, we can again utilize Lemma 6.1, i.e.

$$a(u_k, v_l) \leq C(\frac{1}{\sqrt{2}})^{l-k} \|u_k\| \, 2^l \|v_l\|_0. \tag{7.7}$$

for $k \leq l$ and all functions $u_k \in \mathcal{S}_k$ and $v_l \in \mathcal{S}_l$. With (7.6), one obtains, for functions $v_l \in \mathcal{V}_l$, the strengthened Cauchy–Schwarz inequality

$$a(u_k, v_l) \leq \widehat{C}(\frac{1}{\sqrt{2}})^{l-k} \|u_k\| \, \|v_l\|. \tag{7.8}$$

With (6.6), (7.8) yields the desired Cauchy–Schwarz type inequality (5.3)

$$a(w_k, v_l) \leq \widetilde{C}(\frac{1}{\sqrt{2}})^{l-k}(B_k w_k, w_k)^{1/2}(B_l v_l, v_l)^{1/2} \qquad (7.9)$$

for $k \leq l$ and the functions $w_k \in \mathcal{W}_k$ and $v_l \in \mathcal{V}_l$.

Thus we have proven the convergence of classical multigrid methods with a convergence rate

$$1 - \mathcal{O}(1)$$

which does not deteriorate with the number of refinement levels and without utilizing regularity properties of the boundary value problem.

If the smoothers $B_k : \mathcal{S}_k \to \mathcal{S}_k$ satisfy the stronger and somewhat more restrictive condition (as compared with the condition (6.6))

$$c_1 4^k \|u_k\|_0^2 \leq (u_k, B_k u_k) \leq c_2 4^k \|u_k\|_0^2 \qquad (7.10)$$

for all $u_k \in \mathcal{S}_k$, the proof of the Cauchy–Schwarz type inequality (5.3) can be based directly on (7.7), and (7.9) holds even for all functions $v_l \in \mathcal{S}_l$. According to the remark in Section 5, for this case, interestingly the optimality of the multigrid method does not depend on the order in which the subspace corrections are performed.

If the coefficient functions of the differential operator in (2.6) are no longer differentiable, or if the derivatives are large, one can still apply Theorem 5.4 and gets a nearly optimal convergence rate.

For nonuniformly refined grids, one can work with L_2-like decompositions which are based on *local projections* as introduced in Dahmen and Kunoth (1991) or Bornemann and Yserentant (1992). Such decompositions can be analysed on the basis of the equivalence of the energy norm to the discrete norm (7.3).

8. Additive multilevel methods

Stimulated by the development of domain decomposition and of hierarchical basis methods, the interest has recently shifted from the recursively defined classical multilevel algorithms to additive multilevel methods. The most prominent new example in this class of algorithms is the multilevel nodal basis method of Bramble, Pasciak and Xu.

A main reason for this development is that additive multilevel algorithms fit much better to nonuniformly refined grids (as they are absolutely necessary for the solution of complicated real-life problems) because these algorithms allow the use of simpler, more natural data structures. Another reason is that the higher flexibility of these algorithms simplifies the use of parallel computers, although this should not be viewed too naively. It is fair to mention that additive methods usually need slightly more iteration steps than their multiplicative counterparts, although the single iteration step tends to be cheaper.

Additive multilevel methods fall into the class of the additive subspace correction methods already introduced in Section 2. In the same way, as the multiplicative subspace correction methods introduced in Section 2 correspond to the Gauß–Seidel method, the additive subspace correction methods discussed in this section are associated with the Jacobi iteration. In the notation of Section 2, for given subspaces $\mathcal{W}_0, \ldots, \mathcal{W}_J$ of $\mathcal{S}$ and given approximations B_k, the additive subspace correction method for the solution of the abstract equation (2.4) is

$$\tilde{u} \leftarrow \tilde{u} + \alpha \sum_{k=0}^{J} B_k^{-1} Q_k (f - A\tilde{u}). \tag{8.1}$$

This means that the single subspace corrections are not applied in a sequential order but in parallel. The iteration (8.1) can be rewritten as

$$\tilde{u} \leftarrow \tilde{u} + \alpha \, C(f - A\tilde{u}) \tag{8.2}$$

with the approximate inverse

$$C = \sum_{k=0}^{J} B_k^{-1} Q_k \tag{8.3}$$

of the operator $A : \mathcal{S} \to \mathcal{S}$.

As in the convergence theory for the multiplicative variant, the convergence theory for the additive subspace correction method is based on splitting $\mathcal{S}$ into subspaces $\mathcal{V}_k$ of the spaces $\mathcal{W}_k$. The convergence estimates are based on two assumptions. The first is again the stability assumption (5.2) that, for all $v_k \in \mathcal{V}_k$,

$$\sum_{k=0}^{J} (B_k v_k, v_k) \leq K_1 \Big\| \sum_{k=0}^{J} v_k \Big\|^2. \tag{8.4}$$

The second assumption is that there exists a (new) constant K_2 with

$$\Big\| \sum_{k=0}^{J} w_k \Big\|^2 \leq K_2 \sum_{k=0}^{J} (B_k w_k, w_k) \tag{8.5}$$

for all $w_k \in \mathcal{W}_k$. The assumption (8.5) can be deduced from the Cauchy–Schwarz type estimate (5.23) which is stronger than the assumption (5.3). The related condition (5.8)

$$\Big\| \sum_{k=0}^{J} v_k \Big\|^2 \leq K_2 \sum_{k=0}^{J} (B_k v_k, v_k)$$

for the elements $v_k \in \mathcal{V}_k$ is *not* sufficient.

The main result for additive subspace correction methods is a generalization of well known theorems from the theory of domain decomposition methods; see Widlund (1989) and Björstad and Mandel (1991).

Theorem 8.1 The operator C is symmetric and positive definite with respect to the inner product (2.2) on S. Therefore the eigenvalues λ of the operator CA are real and positive. They range in the interval

$$1/K_1 \leq \lambda \leq K_2. \tag{8.6}$$

Proof. As the B_k are symmetric operators, one has, for all $u, v \in S$,

$$(Cu, v) = \sum_{k=0}^{J}(B_k^{-1}Q_k u, Q_k v) = (u, Cv)$$

so that C is symmetric and also positive definite. Let $v \in S$ have the decomposition $v = \sum_{k=0}^{J} v_k$ with elements $v_k \in \mathcal{V}_k$. Then, by Lemma 5.2,

$$\|v\|^2 = \sum_{k=0}^{J} a(v_k, v) = \sum_{k=0}^{J} a(v_k, P_k v)$$

$$\leq \sqrt{K_1}\|v\| \left(\sum_{k=0}^{J} a(T_k P_k v, P_k v)\right)^{1/2}$$

$$= \sqrt{K_1}\|v\| \left(\sum_{k=0}^{J} a(T_k v, v)\right)^{1/2}$$

where, as in Section 5, $T_k = B_k^{-1}Q_k A$. Because $\sum_{k=0}^{J} T_k = CA$, one gets

$$a(v, v) \leq K_1\, a(CAv, v).$$

Therefore the eigenvalues of CA cannot be less than $1/K_1$. By the new assumption (8.5), one obtains, for all $v \in S$,

$$\|\sum_{k=0}^{J} T_k v\|^2 \leq K_2 \sum_{k=0}^{J}(B_k T_k v, T_k v) = K_2 \sum_{k=0}^{J} a(T_k v, v)$$

or, again with $\sum_{k=0}^{J} T_k = CA$, the estimate

$$\|CAv\|^2 \leq K_2\, a(CAv, v).$$

Therefore the eigenvalues of CA are not greater than K_2. □

By Theorem 8.1, the iteration (8.2) can be accelerated by the conjugate gradient method. In fact, additive subspace correction methods are nearly exclusively used in this way so that the proper choice of the damping parameter α is no longer a question of practical interest. The quality of C as a preconditioner for A is essentially described by the spectral condition number

$$\kappa = \kappa(C^{1/2}AC^{1/2}) \tag{8.7}$$

which is defined as the ratio of the maximum and the minimum eigenvalue of the operator $C^{1/2}AC^{1/2}$. As this operator is similar to CA, Theorem 8.1 says that

$$\kappa \leq K_1 K_2 \tag{8.8}$$

is an upper bound for this condition number.

For additive multilevel methods for the solution of finite element equations like (2.11), the spaces $\mathcal{W}_k \subseteq \mathcal{S}$ are subspaces of the coarse level spaces $\mathcal{S}_k$; for the *multilevel nodal basis method* of Bramble et al. (1990) and Xu (1989) one has $\mathcal{W}_k = \mathcal{S}_k$, and for the *hierarchical basis method* (Yserentant 1986b, 1990, 1992), the spaces $\mathcal{W}_k$ are the hierarchical complements (2.41). Therefore the multilevel nodal basis method can be seen as the additive version of the multigrid V-cycle whereas the hierarchical basis method is the additive version of the hierarchical basis multigrid method.

For both methods, we again require that $B_0 = A_0$ and that, for $k \geq 1$, the operators $B_k : \mathcal{W}_k \rightarrow \mathcal{W}_k$ satisfy an estimate

$$c_1 4^k \|w_k\|_0^2 \leq (w_k, B_k w_k) \leq c_2 4^k \|w_k\|_0^2 \tag{8.9}$$

for all $w_k \in \mathcal{W}_k$. This is essentially the condition (3.6) which is somewhat more restrictive than the condition (6.6) used in the analysis of the multiplicative variants. However, remember that simple point Jacobi and Gauß–Seidel smoothers are covered by (8.9).

Then, for arbitrary subspaces $\mathcal{W}_k \subseteq \mathcal{S}_k$, the Cauchy–Schwarz type inequality

$$a(u,v) \leq C\left(\frac{1}{\sqrt{2}}\right)^{l-k} \|u\| \, 2^l \|v\|_0 \tag{8.10}$$

from Lemma 6.1 for the functions $u \in \mathcal{S}_k$ and $v \in \mathcal{S}_l$, $k \leq l$, and (8.9) yield the new condition (8.5).

The subspaces $\mathcal{V}_k \subseteq \mathcal{W}_k$ are chosen as for the corresponding multiplicative schemes. Therefore the stability condition (8.4) has already been derived in the last section. With Theorem 8.1, we can conclude that the additive multilevel methods have qualitatively the same convergence behaviour as their multiplicative counterparts.

In order to exhibit the advantages of additive multilevel methods, the approximations B_k for the operators A_k should be chosen as simple as possible. The best possible choice is probably the Jacobi method.

In the following we discuss the realization of the multilevel nodal basis method in conjunction with the Jacobi method. Let $\mathcal{N}_k = \{x_1, \ldots, x_{n_k}\}$ be the set of vertices of the triangles in $\mathcal{T}_k$ not lying on the boundary of Ω. Then $\mathcal{S}_k$ is spanned by the nodal basis functions $\psi_i^{(k)}$, $i = 1, \ldots, n_k$, which are defined by

$$\psi_i^{(k)}(x_l) = \delta_{il}, \quad x_l \in \mathcal{N}_k. \tag{8.11}$$

Then, for the given B_k, the operator

$$C = A_0^{-1}Q_0 + \sum_{k=1}^{j} B_k^{-1}Q_k \tag{8.12}$$

can be written as

$$Cr = A_0^{-1}Q_0r + \sum_{k=1}^{j}\sum_{i=1}^{n_k} \frac{(r, \psi_i^{(k)})}{a(\psi_i^{(k)}, \psi_i^{(k)})}\, \psi_i^{(k)}. \tag{8.13}$$

To realize the iteration

$$\tilde{u} \leftarrow \tilde{u} + \alpha\, Cr, \quad r = f - A\tilde{u}, \tag{8.14}$$

or its conjugate gradient accelerated version efficiently, the functions $\tilde{u}$ and the residuals r have to be represented differently. We store $\tilde{u}$ by the values

$$\tilde{u}(x_i), \quad i = 1,\dots,n, \tag{8.15}$$

whereas r is represented by

$$(r, \psi_i), \quad i = 1,\dots,n, \tag{8.16}$$

where, for simplicity, $n = n_j$ and $\psi_i = \psi_i^{(j)}$. The inner products (8.16) are given by

$$(r, \psi_i) = (f, \psi_i) - \sum_{l=1}^{n} a(\psi_i, \psi_l)\, u(x_l), \tag{8.17}$$

so that only the usual residual has to be computed; an explicit representation of the operator A is not needed. Note that the values $(r, \psi_i^{(k)})$ can be recursively computed beginning with the values $(r, \psi_i) = (r, \psi_i^{(j)})$, and that the summation of the single terms in (8.13) can be formulated as a recursive process, too. The function

$$u_0 = A_0^{-1}Q_0r \in \mathcal{S}_0 \tag{8.18}$$

satisfies the relation

$$a(u_0, v) = (r, v), \quad v \in \mathcal{S}_0. \tag{8.19}$$

To compute u_0, therefore one needs only $(r, \psi_i^{(0)})$, $i = 1,\dots,n_0$, but not Q_0r itself, and one has to solve a linear system with the level 0 discretization matrix.

The appropriate modification of the multilevel nodal basis preconditioner to nonuniformly refined grids is

$$Cr = A_0^{-1}Q_0r + \sum_{k=1}^{j} \sum_{\psi_i^{(k)} \neq \psi_i^{(k-1)}} \frac{(r, \psi_i^{(k)})}{a(\psi_i^{(k)}, \psi_i^{(k)})} \psi_i^{(k)} \qquad (8.20)$$

where the inner sum stands for

$$\sum_{\psi_i^{(k)} \neq \psi_i^{(k-1)}} = \sum_{\substack{i=1 \\ \psi_i^{(k)} \neq \psi_i^{(k-1)}}}^{n_{k-1}} + \sum_{i=n_{k-1}+1}^{n_k}. \qquad (8.21)$$

Only those basis functions $\psi_i^{(k)}$ of S_k are still taken into account in the inner sum which are associated with the new nodes $x_i, i = n_{k-1} + 1, \ldots, n_k$, on the level k and with the neighbours of these nodes. With this modification, the operation count for the single iteration step (8.14) remains strictly proportional to the number of unknowns independent of the distribution of the unknowns among the levels.

As the corresponding multigrid methods based on local smoothing procedures, this version of the multilevel nodal basis method can be analysed utilizing the local L_2-decompositions introduced in Bornemann and Yserentant (1992). It turns out that the condition number (8.7) behaves like $\mathcal{O}(1)$, as in a uniform refinement.

The hierarchical basis method goes one step further. In Xu's formulation (Xu, 1989) it is given by

$$Cr = A_0^{-1}Q_0r + \sum_{k=1}^{j} \sum_{i=n_{k-1}+1}^{n_k} \frac{(r, \psi_i^{(k)})}{a(\psi_i^{(k)}, \psi_i^{(k)})} \psi_i^{(k)}. \qquad (8.22)$$

As every term in the double sum can be associated with a node of the final level, the algorithmic realization of this method becomes extremely simple; see Yserentant (1986b, 1990).

If we introduce the hierarchical basis functions $\widehat{\psi}_i, i = 1, \ldots, n$, of S by

$$\widehat{\psi}_i = \psi_i^{(0)}, \quad x_i \in \mathcal{N}_0, \qquad (8.23)$$

and by

$$\widehat{\psi}_i = \psi_i^{(k)}, \quad x_i \in \mathcal{N}_k \setminus \mathcal{N}_{k-1}, \qquad (8.24)$$

the hierarchical basis preconditioner takes the form

$$Cr = A_0^{-1}Q_0r + \sum_{i=n_0+1}^{n} \frac{(r, \widehat{\psi}_i)}{a(\widehat{\psi}_i, \widehat{\psi}_i)} \widehat{\psi}_i. \qquad (8.25)$$

Thus it is, up to a small block of the dimension n_0 of the initial finite element space S_0, Jacobi's old method, now with respect to the hierarchical basis formulation of the discrete elliptic boundary value problem. In this sense, it is the most simple multigrid method.

REFERENCES

O. Axelsson and I. Gustafsson (1983), 'Preconditioning and two-level multigrid methods of arbitrary degree of approximation', *Math. Comput.* **40**, 219–242.

O. Axelsson and P. Vassilevski (1989), 'Algebraic multilevel preconditioning methods I', *Numer. Math.* **56**, 157–177.

N.S. Bachvalov (1966), 'On the convergence of a relaxation method with natural constraints on the elliptic operator' (in Russian), *USSR Comput. Math. and Math. Phys.* **6.5**, 101–135.

E. Bänsch (1991), 'Local mesh refinement in 2 and 3 dimensions', *Impact of Computing in Sci. Engrg.* **3**, 181–191.

R.E. Bank (1981), 'A comparison of two multilevel iterative methods for nonsymmetric and indefinite elliptic finite element equations', *SIAM J. Numer. Anal.* **18**, 724–743.

R.E. Bank (1990), *PLTMG: A Software Package for Solving Elliptic Partial Differential Equations*, SIAM (Philadelphia).

R.E. Bank and C.C. Douglas (1985), 'Sharp estimates for multigrid convergence rates of convergence with general smoothing and acceleration', *SIAM J. Numer. Anal.* **22**, 617–633.

R.E. Bank and T. Dupont (1980), 'Analysis of a two-level scheme for solving finite element equations', Report CNA-159, Center for Numerical Analysis, University of Texas at Austin.

R.E. Bank and T. Dupont (1981), 'An optimal order process for solving finite element equations', *Math. Comput.* **36**, 35–51.

R.E. Bank and D.J. Rose (1982), 'Analysis of a multilevel iterative method for nonlinear finite element equations', *Math. Comput.* **39**, 453–465.

R.E. Bank and A. Weiser (1985), 'Some a-posteriori estimators for elliptic partial differential equations', *Math. Comput.* **44**, 283–301.

R.E. Bank, T. Dupont and H. Yserentant (1988), 'The hierarchical basis multigrid method', *Numer. Math.* **52**, 427–458.

R.E. Bank, B. Welfert and H. Yserentant (1990), 'A class of iterative methods for solving saddle point problems', *Numer. Math.* **56**, 645–666.

P.E. Björstad and J. Mandel (1991), 'On the spectra of sums of orthogonal projections with applications to parallel computing', *BIT* **31**, 76–88.

F.A. Bornemann and H. Yserentant (1992), 'A basic norm equivalence for the theory of multilevel methods', *Numer. Math.*, to appear.

D. Braess (1981), 'The contraction number of a multigrid method for solving the Poisson equation', *Numer. Math.* **37**, 387–404.

D. Braess and W. Hackbusch (1983), 'A new convergence proof for the multigrid method including the V-cycle', *SIAM J. Numer. Anal.* **20**, 967–975.

J.H. Bramble and J.E. Pasciak (1988), 'A preconditioning technique for indefinite systems resulting from mixed approximations of elliptic problems', *Math. Comput.* **181**, 1–17.

J.H. Bramble and J.E. Pasciak (1991), 'New estimates for multilevel algorithms including the V-cycle', Technical Report '91-47, Mathematical Sciences Institute, Cornell University.

J.H. Bramble, J.E. Pasciak, J. Wang and J. Xu (1991a), 'Convergence estimates for product iterative methods with applications to domain decomposition', *Math. Comput.* **57**, 1–21.

J.H. Bramble, J.E. Pasciak, J. Wang and J. Xu (1991b), 'Convergence estimates for multigrid algorithms without regularity assumptions', *Math. Comput.* **57**, 23–45.

J.H. Bramble, J.E. Pasciak and J. Xu (1990), 'Parallel multilevel preconditioners', *Math. Comput.* **55**, 1–22.

A. Brandt (1977), 'Multi-level adaptive solutions to boundary-value problems', *Math. Comput.* **31**, 333–390.

A. Brandt (1982), 'Guide to multigrid development', in *Multigrid Methods* (W. Hackbusch and U. Trottenberg, eds), Lecture Notes in Mathematics 960, Springer (Berlin, Heidelberg, New York).

T. Chan, R. Glowinski, J. Periaux and O.B. Widlund, eds (1989), *Domain Decomposition Methods, Proceedings, Los Angeles 1988*, SIAM (Philadelphia).

P.G. Ciarlet (1978), *The Finite Element Method for Elliptic Problems*, North-Holland (Amsterdam).

W. Dahmen and A. Kunoth (1992), 'Multilevel preconditioning', *Numer. Math.*, to appear.

P. Deuflhard (1992), *Newton-Techniques for Highly Nonlinear Problems – Theory and Algorithms*, Academic Press (New York) in preparation.

P. Deuflhard, P. Leinen and H. Yserentant (1989), 'Concepts of an adaptive hierarchical finite element code', *Impact of Computing in Sci. Engrg* **1**, 3–35.

M. Dryja and O. Widlund (1991), 'Multilevel additive methods for elliptic finite element problems', in *Parallel Algorithms for Partial Differential Equations, Proceedings, Kiel 1990* (W. Hackbusch, ed.), Vieweg, Braunschweig (Wiesbaden).

R.P. Fedorenko (1964), 'The speed of convergence of an iterative process' (in Russian), *USSR Comput. Math. and Math. Phys.* **4.3**, 227–235.

R.W. Freund, G.H. Golub and N.M. Nachtigal (1992), 'Iterative solution of linear systems', *Acta Numerica 1992*, 57–100.

W. Hackbusch (1976), 'Ein iteratives Verfahren zur schnellen Auflösung elliptischer Randwertprobleme', Report 76-12, Mathematisches Institut der Universität zu Köln.

W. Hackbusch (1981), 'On the convergence of multi-grid iterations', *Beiträge Numer. Math.* **9**, 213–329.

W. Hackbusch (1982), 'Multi-grid convergence theory', in *Multigrid Methods* (W. Hackbusch and U. Trottenberg, eds), Lecture Notes in Mathematics 960, Springer (Berlin, Heidelberg, New York).

W. Hackbusch (1985), *Multigrid Methods and Applications*, Springer (Berlin, Heidelberg, New York).

W. Hackbusch (1989a), 'The frequency decomposition multi-grid algorithm', in *Robust Multi-Grid Methods, Proceedings Kiel 1988* (W. Hackbusch, ed.). Vieweg, Braunschweig (Wiesbaden).

W. Hackbusch (1989b), 'The frequency decomposition multi-grid method. Part I: Application to anisotropic equations', *Numer. Math.* **56**, 229–245.

W. Hackbusch (1991), *Iterative Lösung großer schwachbesetzter Systeme*, Teubner (Stuttgart) (English translation in preparation).

W. Hackbusch and A. Reusken (1989), 'Analysis of a damped nonlinear multilevel method', *Numer. Math.* **55**, 225–246.

W. Hackbusch and U. Trottenberg, eds (1982), *Multigrid Methods, Proceedings, Köln 1981*, Lecture Notes in Mathematics 960, Springer (Berlin, Heidelberg, New York).

J. Mandel, S. McCormick and R. Bank (1987), 'Variational multigrid theory', in *Multigrid Methods* (S. McCormick, ed.), SIAM (Philadelphia).

S. McCormick, ed. (1987), *Multigrid Methods*, SIAM (Philadelphia).

P. Oswald (1990), 'On function spaces related to finite element approximation theory', *Zeitschrift für Analysis und ihre Anwendungen* **9**, 43–64.

P. Oswald (1991), 'On discrete norm estimates related to multilevel preconditioners in the finite element method', in *Proceedings of the International Conference on the Constructive Theory of Functions, Varna 1991.*

P. Oswald (1992), 'Stable splittings of Sobolev spaces and fast solution of variational problems', Forschungsergebnisse der Friedrich-Schiller-Universität Jena, Math/92/5.

A. Reusken (1992), 'On maximum norm convergence of multigrid methods for elliptic boundary value problems', Report RANA 92-03, Department of Mathematics and Computer Science, Eindhoven University.

M.C. Rivara (1984), 'Algorithms for refining triangular grids suitable for adaptive and multigrid techniques', *Int. J. Numer. Meth. Engrg* **20**, 745–756.

A. Schatz (1974), 'An observation concerning Ritz–Galerkin methods with indefinite bilinear forms', *Math. Comput.* **28**, 959–962.

R. Stevenson (1992), 'New estimates of the contraction number of V-cycle multigrid with applications to anisotropic equations', Preprint Nr. 713, Department of Mathematics, University of Utrecht.

K. Stüben and U. Trottenberg (1982), 'Multigrid methods: fundamental algorithms, model problem analysis and applications', in *Multigrid Methods* (W. Hackbusch and U. Trottenberg, eds), Lecture Notes in Mathematics 960, Springer (Berlin, Heidelberg, New York).

R.S. Varga (1962), *Matrix Iterative Analysis*, Prentice Hall (Englewood Cliffs, NJ).

P. Vassilevski (1992), 'Preconditioning nonsymmetric and indefinite finite element matrices', *J. Numer. Linear Alg. with Appl.* **1**, 59–76.

P. Wesseling (1992), *An Introduction to Multigrid Methods*, John Wiley (Chichester).

O. Widlund (1989), 'Optimal iterative refinement methods', in *Domain Decomposition Methods* (T. Chan et al., eds), SIAM (Philadelphia).

G. Wittum (1989a), 'Linear iterations as smoothers in multigrid methods: theory with applications to incomplete decompositions', *Impact of Computing in Sci. Engrg* **1**, 180–215.

G. Wittum (1989b), 'On the robustness of ILU smoothing', *SIAM J. Sci. Statist. Comput.* **10**, 699–717.

J. Xu (1989), 'Theory of multilevel methods', Report AM48, Department of Mathematics, Pennsylvania State University.

J. Xu (1992a), 'A new class of iterative methods for nonsymmetric and indefinite problems', *SIAM J. Numer. Anal.* **29**, 303–319.

J. Xu (1992b), 'Iterative methods by space decomposition and subspace correction', *SIAM Rev.* **34**.

D.M. Young (1950), 'Iterative methods for solving partial differential equations of elliptic type', Thesis, Harvard University.

H. Yserentant (1983), 'On the convergence of multi-level methods for strongly nonuniform families of grids and any number of smoothing steps per level', *Computing* **30**, 305–313.

H. Yserentant (1986a), 'The convergence of multi-level methods for solving finite element equations in the presence of singularities', *Math. Comput.* **47**, 399–409.

H. Yserentant (1986b), 'On the multi-level splitting of finite element spaces', *Numer. Math.* **49**, 379–412.

H. Yserentant (1986c), 'On the multi-level splitting of finite element spaces for indefinite elliptic boundary value problems', *SIAM J. Numer. Anal.* **23**, 581–595.

H. Yserentant (1988), 'Preconditioning indefinite discretization matrices', *Numer. Math.* **54**, 719–734.

H. Yserentant (1990), 'Two preconditioners based on the multi-level splitting of finite element spaces', *Numer. Math.* **58**, 163–184.

H. Yserentant (1992), 'Hierarchical bases', in *ICIAM 91* (R.E. O'Malley, ed.), SIAM (Philadelphia).

S. Zhang (1990), 'Optimal order non-nested multigrid methods for solving finite element equations II: on non-quasiuniform meshes', *Math. Comput.* **55**, 439–450.

X. Zhang (1991), 'Multilevel additive Schwarz methods', Technical Report 582, Courant Institute, New York.

X. Zhang (1992), 'Multilevel Schwarz methods', *Numer. Math.*, to appear.